AF411834

Solid State Physics of Finite Systems

Advanced Texts in Physics

This program of advanced texts covers a broad spectrum of topics which are of
current and emerging interest in physics. Each book provides a comprehensive and
yet accessible introduction to a field at the forefront of modern research. As such,
these texts are intended for senior undergraduate and graduate students at the MS
and PhD level; however, research scientists seeking an introduction to particular
areas of physics will also benefit from the titles in this collection.

Springer

Berlin
Heidelberg
New York
Hong Kong
London
Milan
Paris
Tokyo

R.A. Broglia   G. Colò
G. Onida   H.E. Roman

Solid State Physics of Finite Systems

Metal Clusters, Fullerenes, Atomic Wires

With 54 Figures and 31 Tables

Springer

Prof. Ricardo A. Broglia
Dipartimento di Fisica
Università di Milano
and INFN Sez. Milano
Via Celoria 16
20133 Milano, Italy
and
Niels Bohr Institute
University of Copenhagen
Blegdamsvej 17
2100 Copenhagen
Denmark
e-mail : Ricardo.Broglia@mi.infn.it

Dr. Gianluca Colò
Dipartimento di Fisica
Università di Milano
and INFN Sez. Milano
Via Celoria 16
20133 Milano, Italy
e-mail : colo@mi.infn.it

Prof. Giovanni Onida
Dipartimento di Fisica
Università di Milano
and INFM
Via Celoria 16
20133 Milano, Italy
e-mail : onida@mi.infn.it

Dr. H. Eduardo Roman
Dipartimento di Fisica
Università di Milano
and INFN Sez. Milano
Via Celoria 16
20133 Milano, Italy
e-mail : roman@mi.infn.it

ISSN 1439-2674

ISBN 3-540-21035-0   Springer-Verlag Berlin Heidelberg New York

Library of Congress Control Number: 2004105249

Springer-Verlag is a part of Springer Science+Business Media

springeronline.com

Typesetting: Data prepared by the author using a Springer $\TeX$ macro package
Cover design: *design & production* GmbH, Heidelberg

Printed on acid-free paper   SPIN 108992970   57/3141/tr   5 4 3 2 1 0

for Angela, Donatella, Gianandrea and Bettina - RAB
for Claudia, Leonardo and Valentina - GC
for Gioele, Alexander, Anna, Matilde, Giada, and Mattia - GO
for Fulvia, Claudia and Angela - HER

Preface

Quantum mechanics is the set of laws of physics which, to the best of our knowledge, provides a complete account of the microworld. One of its chapters, quantum electrodynamics (QED), is able to account for the quantal phenomena of relevance to daily life (electricity, light, liquids and solids, etc.) with great accuracy. The language of QED, field theory, has proved to be universal providing the theoretical basis to describe the behaviour of many-body systems. In particular finite many-body systems (FMBS) like atomic nuclei, metal clusters, fullerenes, atomic wires, etc. That is, systems made out of a small number of components.

The properties of FMBS are expected to be quite different from those of bulk matter, being strongly conditioned by quantal size effects and by the dynamical properties of the surface of these systems. The study of the electronic and of the collective behaviour (plasmons and phonons) of FMBS and of their interweaving, making use of well established first principle quantum (field theoretical) techniques, is the main subject of the present monograph.

The interest for the study of FMBS was clearly stated by Feynman in his address to the American Physical Society with the title "There is plenty of room at the bottom". On this occasion he said among other things: "When we get to the very, very small world - say circuits of seven atoms - we have a lot of new things that would happen that represent completely new opportunities for design" [1].

Furthermore, FMBS can be used as building blocks to construct new (nanostructured) materials, with properties which are controlled by those of the individual atomic aggregates, in particular by their phonon spectrum, and by the electronic density of levels and plasmons of the solid as a whole.

The material presented here is an outgrowth of lectures given to fourth-years students at the University of Milan, based on research carried out by our group. We have, therefore, placed special emphasis on the role played by the general concepts which are at the basis of the physics of FMBS in general, and of atomic and molecular aggregates in particular. We have supplemented this basic subject matter in an organic fashion, with material taken from work going on at the forefront of research on the structure of atomic and molecular clusters. This monograph should thus be useful to advanced students as well as to young researchers alike, both theorists and experimentalists.

We want to thank F. Alasia, P. Arcagni, F. Barranco, G. Benedek, M. Bernasconi, M. Bianchetti, P. F. Buonsante, N. Breda, F. Ginelli, A. Lorenzoni, F. Marini, C. Milani, J. Pacheco, D. Provasi, Ll. Serra, E. Vigezzi and C. Yannouleas, for many discussions and fruitful collaboration. Discussions with G. F. Bertsch, S. Bjørnholm, P. F. Bortignon, O. Gunnarson, P. Milani, B. R. Mottelson and R. Smalley are gratefully acknowledged. We also express our gratitude for the inspiring atmosphere of the Department of Physics of the University of Milan and of the Niels Bohr Institute of the University of Copenhagen, institutions where the research which is at the basis of the present monograph was carried out. The financial support of the Istituto Nazionale di Fisica Nucleare (INFN) Sez. di Milano, and Istituto Nazionale di Fisica della Materia (INFM) Sez. G, is gratefully acknowledged.

Milan,
April 2004

R. A. Broglia
G. Colò
G. Onida
H. E. Roman

Contents

1. Introduction 1

2. Overview 7
   2.1 k–Mass 10
   2.2 Electrons in Metals 12
        2.2.1 Jellium Model 13
   2.3 Electron Fluctuations (Plasmons) 16
   2.4 Electron–Plasmon Coupling 16
   2.5 ω–Mass 20
   2.6 Total Effective Mass 21
   2.7 Induced Interaction 21
   2.8 Examples 23
   2.9 Minimal Mean Field Theory 23
        2.9.1 Non-locality in Space of the Exchange Contribution ... 24
        2.9.2 Non-locality in Time of the Correlation Contribution . 26
  2.10 Phonons 26
  2.11 Appendix 28
        2.11.1 Second Quantization 28
        2.11.2 Single-Particle Self-Energy 32

3. Electronic Structure 35
   3.1 Born–Oppenheimer Approximation 35
   3.2 Density Functional Theory: Generalities 38
   3.3 The Local Density Approximation 41
        3.3.1 Discussion of Kohn–Sham Method 43
        3.3.2 Pseudopotentials 44
   3.4 Kohn–Sham Equations in a Spherical Basis 45
        3.4.1 The Expansion of V_{ext} in Spherical Harmonics 47
        3.4.2 The Expansion of $\varrho(\boldsymbol{r})$, v_{H} and v_{xc} 50
   3.5 Kohn–Sham Equations in a Cylindrical Basis 51
   3.6 Appendices 54
        3.6.1 Basics on Pseudopotentials 54
        3.6.2 Matrix Elements in a Spherical Basis 57

4. Electronic Response to Time-Dependent Perturbations ... 61
4.1 Linear Response: RPA and TDLDA 61
4.2 Strength Function and Sum Rules 65
4.3 Transition Density and Plasmons 68
4.4 Appendices 70
    4.4.1 Linear Response................................. 70
    4.4.2 TDLDA or RPA in the Configuration Space 72
    4.4.3 Plasmon Wavefunction Components
        for a Separable Interaction 73
    4.4.4 Spherical TDLDA 74
    4.4.5 Cylindrical TDLDA 76

5. Applications to Carbon Structures and Metal Clusters ... 79
5.1 C_{60} Fullerene 79
    5.1.1 Kohn–Sham States for C_{60} 79
    5.1.2 The Optical Absorption of C_{60} 81
5.2 Carbon Nanotubes 85
    5.2.1 Electronic Structure 86
    5.2.2 Electromagnetic Response and Static Polarizabilities .. 87
5.3 Linear Carbon Chains................................ 89
    5.3.1 Electronic Structure 89
    5.3.2 Optical Response and Plasmon Resonances 93
5.4 Metal Clusters: the Case of Na_8 98
5.5 Appendices 100
    5.5.1 Role of Symmetries 100

6. Phonons: Harmonic Approximation 105
6.1 Harmonic Approximation.............................. 106
    6.1.1 A Practical Example: a Linear Triatomic Molecule ... 111
6.2 Phonon Calculations 112
    6.2.1 Introduction 112
    6.2.2 Bond Charge Models 114
    6.2.3 Phonons from ab initio Calculations 128

7. The Car–Parrinello Method 131
7.1 An Efficient Tool for Solving the Electronic Problem 131
7.2 Car–Parrinello Molecular Dynamics...................... 133
7.3 Ionic Motion: Parameter-Free Molecular Dynamics Simulation 136
    7.3.1 Frozen-Phonon Calculations 141

8. Coupling of Electrons to Phonons
and to Plasmons 145
8.1 Definition of the Electron–Phonon Interaction 146
8.2 Diagonal Matrix Elements 148
8.3 Electron–Phonon Coupling in C_{60}^- 149

8.4   Signatures of the Electron–Phonon Interaction:
Photoemission in C_{60}^- 154
     8.4.1   Electron–Phonon Coupling in Other Fullerenes 160
     8.4.2   Electron–Phonon Coupling in Sodium Clusters 163
8.5   The Effective Electron–Electron Interaction 165
8.6   Coupling of Electrons to Plasmons 166
8.7   TDLDA with Electron–Plasmon
and Electron–Phonon Couplings 169
8.8   An Example: Electron–Phonon and
Electron–Plasmon Couplings in Na_9^+ 171
8.9   Appendices 173
     8.9.1   Terms Neglected in the Born–Oppenheimer
Approximation 173
     8.9.2   The Electron–Phonon Coupling Constant λ 175
     8.9.3   Outline of the Derivation of (8.50) 177
     8.9.4   Equation (8.50) Using Second-Order
Perturbation Theory 178

9.   Pairing in Atomic Aggregates 183
9.1   The BCS Solution 184
9.2   Superconductivity in Metals 190
     9.2.1   Finite Temperature 190
9.3   Superconductivity in Fullerides 194
     9.3.1 C_{60} Fullerene Based Superconductors 195
     9.3.2   Nanometer Superconductors Based on Small Fullerenes 197
     9.3.3   Thin Carbon Nanotubes 200
9.4   Selected Open Questions 201
     9.4.1   Hole Doped Fullerites 202
     9.4.2   Screening Effects 204
9.5   Appendices 204
     9.5.1   Simple Estimate of μ 204
     9.5.2   Solution of the Pairing Hamiltonian 206
     9.5.3   Temperature Dependent BCS Solution 213

10. Discussion 221

References 224

Index 233

1. Introduction

Nanostructured materials are made out of atoms in their more common forms, but the atoms are arranged in nanometer or sub-nanometer size units, which become the building blocks of the new materials. A prototype of these building blocks is C_{60} fullerene [2, 3]. These tiny grains, atomic aggregates or clusters, in which the surface plays a paramount role (see e.g. [4]), respond to external stimuli such as light, mechanical stress and electric fields quite differently from bulk matter [5, 6, 7, 8]. Hence nanostructured materials display an array of novel attributes and properties.

To learn how to produce customer tailored nanometer materials, one needs to have a thorough knowledge of the physical properties of the associated building blocks, that is, of atomic aggregates. For example, small silver particles dissolved in the stained glasses of medieval cathedrals led to the blues, the distinct ruby being produced by the presence of gold particles.

When trying to understand the colour of these particles, it is not enough to make reference to the optical response of either the individual atoms or bulk metal crystals. This is because the way that the properties of a solid gradually evolve as atoms are brought together to form increasingly larger units is far from trivial. We understand reasonably well the end points of such evolution, but have still a poor knowledge of the intermediate stages. A sodium atom has a very simple absorbtion spectrum, as shown schematically in Fig. 1.1. It consists of essentially one line in the visible, the well-known yellow light. In quantum physics this phenomenon is well described as a one-electron transition from a quantal state, known as 3s, to an excited state, denoted by 3p. A sodium crystal, on the other hand, has a completely different spectrum. The absorbtion is strong in the infrared, goes through a minimum in the visible, and then rises again in the ultraviolet. The reason for this is that very low-energy photons can excite electrons from the continuum of states just below the energy of the last occupied state to states just above the Fermi energy. The strong ultraviolet absorbtion is caused by interband transitions.

Suppose we chip off a small corner of the crystal, producing a micro-crystal that contains only 8 atoms. Again, the absorbtion spectrum changes completely. A relatively broad absorbtion maximum displaying different lines (resonances) appears in the visible (see also Chap. 8 and Fig. 8.8). This absorbtion is due to a collective excitation of outer electrons called a plasmon excitation (collective excitation of valence electrons). It can be viewed as a collective sloshing motion of the electrons from one side of the microcrystal to the other side. Such a motion has a strong dipole moment, behaving as an antenna which gives rise to a strong absorbtion band in the visible (see Fig. 1.2).

The absorbtion of light by a sodium atom is well described within the picture of independent particle motion. In this picture each electron moves in the atom as if it was alone. It is only when an electron tries to leave the atom that it feels that its motion is confined due to the presence of a wall produced by the pullings and pushings of all the other electrons and of the positively charged atomic nuclei. That is, in the independent particle model,

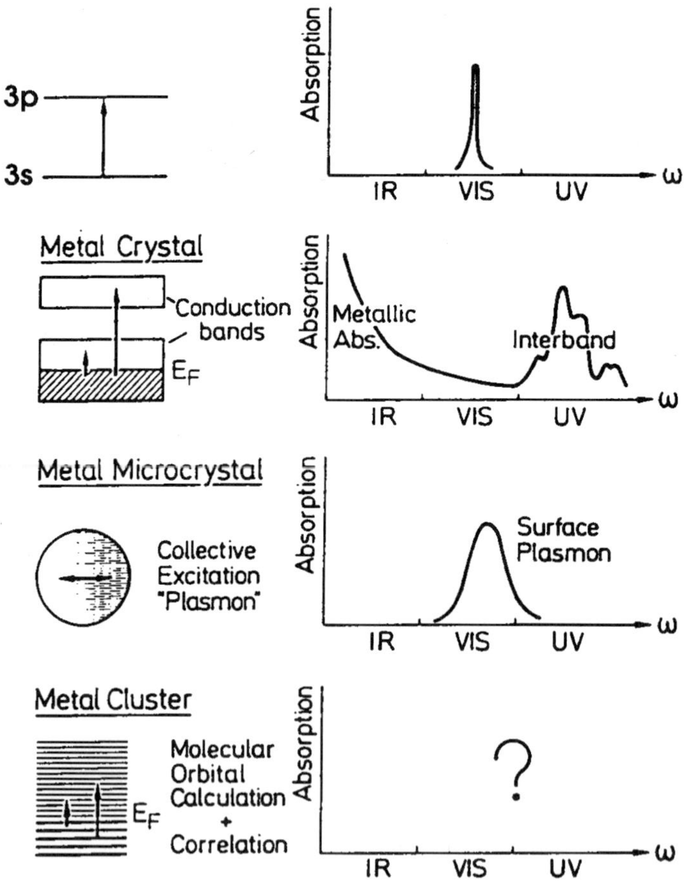

Fig. 1.1. Absorption of light by sodium in different stages of aggregation (after [9])

the electrons move in the average potential generated by the interaction with the other electrons and with the ions. This picture, however, cannot describe even qualitatively the resonance phenomenon (plasmon excitation) that dominates the optical response of an 8–atom cluster. Between one atom and 8 atoms we must completely change our way of looking at the optical response of the system [9, 10].

Because the response to light is quite different in Na_8 than in bulk Na, one would expect that a material made out of small sodium aggregates will display quite a different coulour than a sodium crystal (see Chap. 5). It is likely that it will also display a number of other differences, in particular regarding specific heat, conductivity, elasticity etc. (see Chaps. 7, 8). This is also expected to be true for other metallic aggregates (e.g. Ag- and Au- clusters; see also [11]).

Some of the most promising building blocks of nanophase materials are fullerenes (see Fig. 1.3), molecules which have been found to maintain most of their intrinsic characteristics when placed inside an infinite crystalline lattice, whether they form van der Waals or covalent solids [2, 3, 12]. Colour, strength, transport properties and other features of these materials depend

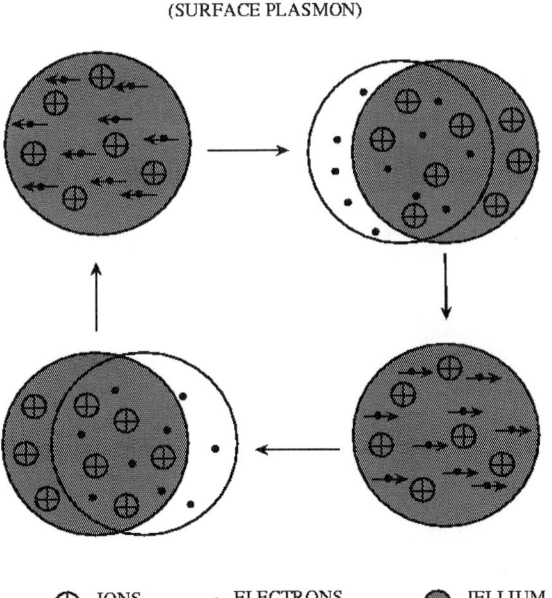

Fig. 1.2. Schematic representation of the collective motion of valence electrons (plasmons) in a metal cluster (after [10])

on the properties of the isolated molecules, in particular on the strength
with which electrons couple to ionic vibrations (phonons). The mechanisms
which are at the basis of this coupling are also important in understanding
the properties of carbon nanotubes and of linear carbon chains, remarkable
examples of molecular quantum wires. This is in keeping with the fact that
nanotubes can be obtained by bisecting a fullerene molecule at the equator,
joining the resulting hemispheres with a cylindrical tube made out of an arbitrary number of belts, each built out of benzoid rings [13]. While single-wall
C_{20}-derived nanotubes [14, 15] (see also Sect. 9.3.3) are the thinnest capped
nanotubes which can be thought of, linear chains made out of carbon atoms
are the ultimate examples of atomic quantum wires, playing a role in such
different phenomena as the presence of diffuse interstellar bands (DIF) in the
spectra of distant stars [16, 17, 18], and, arguably, in the field emission of
open-end carbon nanotubes [19, 20, 21, 22, 23, 24] (see also [25]).

The basic starting point in the study of atomic aggregates like metal
clusters, fullerenes, nanotubes, linear carbon chains etc., is mean field theory,
one of the most useful approximations in all of physics and chemistry (see,
e.g., [10]). In it we replace the many-particle Schrödinger equation by a single-

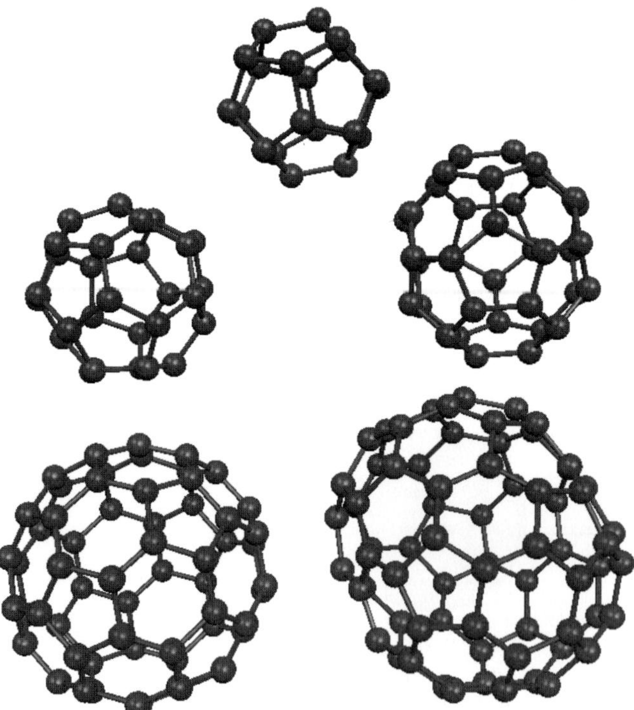

Fig. 1.3. Selected fullerenes: molecular structure for (from top to bottom and from
left to right) C_{20}, C_{28}, C_{36}, C_{60} and C_{70}

particle Scrödinger equation. What started out as a problem too complicated to solve for all but the smallest systems becomes quite manageable when one deals with one particle at a time. The many-body effects come in via the single-particle potential which is generated from the particles themselves. The resulting self-consistent mean field theory has been enormously successful in many domains of quantum mechanics. Whenever a many-particle system exhibits single-particle behaviour, which happens quite commonly, mean field theory is likely to be useful.

In the case of metal clusters, fullerenes, nanotubes, linear carbon chains etc., there are a number of observations which testify to the validity of the independent particle picture. In fact, the dependence of quantities like mass abundance and ionization potential vary strongly with the number of atoms (see Figs. 1.4 and 1.5 and Refs. [5, 26]). In particular, the energy needed to free one electron from the cluster exhibits a distinct shell structure with

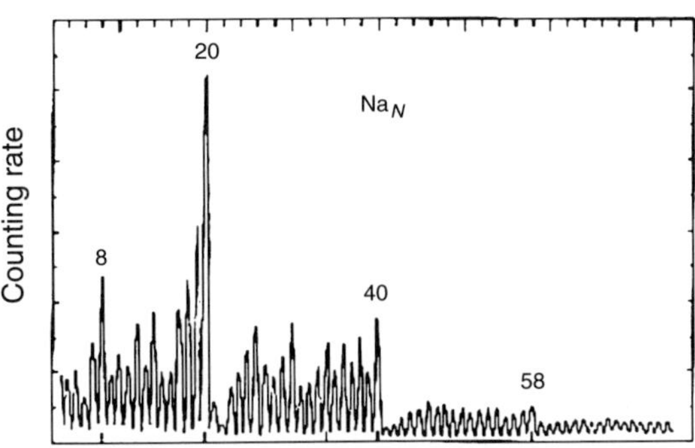

Fig. 1.4. Na clusters abundance (after [5])

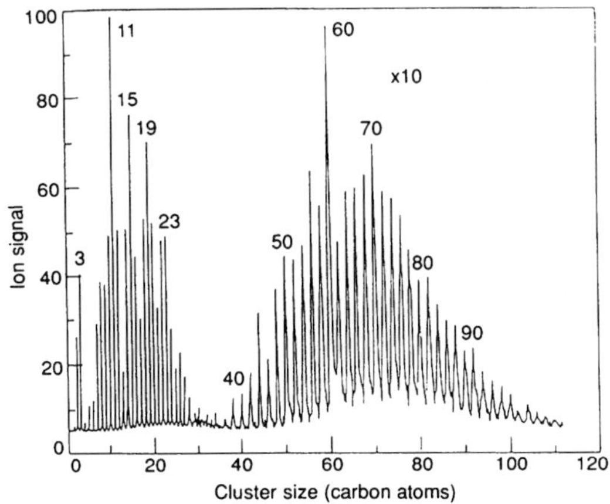

Fig. 1.5. Carbon clusters abundance (after [3])

extreme values at prescribed number of electrons. These discontinuities are strongly connected with the independent particle picture already mentioned. In fact, the orbits in which an electron moves in an average potential display a bunchiness as a function of the number of electrons which can be viewed as layers envelopping the atomic nucleus, much like the configuration of an onion. These layers or shells each contain a number of orbits and thus a given number of electrons, the so-called magic numbers. Special stability is adscribed to the filling of each of these shells. In particular in the case of C_{60}, the single-particle energy gap between the highest occupied molecular orbital (HOMO) and the lowest unoccupied molecular orbital (LUMO) is $1.8{\pm}0.1$ eV [27], a quantity much larger than room temperature (≈ 25 meV), a fact which explains the stability of this system.

In the following Chapter we shall provide an overview of the main contents of the present monography, while in Chap. 3 the tools to carry out detailed *ab initio* calculations (Density Functional Theory (DFT) and Local Density Approximation (LDA)) of the electronic motion will be developed. In Chap. 4, we discuss the time-dependent extension of LDA (TDLDA). Applications of these techniques are carried out in Chap. 5 to the case of the optical response of fullerenes, linear carbon chains, nanotubes and small metal clusters. In Chap. 6 we illustrate how to calculate the phonon spectrum and in Chap. 7 we briefly review the unification of the DFT and molecular dynamics techniques achieved by the Car–Parrinello scheme. As examples of the theory developed in these Chapters, we calculate the phonon spectrum of fullerenes, linear chains and metal clusters. The electron–phonon interaction and the first principle techniques used to calculate it are discussed in Chap. 8. Also discussed in this Chapter is the electron–plasmon coupling. In particular, the consequences this coupling has on the optical response of metallic clusters. In Chap. 9, we apply the techniques developed in previous Chapters to calculate the transition temperature between the normal and the superconducting phase in doped fullerides. We also touch upon the possibility of creating high T_c fullerides based on the small fullerenes and on small metallic aggregates.

2. Overview

We start with the many-body Hamiltonian of a system of N-particles[1]

$$H = -\frac{\hbar^2}{2m_\mathrm{e}} \sum_{l=1}^{N} \nabla_l^2 + \frac{1}{2} \sum_{l \neq m=1}^{N} v(|\boldsymbol{r}_l - \boldsymbol{r}_m|). \tag{2.1}$$

The quantities which one would like to obtain by solving this Hamiltonian, and which are useful for the interpretation of experimental results, are related to both the ground state and the excited states of the many-electron system. For example, calculation of the cohesive energy will require knowledge of the ground state total energy, while the interpretation of a spectroscopic measurement (e.g., photoemission) will also require knowledge of excited state energies. Such excitations can be, in general, of a collective character (i.e., involving the correlated motion of a large number of particles); however, there is strong evidence which testifies to the fact that, in many-fermion systems, a large class of excitations can be described in term of "quasiparticles", i.e., in terms of one single particle[2] moving in the *mean field* produced by the action of all other particles. To extract this field from the above Hamiltonian, one has to eliminate one coordinate from the two-body interaction. The simplest way to do this is in a classical approximation, that is, to freeze it,

$$\boldsymbol{r}_m \to \boldsymbol{r}_m^{\,0} \text{ (fixed value of the coordinate).}$$

Thus

$$H' = \sum_l H_l,$$

$$H_l = -\frac{\hbar^2}{2m_\mathrm{e}} \nabla_l^2 + \sum_m v(|\boldsymbol{r}_l - \boldsymbol{r}_m^{\,0}|),$$

where $\boldsymbol{r}_l$ is a generic coordinate which we shall call $\boldsymbol{r}$. Consequently,

---

[1] In what follows m_e and M_n will denote the electron and the ion masses, respectively. Moreover, we neglect for simplicity the possible presence of an external potential $V_\mathrm{ext}(r)$, as that which will be included, e.g., in (2.18) or (3.1).

[2] The properties of these single particles, however, will deviate from those of truly free particles, since the response of the whole system is reflected in their properties.

$$H_l \to H_{\mathrm{sp}} = -\frac{\hbar^2}{2m_{\mathrm{e}}}\nabla_r^2 + U(r),$$

$$U(r) = \sum_m v(|\boldsymbol{r} - \boldsymbol{r}_m^0|) \approx N v(|\boldsymbol{r} - \boldsymbol{r}^{\,0}|), \tag{2.2}$$

where $\boldsymbol{r}^0$ indicates the center-of-mass coordinate of the system and $U(r)$ the mean field potential. Now, in quantum mechanics the position of a particle is determined by the squared modulus of the wavefunction describing its motion. Thus, instead of the replacement $\boldsymbol{r}_m \to \boldsymbol{r}_m^0$ we introduce

$$\sum_m v(|\boldsymbol{r} - \boldsymbol{r}_m|) \to \sum_{i \in \{\mathrm{occ}\}} \int \mathrm{d}^3 r' \, |\varphi_i(\boldsymbol{r}')|^2 \, v(|\boldsymbol{r} - \boldsymbol{r}'|),$$

where the sum is over the occupied single-particle states. In this case

$$H_l = -\frac{\hbar^2}{2m_{\mathrm{e}}}\nabla_l^2 + \sum_{i \in \{\mathrm{occ}\}} \int \mathrm{d}^3 r' |\varphi_i(\boldsymbol{r}')|^2 v(|\boldsymbol{r}_l - \boldsymbol{r}'|). \tag{2.3}$$

One can then write (see Fig. 2.1, (1) and (3))

$$H_{\mathrm{sp}} = -\frac{\hbar^2}{2m_{\mathrm{e}}}\nabla^2 + v_{\mathrm{H}}(r), \tag{2.4}$$

where

$$v_{\mathrm{H}}(\boldsymbol{r}) = \int \mathrm{d}^3 r' \varrho(\boldsymbol{r}')v(|\boldsymbol{r} - \boldsymbol{r}'|), \tag{2.5}$$

and

$$\varrho(\boldsymbol{r}) = \sum_{i \in \{\mathrm{occ}\}} |\varphi_i(\boldsymbol{r})|^2, \tag{2.6}$$

leading to the so called Hartree equations

$$\left[-\frac{\hbar^2}{2m_{\mathrm{e}}}\nabla^2 + \sum_{i \in \{\mathrm{occ}\}} \int \mathrm{d}^3 r' \varphi_i^*(\boldsymbol{r}')v(|\boldsymbol{r} - \boldsymbol{r}'|)\varphi_i(\boldsymbol{r}') \right] \varphi_j(\boldsymbol{r}) = \varepsilon_j \varphi_j(\boldsymbol{r}). \tag{2.7}$$

The quantity $\varrho(\boldsymbol{r})$ is the density of the system, while v_{H} is the mean field Hartree potential. The N-particle solution of the system can then be written as

$$\Psi(\boldsymbol{r}_1 \ldots \boldsymbol{r}_N) = \prod_{m=1}^{N} \varphi_m(\boldsymbol{r}_m). \tag{2.8}$$

Because the particles we are dealing with are fermions, no two of them can occupy the same quantum state. Consequently, product wavefunctions have to be antisymmetrized. In particular, one has to replace the product $\varphi_i(\boldsymbol{r}')\varphi_j(\boldsymbol{r})$ appearing in the second term of the left hand side of (2.7) by

$$\varphi_i(\boldsymbol{r}')\varphi_j(\boldsymbol{r}) - \varphi_j(\boldsymbol{r}')\varphi_i(\boldsymbol{r}).$$

In this way, a particle does not interact with itself and all fermions become indistiguishable. By the same token, the wavefunction introduced in (2.8) becomes

$$\Psi(\boldsymbol{r}_1\dots\boldsymbol{r}_N) = \frac{1}{\sqrt{N!}}\det[\varphi_1(\boldsymbol{r}_1)\dots\varphi_N(\boldsymbol{r}_N)], \tag{2.9}$$

that is, a normalized determinant. One then gets

$$\left[-\frac{\hbar^2}{2m_{\mathrm{e}}}\nabla^2 + v_{\mathrm{H}}(r)\right]\varphi_j(\boldsymbol{r}) + \int \mathrm{d}^3 r'U_{\mathrm{x}}(\boldsymbol{r},\boldsymbol{r}')\varphi_j(\boldsymbol{r}') = \varepsilon_j\varphi_j(\boldsymbol{r}), \tag{2.10}$$

i.e., Hartree–Fock equations (HF equations), where (see Fig. 2.1, (2) and (4))

$$U_{\mathrm{x}}(\boldsymbol{r},\boldsymbol{r}') = -\sum_{i\in\{\mathrm{occ}\}}\varphi_i^*(\boldsymbol{r}')v(|\boldsymbol{r}-\boldsymbol{r}'|)\varphi_i(\boldsymbol{r}), \tag{2.11}$$

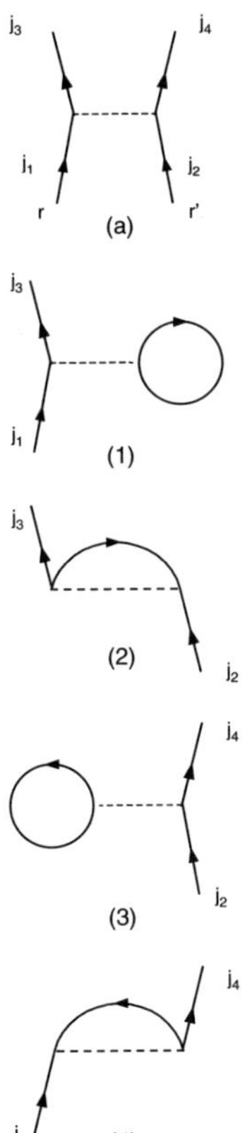

Fig. 2.1. (a): scattering of two electrons through the Coulomb interaction. (1) and (3): contributions to the (direct) Hartree potential (see (2.5) and (2.6)). (2) and (4): contributions to the (exchange) Fock-potential (see (2.11))

is the exchange (Fock) potential. The wavefunction introduced in (2.9) is an approximation to the exact many-body wavefunction describing the quantum behaviour of N interacting electrons. More precisely, in this approximation one neglects electron–electron correlations beyond those induced by the Pauli principle.

From the computational point of view, a crucial reduction of the complexity of the problem is achieved, by reducing an N-body problem to N (coupled) one-body problems. A solution of the HF equations yields hence both an approximate value for the total energy of the system, and also an approximate solution for the quasiparticle states, which can be identified with the one-particle HF orbitals. A formal and rigourous demonstration can be given within the framework of Green's functions theory (see, e.g., [28]) making use of Koopman's theorem[3] (see e.g. [29]). While Hartree–Fock theory gives, as a rule, a quite accurate ground state energy[4] and the correct sequence of single-particle levels, it produces, as a rule, a too low density of levels around the Fermi energy ϵ_F (see, e.g., [30] and Fig. 2.2).

Another strategy to treat the ground-state problem of an N-electron system comes from Density Functional Theory (DFT), which will be introduced in the next Chapter. In that case, the mapping of the N-body problem into a set of (coupled) one-body problems is obtained as a consequence of the Kohn–Sham (KS) approach to the DFT. In a first instance, this mapping bears only a formal resemblance to the mean field equations described above[5]. However, once the problem of the determination of ground and excited states is considered in the framework of Green's function theory (see e.g. [33]), it is possible to identify the KS-DFT effective potential (see (3.23) below) with an approximate local and energy-independent self-energy (see (2.49)), or, to say it more precisely, one can consider the exchange and correlation potential of the Density Functional Theory (see (3.23)) as an approximation to $\tilde{U}(r)$ and m^* in (2.53). This will be discussed in more detail in Sect. 2.9.

2.1 k–Mass

In what follows we shall elaborate on the meaning and consequences of the non-local potential U_{x}, making use of the time-dependent Schrödinger equation

$$i\hbar\frac{\partial\varphi(x,t)}{\partial t} = -\frac{\hbar^2}{2m_{\mathrm{e}}}\nabla^2\varphi(x,t) + \int \mathrm{d}x' U_{\mathrm{x}}(x,x')\varphi(x',t) \qquad (2.12)$$

---

[3] This theorem states that the energy ε of the highest occupied state is equal to the difference between the ground state energies of the N- and $(N-1)$-particle systems (aside from correction of the order of ε/N).

[4] Given by $\sum_i \varepsilon_i - \frac{1}{2}\int \mathrm{d}r v_{\mathrm{H}}(r)\varrho(r) - \frac{1}{2}\int \mathrm{d}r\mathrm{d}r' U_{\mathrm{x}}(r,r') \sum_{i\in\{\mathrm{occ}\}} \varphi_i^*(r')\varphi_i(r)$.

[5] In principle, the solution of the KS equations yields only the ground state properties [31], and no Koopman's theorem applies to the KS-DFT [32].

(we treat a one-dimensional system for the sake of simplicity). Let us assume a uniform system of volume V (which eventually goes to infinity). In such a case

$$\varphi(x,t) \sim \frac{1}{\sqrt{V}} \mathrm{e}^{\mathrm{i}(kx-\omega t)}.$$

Inserting this wavefunction in (2.12) one obtains

$$\mathrm{i}\hbar(-\mathrm{i}\omega)\varphi(x,t) = -\frac{\hbar^2}{2m_\mathrm{e}}(\mathrm{i}k)^2\varphi(x,t) + \int \mathrm{d}x' U_\mathrm{x}(x,x')\varphi(x',t).$$

Multiplying from the left by $\varphi^*(x,t)$ and integrating over x leads to

$$\hbar\omega = \frac{\hbar^2 k^2}{2m_\mathrm{e}} + \int \frac{\mathrm{d}x\,\mathrm{d}x'}{V}\, U_\mathrm{x}(x,x')\mathrm{e}^{-\mathrm{i}k(x-x')}. \tag{2.13}$$

Making use of the coordinate transformation connecting the set (x,x') with the relative and center-of-mass set $(y = x - x',\, Y = (x+x')/2)$, together with $|\partial(y,Y)/\partial(x,x')| = 1$ and the fact that U_x does not depend on Y, one can write

$$\hbar\omega = \frac{\hbar^2 k^2}{2m_\mathrm{e}} + \tilde{U}_\mathrm{x}(k), \tag{2.14}$$

where $\tilde{U}_\mathrm{x}(k)$ is the Fourier transform of U_x. We now impose the requirement that the system displays a single-particle dispersion relation like that of a free particle. This condition can, in principle, be fulfilled by introducing an effective mass (the so-called k-mass), that is,

$$\hbar\omega = \frac{\hbar^2 k^2}{2m_k}.$$

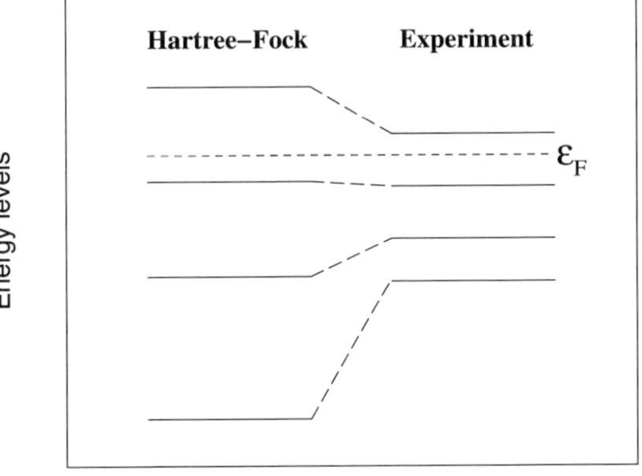

Fig. 2.2. Schematic representation of the single-particle energy levels of a finite system around the Fermi energy ε_F predicted by Hartree–Fock theory in comparison with experiment (see [30] and Refs. therein)

Making use of the relations

$$\frac{\mathrm{d}}{\mathrm{d}k}\hbar\omega = \frac{\hbar^2 k}{m_k}, \qquad \text{and} \qquad \frac{\mathrm{d}}{\mathrm{d}k}\hbar\omega = \frac{\hbar^2 k}{m_e} + \frac{\partial \tilde{U}_{\mathrm{x}}}{\partial k},$$

in keeping with the fact that we are calculating an inertia, one obtains

$$\frac{\hbar^2 k}{m_k} = \frac{\hbar^2 k}{m_e} + \frac{\partial \tilde{U}_{\mathrm{x}}}{\partial k},$$

leading to

$$m_k = m_e \left(1 + \frac{m_e}{\hbar^2 k}\frac{\partial \tilde{U}_{\mathrm{x}}}{\partial k}\right)^{-1}. \tag{2.15}$$

Summing up one can, for many purposes, take into account approximately some of the effects of the non-local exchange potential, in terms of an effective mass (k-mass) [30],

$$\left(-\frac{\hbar^2}{2m_k}\nabla^2 + U'(r)\right)\varphi_j(\boldsymbol{r}) = \varepsilon_j\varphi_j(\boldsymbol{r}), \tag{2.16}$$

where

$$U'(\boldsymbol{r}) = \frac{m_e}{m_k}U(\boldsymbol{r}). \tag{2.17}$$

Because, as a rule, $m_k < m_e$, one can understand from the above relations the result schematically displayed in Fig. 2.2 (see also Sect. 2.2.1). In what follows, we shall illustrate the fact that $m_k < m_e$ by calculating explicitly the k-mass in the case of electrons in a metal, making use of a simplified treatment (i.e. the *jellium* model) to describe the role played by the ions.

2.2 Electrons in Metals

In the electron gas, the ions are assumed to be frozen and their charge is smeared out to form the positively charged background. In real solids, the background is neither uniform nor frozen. The band structure renders the Fermi surface non spherical: it also introduces a contribution to the effective mass of the electrons which reflects the curvature of the energy bands. Since we consider a uniform system here, we do not discuss this effect: it is indeed quite small in e.g. Na [34].

The Schrödinger equation describing the motion of electrons in a metal, leaving the ions fixed (see Sect. 3.1) in their spatial positions, can be written as

$$H\Psi = -\sum_i \frac{\hbar^2}{2m_e}\nabla_i^2\Psi - Ze^2\sum_n \frac{1}{|\boldsymbol{r} - \boldsymbol{R}_n|}\Psi + \frac{1}{2}\sum_{i \neq j}\frac{e^2}{|\boldsymbol{r}_i - \boldsymbol{r}_j|}\Psi$$

$$= E\Psi, \tag{2.18}$$

where

$$\Psi = \Psi(\boldsymbol{r}_1, s_1, \ldots, \boldsymbol{r}_N, s_N).$$

The set of quantities $\{\boldsymbol{R}_n\} \equiv (\boldsymbol{R}_1, \boldsymbol{R}_2 \ldots \boldsymbol{R}_{N_{\mathrm{at}}})$ denote the positions of the N_{at} ions, while $\boldsymbol{r}_i$ and s_i stands for the spatial coordinate and the spin of the i-th electron, respectively. The Hartree equations associated with the Hamiltonian H introduced in (2.18) are

$$\left(-\frac{\hbar^2}{2m_{\mathrm{e}}} \nabla^2 + U_{\mathrm{ion}}(\boldsymbol{r}) + U_{\mathrm{el}}(\boldsymbol{r}) \right) \varphi_j(\boldsymbol{r}) = \varepsilon_j \varphi_j(\boldsymbol{r}), \tag{2.19}$$

where

$$U_{\mathrm{el}}(\boldsymbol{r}) = e^2 \sum_{i \in \{\mathrm{occ}\}} \int \mathrm{d}^3 r' \, |\varphi_i(\boldsymbol{r}')|^2 \frac{1}{|\boldsymbol{r} - \boldsymbol{r}'|}, \tag{2.20}$$

and

$$U_{\mathrm{ion}}(\boldsymbol{r}) = -Ze^2 \sum_n \frac{1}{|\boldsymbol{r} - \boldsymbol{R}_n|}. \tag{2.21}$$

Taking into account the Pauli principle one obtains the Hartree–Fock equations,

$$\left(-\frac{\hbar^2}{2m_{\mathrm{e}}} \nabla^2 + U_{\mathrm{ion}}(\boldsymbol{r}) + U_{\mathrm{el}}(\boldsymbol{r}) \right) \varphi_j(\boldsymbol{r}) + \int \mathrm{d}^3 r' U_{\mathrm{x}} \, \varphi_j(\boldsymbol{r}') = \varepsilon_j \varphi_j(\boldsymbol{r}), \tag{2.22}$$

where

$$U_{\mathrm{x}}(\boldsymbol{r}, \boldsymbol{r}') = -e^2 \sum_{i \in \{\mathrm{occ}\}} \varphi_i^*(\boldsymbol{r}') \frac{1}{|\boldsymbol{r} - \boldsymbol{r}'|} \varphi_i(\boldsymbol{r}),$$

is the exchange (Fock) potential (see (2.11)).

2.2.1 Jellium Model

Let us spread uniformely the positive charge of the ions throughout the system. One can then write

$$U_{\mathrm{ion}}(r) = -Ze^2 \int \frac{\mathrm{d}^3 r'}{V} \frac{1}{|\boldsymbol{r} - \boldsymbol{r}'|}.$$

In the case of a uniform, three dimensional system,

$$\varphi_j(\boldsymbol{r}) = \frac{1}{\sqrt{V}} \mathrm{e}^{\mathrm{i}\boldsymbol{k}_j \cdot \boldsymbol{r}}.$$

Equation (2.20) thus becomes,

$$U_{\mathrm{el}} = e^2 \sum_{i \in \{\mathrm{occ}\}} \int \frac{\mathrm{d}^3 r'}{V} \frac{1}{|\boldsymbol{r} - \boldsymbol{r}'|} = Ze^2 \int \frac{\mathrm{d}^3 r'}{V} \frac{1}{|\boldsymbol{r} - \boldsymbol{r}'|},$$

and

$$U_{\text{ion}} + U_{\text{el}} = 0.$$

Consequently, only the exchange term survives, that is,

$$-\frac{\hbar^2}{2m_{\text{e}}}\nabla^2\varphi_j(\boldsymbol{r}) - \sum_{i\in\{\text{occ}\}}\int \mathrm{d}^3r'\,\frac{e^2}{|\boldsymbol{r}-\boldsymbol{r}'|}\varphi_i^*(\boldsymbol{r}')\varphi_i(\boldsymbol{r})\varphi_j(\boldsymbol{r}') = \varepsilon_j\varphi_j(\boldsymbol{r}). \tag{2.23}$$

In other words, within the jellium approximation, the system is bound because the electrons are fermions.

Let us now work out in detail the exchange term, making again use of the eigenfunctions of the infinite system (plane waves),

$$-\frac{1}{V}\sum_{i\in\{\text{occ}\}}\int \mathrm{d}^3r'\,\frac{e^2}{|\boldsymbol{r}-\boldsymbol{r}'|}\,\mathrm{e}^{\mathrm{i}(\boldsymbol{k}_j-\boldsymbol{k}_i)\cdot\boldsymbol{r}'}\,\varphi_i(\boldsymbol{r})$$

$$= -\frac{1}{V}\sum_{i\in\{\text{occ}\}}\int \mathrm{d}^3r'\,\frac{e^2}{|\boldsymbol{r}-\boldsymbol{r}'|}\,\mathrm{e}^{\mathrm{i}(\boldsymbol{k}_j-\boldsymbol{k}_i)\cdot(\boldsymbol{r}'-\boldsymbol{r})}\,\mathrm{e}^{\mathrm{i}(\boldsymbol{k}_j-\boldsymbol{k}_i)\cdot\boldsymbol{r}}\,\varphi_i(\boldsymbol{r})$$

$$= -\frac{1}{V}\sum_{i\in\{\text{occ}\}}\int \mathrm{d}^3r'\,\frac{e^2}{|\boldsymbol{r}-\boldsymbol{r}'|}\,\mathrm{e}^{\mathrm{i}(\boldsymbol{k}_j-\boldsymbol{k}_i)\cdot(\boldsymbol{r}'-\boldsymbol{r})}\,\varphi_j(\boldsymbol{r}). \tag{2.24}$$

Let us now calculate the integral making use of the change of variables $\boldsymbol{\xi} = \boldsymbol{r}' - \boldsymbol{r}$, $\mathrm{d}^3\xi = \mathrm{d}^3r'$ ($\boldsymbol{r}$ fixed),

$$\int \mathrm{d}^3\xi\,\frac{e^2}{\xi}\,\mathrm{e}^{\mathrm{i}(\boldsymbol{k}_j-\boldsymbol{k}_i)\cdot\boldsymbol{\xi}} = \frac{4\pi e^2}{(\boldsymbol{k}_j-\boldsymbol{k}_i)^2}. \tag{2.25}$$

Consequently, the exchange term becomes

$$-\frac{1}{V}\sum_{i\in\{\text{occ}\}}\frac{4\pi e^2}{(\boldsymbol{k}_j-\boldsymbol{k}_i)^2}\,\varphi_j(\boldsymbol{r})$$

$$= -\frac{1}{V}\sum_{k'<k_{\text{F}}}\frac{4\pi e^2}{(\boldsymbol{k}-\boldsymbol{k}')^2}\,\varphi_j(\boldsymbol{r}) = -\int_{k'<k_{\text{F}}}\frac{\mathrm{d}^3k'}{(2\pi)^3}\,\frac{4\pi e^2}{(\boldsymbol{k}-\boldsymbol{k}')^2}\,\varphi_j(\boldsymbol{r})$$

$$= v_{\text{x}}^{\text{hom}}(k)\,\varphi_j(\boldsymbol{r}), \tag{2.26}$$

where

$$v_{\text{x}}^{\text{hom}}(k) = -\frac{2e^2}{\pi}\,k_{\text{F}}\,F(k/k_{\text{F}}), \tag{2.27}$$

with (see Fig. 2.3)

$$F(x) = \frac{1}{2} + \frac{1-x^2}{4x}\,\ln\left|\frac{1+x}{1-x}\right|. \tag{2.28}$$

Equation (2.23) now reads

$$\left(\frac{\hbar^2 k_j^2}{2m_{\text{e}}} - \frac{2e^2}{\pi}\,k_{\text{F}}\,F(k/k_{\text{F}})\right)\varphi_j(\boldsymbol{r}) = \epsilon_j\varphi_j(\boldsymbol{r}). \tag{2.29}$$

Multiplying from the left with $\varphi_j^*(\boldsymbol{r})$ and integrating over the whole volume one obtains,

$$\varepsilon(k) = \frac{\hbar^2 k^2}{2m_{\mathrm{e}}} - \frac{2e^2}{\pi}\, k_{\mathrm{F}}\, F(k/k_{\mathrm{F}}) = \hbar\omega.$$

Comparing this Equation with (2.14) leads to

$$\tilde{U}_{\mathrm{x}}(k) = -\frac{2e^2}{\pi}\, k_{\mathrm{F}}\, F(k/k_{\mathrm{F}}). \tag{2.30}$$

Because

$$\lim_{x\to 1} \frac{\partial F}{\partial x} = -\infty, \tag{2.31}$$

the associated k-mass vanishes at the Fermi energy, that is (see (3.20.8) of [30]),

$$\lim_{k\to k_{\mathrm{F}}} m_k = 0.$$

This result implies that there are no states at finite energy, in strong disagreement with experimental values of the specific heat which indicates that the effective mass m^* of the electrons at the Fermi energy has the same magnitude as the bare mass m_{e} [35]. As we discuss below, this result does not depend on the poor treatment of the ions (jellium model), but on the fact that the collective degrees of freedom associated with the electrons (plasmons) (see Chaps. 4 and 5) and their coupling to the electrons (see Chap. 8) have been neglected.

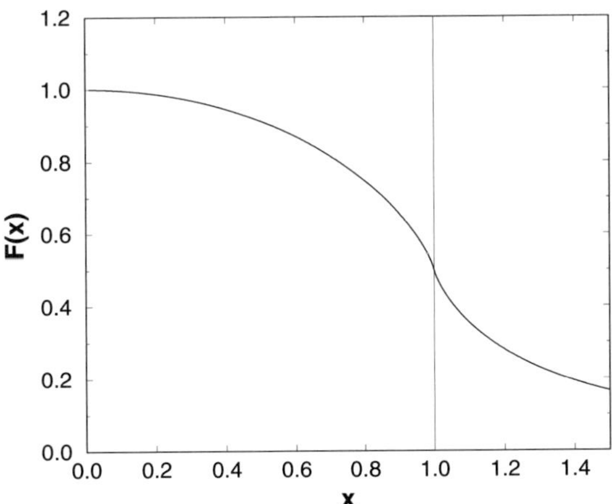

Fig. 2.3. The function $F(x)$ in (2.28) versus x

2.3 Electron Fluctuations (Plasmons)

The fact that

$$\lim_{k \to k_{\mathrm{F}}} m_k = 0,$$

can be traced back to the divergence of the Fourier transform of the Coulomb interaction (see (2.25)) at $q = |\boldsymbol{k}_j - \boldsymbol{k}_i| = 0$. This, in turn, reflects the infinite range of the Coulomb interaction. If this interaction was replaced by a screened interaction of the form $e^2 \mathrm{e}^{-q_0 r}/r$, then its Fourier trasform would be the function $4\pi e^2/(q + q_0)^2$, the $q = 0$ divergence would be eliminated, and the unphysical singularity of $\partial \tilde{U}_{\mathrm{x}}(k)/\partial k$ removed (see Sect. 2.7).

The failure of Hartree–Fock approximation is due to the fact that all many-body correlations beyond the static ones associated with the Pauli principle are neglected. It is known that an interacting many-electron system displays dynamical correlation effects in its ground state. In fact, while an isolated particle only displays the fluctuations associated with Heisenberg's relation $(\Delta x \Delta p \geq \hbar)$, existing between the momentum and the coordinate, a many-body system displays, in addition, the effects of the zero point motion associated with its collective excitations. As an illustration of this point, let us consider the harmonic oscillator as a simple picture of the many-body system. Making use of the Dirac representation (see Appendix 2.11.1)

$$H_{\mathrm{osc}} = \sum_{\alpha} \hbar \omega_\alpha \left(\Gamma_\alpha^\dagger \Gamma_\alpha + \frac{1}{2} \right),$$

$$\Gamma_\alpha^\dagger |0\rangle_{\mathrm{B}} = |n_\alpha = 1\rangle, \quad \Gamma_\alpha |0\rangle_{\mathrm{B}} = 0,$$

where $\Gamma_\alpha^\dagger$ and Γ_α are the creation and annihilation phonon operators, obeying $[\Gamma_\alpha, \Gamma_{\alpha'}^\dagger] = \delta(\alpha, \alpha')$, $\hbar \omega_\alpha$ is the energy of the one-phonon state $|n_\alpha = 1\rangle$, while $|0\rangle_{\mathrm{B}}$ is the vacuum (ground) state. Thus,

$$H_{\mathrm{osc}} |0\rangle_{\mathrm{B}} = \frac{1}{2} \hbar \omega_\alpha |0\rangle_{\mathrm{B}}, \tag{2.32}$$

where $\hbar \omega_\alpha/2$ is the zero point energy of the system, to which it is associated a zero point fluctuation of the coordinate of the system equal to $(\hbar \omega_\alpha/2C_\alpha)^{1/2}$, C_α being the restoring force constant.

2.4 Electron–Plasmon Coupling

In what follows we shall consider correlations arising from the fluctuations of the mean field. In particular, we shall consider the coupling of plasmons to electrons. Let us render explicit the discussion making use of a finite system of radius R_0 (e.g., a C_{60} molecule or a Na_8 cluster considered schematically as a spherical system). Surface fluctuations lead to dynamic deformations which can be parametrized in terms of the multipole expansion [10]

$$R = R_0 \left(1 + \sum_{\lambda\mu} \alpha_{\lambda\mu}(t) Y^*_{\lambda\mu}(\hat{r}) \right), \tag{2.33}$$

where $\alpha_{\lambda\mu}(t)$ is a dynamic variable which in Dirac's notation can be written as

$$\alpha_{\lambda\mu} = \sqrt{\frac{\hbar\omega_\lambda}{2C_\lambda}}\,(\Gamma^\dagger_{\lambda\mu} + \Gamma_{\lambda\mu}), \tag{2.34}$$

being the coordinate of the harmonic oscillator which describes the small amplitude vibrations of the system[6].

Expanding the mean (Hartree) field of the (finite) system to first order in α one obtains (assuming $\alpha^2 \ll \alpha$),

$$U(\boldsymbol{r} - \boldsymbol{R}) = U(\boldsymbol{r} - \boldsymbol{R}_0) + \delta U(\boldsymbol{r}),$$

where

$$\delta U = -R_0 \frac{\partial U}{\partial r} \sum_{\lambda\mu} \alpha_{\lambda\mu} Y^*_{\lambda\mu}. \tag{2.35}$$

Consequently, the Schrödinger equation describing the motion of the electrons becomes

$$(H_0 + \delta U)\varphi_j(\boldsymbol{r}) = \varepsilon_j \varphi_j(\boldsymbol{r}), \tag{2.36}$$

where

$$H_0 = -\frac{\hbar^2}{2m_k}\nabla^2 + U(\boldsymbol{r}). \tag{2.37}$$

Let us now solve $H_0 + \delta U$ in perturbation theory. The matrix element coupling the particles to the collective vibrations of the electrons is (see Fig. 2.4 and Sect. 8.6)

$$\langle n_\alpha = 1,\ j'|\delta U|j\rangle = \sqrt{\frac{\hbar\omega_\alpha}{2C_\alpha}}\,\langle j'|R_0\frac{\partial U}{\partial r}Y^*_{\lambda\mu}(\hat{r})|j\rangle, \tag{2.38}$$

where $\sqrt{\hbar\omega_\alpha/2C_\alpha}$ is the matrix element $\langle n_\alpha = 1|\alpha|0\rangle_{\mathrm{B}}$, between the (boson) ground state (vacuum) and the one phonon state $|n_\alpha = 1\rangle$. This coupling leads, within second-order perturbation theory, to an energy (self-energy) correction (see Fig. 2.5 and Appendix 2.11.2). In what follows we shall be interested in the situation $\hbar\omega \approx \varepsilon_{j'} + \hbar\omega_\alpha$, in which case

$$\Sigma_j(\omega) \approx \sum_{j',\alpha} \frac{|\langle n_\alpha = 1, j'|\delta U|j\rangle|^2}{\hbar\omega - (\varepsilon_{j'} + \hbar\omega_\alpha)}, \tag{2.39}$$

---

[6] The same expansion (2.33) is also valid in the case in which the system displays a static, e.g. quadrupole, deformation. This is measured in the case of atomic nuclei and for axially symmetric systems by $\beta_2 = \sqrt{5}\alpha_{20} = (R_\| - R_\perp)/R_\perp$ (see e.g. [10] and refs. therein). $R_\|$ and $R_\perp$ are the largest and the smallest radii (parallel and perpendicular with respect to the symmetry axis), while β_2 is the reduced matrix element of the $\alpha_{2\mu}$ tensor ($\sqrt{5}\langle 2\mu|\alpha_{2\mu}|0\rangle$).

associated with the (virtual) excitation, and eventual reabsorption, of vibrational modes (plasmons) by an electron. Σ_j is also known as the mass operator [30].

Noted that (2.39) is not well defined, as the energy denominator may vanish. This problem can be circumvented by making an extension into the complex plane, that is, by replacing

$$\varepsilon_{j'} + \hbar\omega_\alpha \to E_{\alpha'} + i\frac{\Delta}{2}, \tag{2.40}$$

(which amounts to averaging the energy of the state by means of a Lorentzian function having width equal to Δ). We then obtain

$$\Sigma_j(\omega) = \Delta E_j(\omega) - \frac{i}{2}\Gamma_j(\omega) \tag{2.41}$$

where

$$\Delta E_j(\omega) = \lim_{\Delta \to 0} \sum_{j',\alpha'} \frac{(\hbar\omega - E_{\alpha'})|\langle n_\alpha = 1, j'|\delta U|j\rangle|^2}{(\hbar\omega - E_{\alpha'})^2 + (\Delta/2)^2}, \tag{2.42}$$

and

$$\Gamma_j(\omega) = \lim_{\Delta \to 0} \sum_{\alpha'} \frac{\Delta|\langle n_\alpha = 1, j'|\delta U|j\rangle|^2}{(\hbar\omega - E_{\alpha'})^2 + (\Delta/2)^2}, \tag{2.43}$$

are the real and imaginary part of the self energy $\Sigma_j(\omega)$.

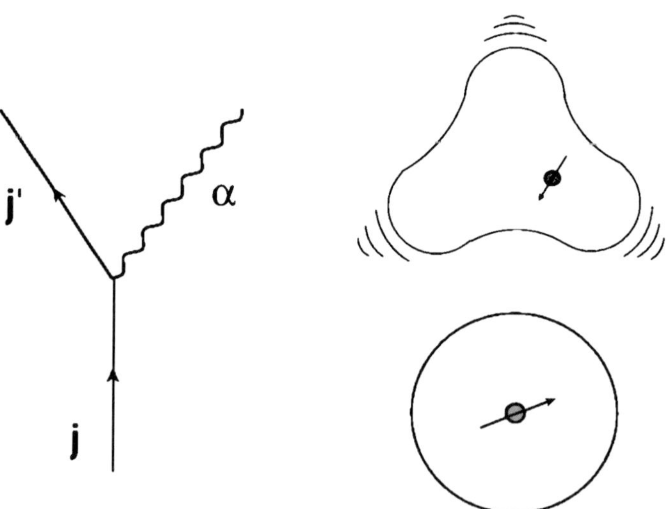

Fig. 2.4. Particle–vibration coupling vertex. The arrowed line represents the fermion while the wavy line the boson (plasmon in this case). To the right, we schematically represent the bouncing inelastically off of the electron on the cluster surface which is set into vibration

Let us first discuss this latter contribution. Making use of the fact that

$$\lim_{\Delta \to 0} \frac{\Delta}{(E - E_0)^2 + (\Delta/2)^2} = 2\pi\delta(E - E_0), \tag{2.44}$$

and replacing $|\langle n_\alpha = 1, j'|\delta U|j\rangle|^2$ by the average value of the different squared matrix elements which we shall call v^2, (2.43) becomes

$$\Gamma_j(\omega) = 2\pi v^2 \sum_{\alpha'} \delta(\hbar\omega - E_{\alpha'}). \tag{2.45}$$

This is the Golden rule for the decay of a simple state j into a set of more complicated states α' lying at the same energy (real, on-the-energy-shell transition).

The single-particle Schrödinger equation now reads

$$H_0\varphi_j(\boldsymbol{r}) = (\tilde{\varepsilon}_j - \frac{\mathrm{i}}{2}\Gamma_j(\omega))\varphi_j(\boldsymbol{r}), \tag{2.46}$$

where H_0 has been defined in (2.37) and where $\tilde{\varepsilon}_j = \varepsilon_j + \Delta E_j(\omega)$. Making use of the time-dependent Schrödinger equation $H_0\varphi_j(\boldsymbol{r},t) = \mathrm{i}\hbar\partial\varphi_j(\boldsymbol{r},t)/\partial t$ one obtains

$$\varphi_j(\boldsymbol{r},t) = \mathrm{e}^{-\mathrm{i}(\tilde{\varepsilon}_j - \mathrm{i}\Gamma_j(\omega)/2)t/\hbar}\varphi_j(\boldsymbol{r}), \tag{2.47}$$

yielding

$$|\varphi_j(\boldsymbol{r},t)|^2 = \mathrm{e}^{-\Gamma_j(\omega)t/\hbar}. \tag{2.48}$$

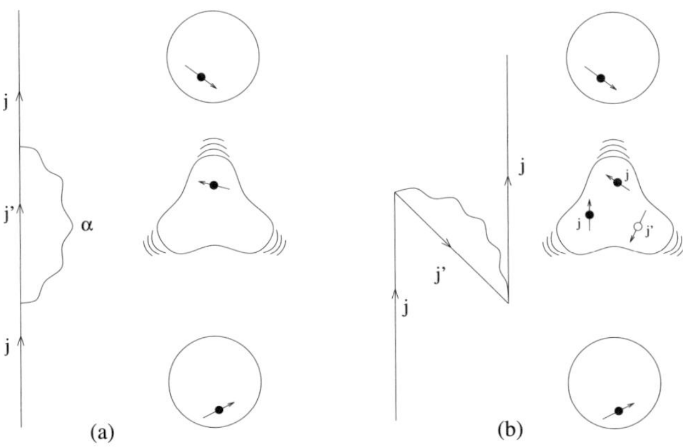

Fig. 2.5. The lowest order process by which the motion of a fermion is renormalized by the coupling to a vibration (boson). In (a), the particle excites the vibration by bouncing inelastically off from the surface. Particles are represented by an arrowed line (solid arrowed dot). In (b), the vibration is excited by a virtual process (vacuum fluctuation). The surface vibration is drawn as a wavy line. Time increases moving upwards

The quantity $\tau = \hbar/\Gamma_j(\omega)$ is the quasi-particle lifetime: in other words, if at $t = 0$ the particle was with probability 1 in the state j, at time τ this probability has more than halved (i.e., $|\varphi(\boldsymbol{r}, t)|^2 = 1/e$) due to the coupling to states containing an electron and a plasmon (see Fig. 2.5).

2.5 ω–Mass

In what follows we concentrate our attention on single-particle states with energies close to the Fermi energy. Thus, energy conserving transitions of the type required by the relation (2.45) are not available for single-particle states and the associated Schrödinger equation reads

$$\left(-\frac{\hbar^2}{2m_e}\nabla^2 + U(r) + \Delta E_j(\omega)\right)\varphi_j(\boldsymbol{r}) = \varepsilon_j\varphi_j(\boldsymbol{r}), \tag{2.49}$$

where $\Delta E_j(\omega)$ is known as the self-energy (real part). Let us again play the same game as with m_k, i.e., consider an infinite three dimensional system and assume $\varphi_j(\boldsymbol{r}) \sim e^{i(\boldsymbol{k}\cdot\boldsymbol{r}-\omega t)}$. Then, the time dependent Schrödinger equation leads to

$$\hbar\omega = \frac{\hbar^2 k^2}{2m_e} + \Delta E(k, \omega). \tag{2.50}$$

If we still want to conserve the independent-particle description we can do it again in terms of an effective mass, the so-called ω–mass m_ω, given by the relation [30]

$$\hbar\omega = \frac{\hbar^2 k^2}{2m_\omega}.$$

Equating the first derivative of this equation and of (2.50), that is,

$$\frac{\partial(\hbar\omega)}{\partial k} = \frac{\hbar^2 k}{m} + \frac{\partial\Delta E(k, \omega)}{\partial(\hbar\omega)}\frac{\partial(\hbar\omega)}{\partial k},$$

or equivalently,

$$\frac{\partial(\hbar\omega)}{\partial k}\left(1 - \frac{\partial\Delta E(k, \omega)}{\partial(\hbar\omega)}\right) = \frac{\hbar^2 k}{m_e},$$

and

$$\frac{\partial(\hbar\omega)}{\partial k} = \frac{\hbar^2 k}{m_\omega}, \tag{2.51}$$

one obtains

$$m_\omega = m_e\left(1 - \frac{\partial\Delta E(k, \omega)}{\partial(\hbar\omega)}\right). \tag{2.52}$$

This relation defines the ω-mass. Note that the left-hand side of (2.51) can be interpreted as the rate of change in energy when the momentum changes or, equivalently, when the number of nodes allows for length changes. Since the latter can be used to label the single-particle states, the energy spacings between levels decreases for increasing values of the effective mass. Thus, the effective mass is proportional to the density of single-particle levels.

2.6 Total Effective Mass

Parametrizing simultaneously the Fock potential and the effect arizing from the coupling of the single-particle degrees of freedom to fluctuations of the mean field (plasmons) in terms of an effective mass, one obtains

$$\left(-\frac{\hbar^2}{2m^*}\nabla^2 + \tilde{U}(r) \right) \varphi_j(\boldsymbol{r}) = \varepsilon_j \varphi_j(\boldsymbol{r}), \tag{2.53}$$

where

$$\tilde{U}(r) = \frac{m_e}{m^*}U(r)$$

and where

$$m^* = \frac{m_k m_\omega}{m_e}. \tag{2.54}$$

Note that m^* can be of the order of or larger than m_e, even if $m_k < m_e$. This is because m_ω can be considerably larger than m_e. Consequently, (2.53) gives, as a rule, not only the right sequence of single-particle levels but also the correct order of magnitude of the density of states around the Fermi energy (related to the specific heat, see Sect. 2.10).

While (2.53), with the definition of m^* given by (2.54), looks very much like the equation describing the independent particle motion of electrons with bare mass m_e, it differs from it in at least one fundamental way. In fact, the occupation probability of single-particle states around the Fermi energy, is given by [30],

$$Z_\omega = \frac{m_e}{m_\omega}. \tag{2.55}$$

Because $m_\omega > m$, $Z_\omega < 1$, reflecting the fact that the particle spends only a fraction of the time in the state in which it started with probability 1 at $t = 0$ (see Fig. 2.5).

2.7 Induced Interaction

When an isolated particle, bouncing inelastically off the surface excites a plasmon, there is only one way for the particle to make the process possible: to reabsorb the plasmon at a later stage (virtual process, see Fig. 2.5). In the presence of another particle, the plasmon can be absorbed by this second particle, leading to an induced interaction with matrix elements (see Fig. 2.6(b))

$$M_{\mathrm{ind}}(\nu_3\nu_4,\nu_1\nu_2) = \langle \nu_3\nu_4|\delta U|\nu_1\nu_2\rangle$$
$$= \sum_\alpha \frac{\langle \nu_4|\delta U|\nu_2, n_\alpha = 1\rangle\langle \nu_3, n_\alpha = 1|\delta U|\nu_1\rangle}{E - (\varepsilon_{\nu_3} + \varepsilon_{\nu_2} + \hbar\omega_\alpha)}. \tag{2.56}$$

Because the sign of $\langle \nu_3, n_\alpha = 1|\delta U|\nu_1\rangle$ is, as a rule, equal to the sign of $\langle \nu_2, n_\alpha = 1|\delta U|\nu_4\rangle$, and $E - (\varepsilon_{\nu_3} + \varepsilon_{\nu_2} + \hbar\omega_\alpha) < 0$ for particles close to the Fermi energy, $M_{\mathrm{ind}} < 0$.

Consequently, the sum of the Coulomb interaction and of the induced interaction given in (2.56) leads to a *screened* Coulomb potential (see Fig. 2.6(c)),

$$U_{\mathrm{C}} = \frac{V(q)}{\kappa_0(q,\omega)},$$

where

$$V(q) = \frac{4\pi e^2}{q^2},$$

is the bare Coulomb interaction (see (2.25) and Fig. 2.6(a)), while $\kappa_0(q,\omega)$ is the dielectric function which turns out to be larger than 1. In the static limit $(\omega \to 0)$ of the Thomas–Fermi approximation [35], one can write

$$\kappa_0(q,0) = 1 + \frac{k_{\mathrm{s}}^2}{q^2},$$

where the square of the screening wavenumber is $k_{\mathrm{s}}^2 = 6\pi\varrho e^2/\varepsilon_{\mathrm{F}}$, and ϱ is the density of the system. Consequently,

$$U_{\mathrm{C}} = \frac{4\pi e^2}{q^2 + k_{\mathrm{s}}^2}. \tag{2.57}$$

Summing up, the coupling of electrons to plasmons has, among other things, the effect of renormalizing the heat capacity of the system, by changing the (effective) mass of the electrons, and of reducing the effective range of the Coulomb interaction.

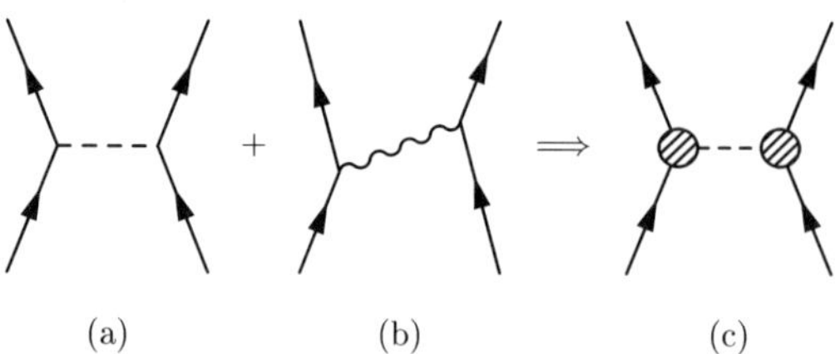

(a)   (b)   (c)

Fig. 2.6. The sum of the bare Coulomb interaction (a) and of the induced plasmon interaction (b), leads to a screened Coulomb potential (c)

2.8 Examples

The electronic density of simple metals can be expressed in terms of the
Wigner–Seitz radius $r_0 = r_s a_B$ defined as the radius of a sphere whose volume
is equal to the volume per conduction electron. Thus

$$\frac{V}{N} = \frac{1}{\varrho} = \frac{4\pi r_0^3}{3} \quad \text{and} \quad r_0 = \left(\frac{3}{4\pi\varrho}\right)^{1/3}. \tag{2.58}$$

Typical values of r_s for simple metals range from 2 to 5 (see e.g. [35]).

In Fig. 2.7 we display both $\tilde{U}_x$ (see (2.30)) as well as $\Delta E(k,\omega)$ (see (2.42))
calculated in an approximation called the Random Phase Approximation
(RPA) in which the plasmons are treated as harmonic vibrations (see Chap.
4). $\tilde{U}_x$ displays a vertical slope at $k = k_F$, in keeping with (2.28) and (2.31).
This feature is eliminated by the screening induced by the coupling to plasmons. In fact $\Delta E(k,\omega)$, which is the correction to the potential energy of
the electrons, is quite flat at $k \approx k_F$. Making use of the relations given in
(2.15) and (2.52), one obtains the result displayed in Fig. 2.8. Also displayed
in this figure, is the total effective mass defined in (2.54), while in Fig. 2.9 we
display the associated momentum distribution, and in Fig. 2.10 we compare
the theoretical results with experiment. To be noted that in the case of the
electron gas, the screening corrections are so important that there exists no
enhancement on m^* near the Fermi momentum.

2.9 Minimal Mean Field Theory

The simplest mean field equations needed to describe quantitatively the experimental findings for levels close to the Fermi energy ($\Gamma_j \approx 0$, see (2.45)),

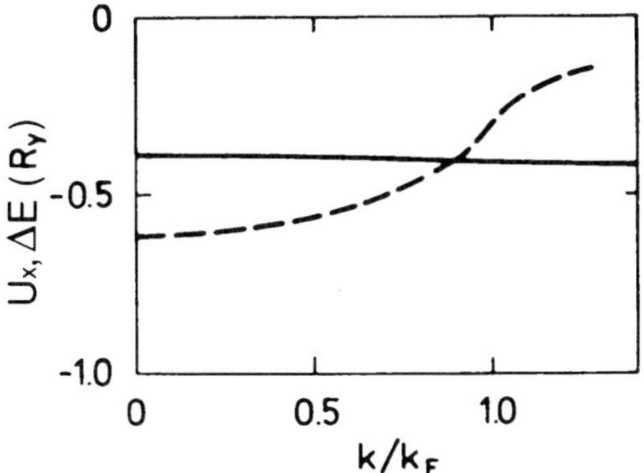

Fig. 2.7. Adapted from [28]. Dependence upon k/k_F of the real part $\Delta E(k,\omega)$ of
the mass operator for an electron gas with $r_s = 4$. The dashed curve represents the
Fock contribution U_x (2.26), and the full line gives $\Delta E(k,\omega)$ in the Random Phase
Approximation (RPA, see Chap. 4), all in units of Ry ($=13.6058$ eV)

namely,

$$\left[-\frac{\hbar^2}{2m_{\mathrm{e}}}\nabla^2 + U(r) + \Delta E_j(k,\omega)\right]\varphi_j(\boldsymbol{r}) + \int \mathrm{d}^3 r' U_{\mathrm{x}}(\boldsymbol{r},\boldsymbol{r}')\varphi_j(\boldsymbol{r}')$$
$$= \varepsilon_j \varphi_j(\boldsymbol{r})\,, \tag{2.59}$$

are non-local both in space (see Fig. 2.1, (2) and (4)) and in time (see Fig. 2.5). In the previous Sections we have discussed a possible strategy to make the above equation local, through the introduction of an effective mass m^*, product of the k- and of the ω-mass (see (2.53) and (2.54)), and of the associated occupation discontinuity Z at the Fermi energy (see (2.55)). While this is the main strategy used in dealing with the single-particle spectrum of atomic nuclei [30], we shall in the next Section discuss different approximations, largely employed in condensed matter physics, to make (2.59) local (see also Chap. 3).

2.9.1 Non-locality in Space of the Exchange Contribution

To make (2.59) local in space, one can assume that

$$\int \mathrm{d}^3 r' U_{\mathrm{x}}(\boldsymbol{r},\boldsymbol{r}')\varphi_j(\boldsymbol{r}') = -\int \mathrm{d}^3 r' v(|\boldsymbol{r}-\boldsymbol{r}'|) \sum_{i \in \{\mathrm{occ}\}} \varphi_i^*(\boldsymbol{r}')\varphi_i(\boldsymbol{r})\varphi_j(\boldsymbol{r}')$$

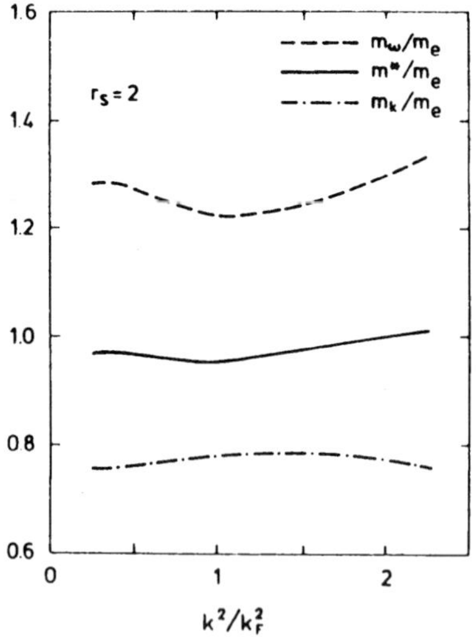

Fig. 2.8. Adapted from [36]. Dependence upon k^2/k_{F}^2 of the effective mass m^*/m_{e} (*full line*), of the ω-mass m_ω/m_{e} (*long dashes*) and of the $k-$mass m_k/m_{e} (*dash-and-dot*), as calculated from the Random Phase Approximation of the correlated basis function approach, for an electron gas with $r_{\mathrm{s}} = 2$. To be noted that in this case the definition is $m_k = m(1 - \frac{m}{\hbar^2 k}\frac{\partial \Sigma(k,E)}{\partial k})^{-1}$ with $\Sigma(k,E) = \tilde{U}_{\mathrm{x}}(k) + \Delta E(k,E)$

$$\approx v_{\mathrm{x}}(\boldsymbol{r})\,\varphi_j(\boldsymbol{r}). \qquad (2.60)$$

A useful parametrization for the local potential $v_{\mathrm{x}}(\boldsymbol{r})$ is that obtained within the Local Density Approximation (LDA) of the Density Functional Theory (DFT) (see also [37]),

$$v_{\mathrm{x}}^{\mathrm{LDA}}(\boldsymbol{r}) = -e^2\,\frac{1}{\pi}\,\left(3\pi^2\varrho(\boldsymbol{r})\right)^{1/3} \approx -0.9848\,e^2\,\varrho^{1/3}, \qquad (2.61)$$

which is discussed in detail in Chap. 3 (see Sect. 3.3).

Typically, $\varrho \approx 10^{22}$ cm^{-3} in a metal (see Table 1.1 in [35]). Using (2.58) one obtains $r_0 = (4\pi\varrho/3)^{-1/3} \approx 3$ Å. From the relation $e^2 = 2\mathrm{Ry}\,a_{\mathrm{B}} \approx 14.4$ Å eV, one obtains,

$$v_{\mathrm{x}}(\boldsymbol{r}) \approx -3.1 \text{ eV}, \qquad (2.62)$$

a number which can be compared with the bare Coulomb interaction between two electrons at a distance of 3 Å,

$$\frac{e^2}{3\text{Å}} \approx 4.8 \text{ eV}. \qquad (2.63)$$

In the above expression of e^2 use was made of Bohr's radius $a_{\mathrm{B}} = \hbar^2/m_{\mathrm{e}}e^2$ (=0.529 Å) and of the energy unit Ry$= e^2/2a_{\mathrm{B}} = \hbar^2/2m_{\mathrm{e}}a_{\mathrm{B}}^2$ (=13.6058 eV).

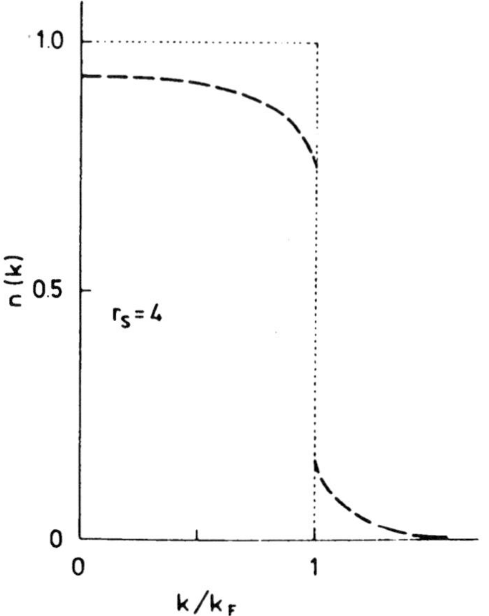

Fig. 2.9. Adapted from [38]. Momentum distribution in the electron gas for $r_{\mathrm{s}} = 4$. The *dashed curve* corresponds to the random phase approximation [39]

2.9.2 Non-locality in Time of the Correlation Contribution

The approximation is made that

$$\Delta E_j(k,\omega) \approx \frac{\partial E_{\text{corr}}(\varrho)}{\partial \varrho}, \tag{2.64}$$

where $E_{\text{corr}}(\varrho)$ is the correlation energy of the interacting electron gas. This is often calculated parametrizing the results of the infinite, homogeneous system as in the LDA described in the next Chapter (see Sect. 3.3). A possible simple parametrization of $E_{\text{corr}}(\varrho)$ leads to [41, 42]

$$v_{\text{c}} = \frac{\partial E_{\text{corr}}(\varrho)}{\partial \varrho} = -0.91 \ln\left(1 + \frac{11.41}{r_{\text{s}}}\right) \text{eV}, \tag{2.65}$$

which for $r_{\text{s}}=4$ leads to $v_{\text{c}} = -1.23$ eV.

2.10 Phonons

In the previous Sections we have essentially dealt with the electronic degrees of freedom and have frozen the degrees of freedom of the ions. It is well known that the lattice can vibrate leading to phonons (see Chap. 6). The coupling of the phonons to electrons (see Chap. 8) renormalize their properties in an important way as testified by the specific heat (associated with processes of the type displayed in Fig. 2.5 where now the *wavy line* represents the phonons) and by the phenomenon of superconductivity (associated with processes similar to those displayed in Fig. 2.6(b), where again the *wavy line* now represents the phonon).

In fact, at low temperature the total specific heat per mole of a (non magnetic) metal can be written as

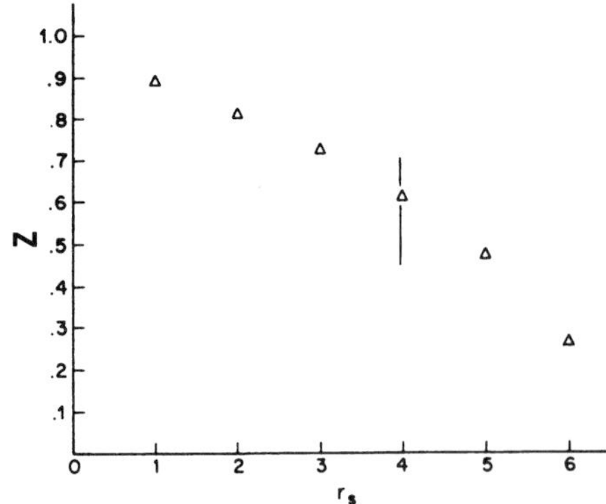

Fig. 2.10. Taken from [40]. Dependence upon r_{s} of the quantity $Z(k_{\text{F}})$ (see (2.55)). The calculated values (open triangles) are from [39]. The vertical bar shows the value deduced from the analysis of the X–ray Compton profile of Na [40]

$$C = C_\mathrm{e} + C_\mathrm{ph}, \tag{2.66}$$

sum of an electron and of a phonon contribution. The electron term can be written as $C_\mathrm{e} = \gamma T$, since for the free electron gas

$$\gamma^{(0)} = \frac{\pi^2 k_\mathrm{B}^2 N_0 Z}{3\varrho} N^{(0)}(0), \tag{2.67}$$

where k_B is the Boltzmann's constant, N_0 Avogadro's number, while $N^{(0)}(0) = 3\varrho/2\varepsilon^{(0)}\mathrm{F}$ is the density of levels at the Fermi energy. It is convenient to express γ for a real metal as

$$\gamma = \frac{N(0)}{N^{(0)}(0)}\gamma^{(0)} = \frac{\tilde{m}}{m^{(0)}}\gamma^{(0)}. \tag{2.68}$$

In the above expressions the superscript $^{(0)}$ denotes the free electron value. In particular $m^{(0)}$ stands for m^* (see Sects. 2.1 and 2.5). Assuming this effective mass to be $\approx m_\mathrm{e}$ (i.e. $m^{(0)} \approx m_\mathrm{e}$) one can write $\varepsilon^{(0)} = \hbar^2 k_\mathrm{F}^2/2m_\mathrm{e}$.

Consequently, since the theoretical value $\gamma^{(0)}$ is proportional to the density of levels $N^{(0)}(0)$ at the Fermi energy, which in turn is proportional to m_e, it proves convenient to define a specific heat effective mass[7] $\tilde{m} = (1 + \lambda)m_\mathrm{e}$, so that $\tilde{m}/m_\mathrm{e}$ is the ratio of the measured γ to the free electron gas value $\gamma^{(0)}$, that is,

$$\frac{\gamma}{\gamma^{(0)}} = 1 + \lambda \tag{2.69}$$

with

$$\gamma^{(0)} = \frac{\pi^2 k_\mathrm{B}^2 N_0 m}{\hbar^2 (3\pi^2 \varrho)^{2/3}}, \tag{2.70}$$

(see also Sect. 2.5).

The quantity λ is the so called mass enhancement factor (see Sects. 8.4 and 8.9.2). It arizes from processes where the phonon dresses the electron (see Fig. 2.5). The expression of λ is $-\partial(\Sigma_j)_\mathrm{phon}/\partial(\hbar\omega)$, formally identical to that appearing in (2.52) but where the coupling is now to the phonon and not to the plasmon. Typical values of $(1 + \lambda)$ arizing from specific heat measurements normalized with respect to the free electron estimates are of the order of 1.3 for Na, Mg, Al and ≈ 2 for Pb and Hg (see Table 2.3 of [35], Table 1.1 of [43], and also Table 9.2).

The exchange of phonons between the electrons in processes like the one displayed in Fig. 2.6(b), can lead to an attraction between pairs of electrons moving close to the Fermi energy, giving rise to the so called Cooper pairs,

---

[7] To be noted that in other finite many-body systems like e.g. the atomic nucleus, where there are not two types of vibrations (plasmons and phonons) as in the case of molecular aggregates, but only one type, associated with the collective response of the nucleons and thus equivalent to the plasmon type of oscillations, m^* is the total effective mass of the system (see Chaps. 8, 9 and [45] and references therein).

which behave as (quasi) bosons. The associated binding energy in the weak coupling situation ($\lambda \ll 1$) is

$$E = 2\hbar\omega_{\mathrm{D}}\,e^{-1/\lambda}, \qquad (2.71)$$

where ω_{D} is the Debye frequency of the system [44]. Cooper pair formation leads, at low temperatures to a condensation, and is at the basis of the phenomenon of superconductivity. As discussed in Chap. 9, a better aproximation to E is provided by the expression

$$E = 2\hbar\omega_{\mathrm{D}}e^{-1/(\lambda-\mu^*)}, \qquad (2.72)$$

where μ^* is a typical screened Coulomb potential (see Sect. 9.2 and Appendix 9.5.1).

2.11 Appendix

2.11.1 Second Quantization

Second quantization provides an economic representation of quantum mechanics which includes automatically the statistics fullfilled by the particles composing the system. This appendix summarizes some of the basic results for fermions and bosons.

Fermions

Let us consider a system of n identical fermions and let $\Psi(\boldsymbol{r}_1, \boldsymbol{r}_2, \ldots \boldsymbol{r}_n)$ denote the exact wavefunction of the system. The state $\Phi(\boldsymbol{r}_1, \boldsymbol{r}_2, \ldots \boldsymbol{r}_n)$ is a member of a complete set of n-particle wavefunctions. It is constructed as a properly antisymmetrized product of one-particle wavefunctions $\varphi_\nu(\boldsymbol{r})$, which form a complete orthonormal set

$$\int \mathrm{d}^3 r \; \varphi_\nu^*(\boldsymbol{r})\, \varphi_{\nu'}(\boldsymbol{r}) = \delta(\nu, \nu') \qquad (2.73)$$

$$\sum_\nu \varphi_\nu^*(\boldsymbol{r}')\, \varphi_\nu(\boldsymbol{r}) = \delta(\boldsymbol{r} - \boldsymbol{r}'). \qquad (2.74)$$

The function $\Phi(\boldsymbol{r}_1, \boldsymbol{r}_2, \ldots \boldsymbol{r}_n)$, in the case of fermions, is given by the determinant of the single particle wavefunctions

$$\Phi(\boldsymbol{r}_1, \boldsymbol{r}_2, \ldots \boldsymbol{r}_n) = \frac{1}{\sqrt{n!}} \det(\varphi_{\nu_1}(\boldsymbol{r}_1)\varphi_{\nu_2}(\boldsymbol{r}_2)\cdots\varphi_{\nu_n}(\boldsymbol{r}_n)). \qquad (2.75)$$

The function Ψ is thus a linear combination of determinants. Taking only one of them means to neglect all correlation effects beyond those arising from the Pauli principle.

We now introduce the creation and annihilation fermion operator $a_\nu^\dagger$ and a_ν respectively, acting on the fermion vacuum state $|0\rangle_{\rm F}$. These operators satisfy the anticommutation relation

$$\{a_\nu, a_{\nu'}^\dagger\} = a_\nu a_{\nu'}^\dagger + a_{\nu'}^\dagger a_\nu = \delta(\nu, \nu'), \tag{2.76}$$

and

$$\{a_\nu, a_{\nu'}\} = \{a_\nu^\dagger, a_{\nu'}^\dagger\} = 0. \tag{2.77}$$

This choice restricts the occupation number of the states ν to 0 or 1 as required by Fermi statistics and to antisymmetric normalized states. Acting with the creation operator $a_j^\dagger$ on the vacuum one creates a single particle state

$$a_j^\dagger |0\rangle_{\rm F} = |j\rangle, \tag{2.78}$$

whose r-representation coincides with the single-particle wavefunction

$$\langle \boldsymbol{r}|j\rangle = \varphi_j(\boldsymbol{r}).$$

The orthonormalization condition is

$$\langle j|j'\rangle = {}_{\rm F}\langle 0|a_j a_{j'}^\dagger|0\rangle_{\rm F} = {}_{\rm F}\langle 0|\delta(j, j') - a_{j'}^\dagger a_j|0\rangle_{\rm F}$$
$$= \delta(j, j') \equiv {}_{\rm F}\langle 0|\overline{a_j a_{j'}^\dagger}|0\rangle_{\rm F}, \tag{2.79}$$

where the relation given by (2.76) has been used together with

$$a_j|0\rangle_{\rm F} = 0, \tag{2.80}$$

and

$${}_{\rm F}\langle 0|0\rangle_{\rm F} = 1. \tag{2.81}$$

The symbol in the last term of (2.79) denotes a contraction. According to Wick's theorem, to calculate overlaps or matrix elements involving $a^\dagger$, and a, one should carry out all possible contractions between creation and annihilation operators, introducing a minus sign each time when in the contraction one jumps over an odd number of operators and a plus sign otherwise.

A two-particle wavefunction in this representations reads

$$a_j^\dagger a_{j'}^\dagger |0\rangle_{\rm F} = |j, j'\rangle_{\rm F}. \tag{2.82}$$

Making use of the anticommutation relation given in (2.77) one can show that

$$|j, j'\rangle = -|j', j\rangle, \tag{2.83}$$

that is, the two-particle state is antisymmetric. Consequently,

$$|j, j\rangle = 0, \tag{2.84}$$

that is, no two fermions can occupy the same quantal states, as required by the Pauli principle.

The orthonormalization condition of the state $|j, j'\rangle$ is given by the relation

$$\langle j_1, j_2 | j_1', j_2' \rangle = {}_{\text{F}}\langle 0 | a_{j_2} a_{j_1} a_{j_1'}^\dagger a_{j_2'}^\dagger | 0 \rangle_{\text{F}}$$

$$= \delta(j_2, j_2')\,\delta(j_1, j_1') - \delta(j_1, j_2')\,\delta(j_1', j_2). \tag{2.85}$$

This result can also be obtained directly without using Wick's theorem by making repeated use of the anticommutation relation given in (2.76). Equations (2.83), (2.84) and (2.85) indicate that

$$\langle \boldsymbol{r}, \boldsymbol{r}' | j_1 \, j_2 \rangle = \frac{1}{\sqrt{2}} \begin{vmatrix} \varphi_{j_1}(\boldsymbol{r}) & \varphi_{j_2}(\boldsymbol{r}) \\ \varphi_{j_1}(\boldsymbol{r}') & \varphi_{j_2}(\boldsymbol{r}') \end{vmatrix}. \tag{2.86}$$

Let us now calculate the matrix element of a two-body interaction

$$\langle j_1 \, j_2 | v | j_1', j_2' \rangle_a$$

$$= \frac{1}{2} \int \mathrm{d}^3 r \, \mathrm{d}^3 r' \begin{vmatrix} \varphi_{j_1}(\boldsymbol{r}) & \varphi_{j_2}(\boldsymbol{r}) \\ \varphi_{j_1}(\boldsymbol{r}') & \varphi_{j_2}(\boldsymbol{r}') \end{vmatrix}^* v(|\boldsymbol{r} - \boldsymbol{r}'|) \begin{vmatrix} \varphi_{j_1'}(\boldsymbol{r}) & \varphi_{j_2'}(\boldsymbol{r}) \\ \varphi_{j_1'}(\boldsymbol{r}') & \varphi_{j_2'}(\boldsymbol{r}') \end{vmatrix}$$

$$= \int \mathrm{d}^3 r \, \mathrm{d}^3 r' \, \varphi_{j_1}^*(\boldsymbol{r}) \, \varphi_{j_2}^*(\boldsymbol{r}') \, v(|\boldsymbol{r} - \boldsymbol{r}'|) \, \varphi_{j_1'}(\boldsymbol{r}) \, \varphi_{j_2'}(\boldsymbol{r}')$$

$$- \int \mathrm{d}^3 r \, \mathrm{d}^3 r' \, \varphi_{j_1}^*(\boldsymbol{r}') \, \varphi_{j_2}^*(\boldsymbol{r}) \, v(|\boldsymbol{r} - \boldsymbol{r}'|) \, \varphi_{j_1'}(\boldsymbol{r}') \, \varphi_{j_2'}^*(\boldsymbol{r}). \tag{2.87}$$

Note that this matrix element changes sign each time two particles are exchanged either in the initial or in the final states, that is,

$$\langle j_1, j_2 | v | j_1', j_2' \rangle_a = -\langle j_1, j_2 | v | j_2', j_1' \rangle_a$$

$$= -\langle j_2, j_1 | v | j_1', j_2' \rangle_a = \langle j_2, j_1 | v | j_2', j_1' \rangle_a. \tag{2.88}$$

Bosons

In the case of particles fulfilling Bose–Einstein statistics we introduce the boson operators $\Gamma_\alpha^\dagger$, Γ_α which create and annihilate a boson in a state α, and respect the commutation relations

$$[\Gamma_\alpha, \Gamma_{\alpha'}^\dagger] = \Gamma_\alpha \Gamma_{\alpha'}^\dagger - \Gamma_{\alpha'}^\dagger \Gamma_\alpha = \delta(\alpha, \alpha')\,, \tag{2.89}$$

and

$$[\Gamma_\alpha, \Gamma_{\alpha'}] = [\Gamma_\alpha^\dagger, \Gamma_{\alpha'}^\dagger] = 0. \tag{2.90}$$

Calling $|0\rangle_{\text{B}}$ the normalized boson vacuum state, that is,

$$_{\text{B}}\langle 0 | 0 \rangle_{\text{B}} = 1, \tag{2.91}$$

one obtains, by definition

$$\Gamma_\alpha | 0 \rangle_{\text{B}} = 0. \tag{2.92}$$

The one-phonon state is defined as

$$\Gamma_\alpha^\dagger |0\rangle_{\mathrm B} = |n_\alpha = 1\rangle \,, \tag{2.93}$$

where n_α indicate the number of phonons in the quantal state α. This state is normalized. In fact

$$\langle n_\alpha = 1 | n_{\alpha'} = 1\rangle = {}_{\mathrm B}\langle 0|\Gamma_\alpha \Gamma_{\alpha'}^\dagger |0\rangle_{\mathrm B}$$

$$= {}_{\mathrm B}\langle 0|\big(\delta(\alpha,\alpha') + \Gamma_{\alpha'}^\dagger \Gamma_\alpha\big)|0\rangle_{\mathrm B} = {}_{\mathrm B}\langle 0|\Gamma_\alpha \Gamma_{\alpha'}^\dagger|0\rangle = \delta(\alpha,\alpha'). \tag{2.94}$$

In the last step of the above equation we have set in evidence the (single) possible contractions between creation and annihilation operators. In this case, Wick's theorem applies as in the case of fermions but using a plus sign in all cases, irrespective of how many operators are jumped over in carrying out the contraction. The commutation relation given in (2.90) implies that

$$\Gamma_\alpha^\dagger \Gamma_\alpha^\dagger |0\rangle_{\mathrm B} \neq 0 \,, \tag{2.95}$$

that is, bosons can occupy the same quantal state. Let us now work out the orthonormalization of this two-phonon state by carrying out all contractions

$${}_{\mathrm B}\langle 0|\Gamma_{\alpha'} \Gamma_{\alpha'} \Gamma_\alpha^\dagger \Gamma_\alpha^\dagger|0\rangle_{\mathrm B} = \delta(\alpha,\alpha')\,\delta(\alpha,\alpha') + \delta(\alpha,\alpha')\,\delta(\alpha,\alpha')$$

$$= 2\delta(\alpha,\alpha'). \tag{2.96}$$

Consequently, the two boson state

$$|n_\alpha = 2\rangle = \frac{1}{\sqrt{2}}\Gamma_\alpha^\dagger \Gamma_\alpha^\dagger |0\rangle_{\mathrm B} \,, \tag{2.97}$$

is a normalized state. Note that

$$\Gamma_\alpha^\dagger |n_\alpha = 1\rangle = \Gamma_\alpha^\dagger \Gamma_\alpha^\dagger |0\rangle_{\mathrm B} = \sqrt{2}\,|n_\alpha = 1\rangle \,, \tag{2.98}$$

and, in general,

$$\Gamma_\alpha^\dagger |n_\alpha = N\rangle = \sqrt{N+1}\,|n_\alpha\rangle = |N+1\rangle. \tag{2.99}$$

We shall now write the harmonic oscillator Hamiltonian in second quantization as originally done by Dirac,

$$H = \sum_{\alpha'} \hbar\omega_{\alpha'}\Big(\Gamma_{\alpha'}^\dagger \Gamma_{\alpha'} + \frac{1}{2}\Big). \tag{2.100}$$

The energy of the ground state (vacuum state) is

$$H|0\rangle_{\mathrm B} = E_0, \tag{2.101}$$

where

$$E_0 = \frac{1}{2}\sum_{\alpha'} \hbar\omega_{\alpha'}, \tag{2.102}$$

It receives a $\frac{1}{2}\hbar\omega_\alpha$ contribution (zero point fluctuation) for any degree of freedom of the system.

Because the coordinate associated with the harmonic motion described by the Hamiltonian defined in (2.100) is

$$\alpha = \sum_\alpha \sqrt{\frac{\hbar\omega_\alpha}{2C_\alpha}}\,(\Gamma_\alpha^\dagger + \Gamma_\alpha) \tag{2.103}$$

the zero point fluctuation associated with the energy (2.102) is

$$_{\rm B}\langle 0|\alpha_\alpha^2|0\rangle_{\rm B}^{1/2} = \left(\sum_\alpha \frac{\hbar\omega_\alpha}{2C_\alpha}\right)^{1/2}. \tag{2.104}$$

The one-phonon state has an energy

$$\begin{aligned}
H|n_\alpha = 1\rangle &= \sum_{\alpha'} \hbar\omega_{\alpha'}\left(\Gamma_{\alpha'}^\dagger \Gamma_{\alpha'} + \frac{1}{2}\right)\Gamma_\alpha^\dagger|0\rangle_{\rm B} \\
&= \sum_\alpha \hbar\omega_{\alpha'}\Gamma_{\alpha'}^\dagger \Gamma_{\alpha'}\Gamma_\alpha^\dagger\,|0\rangle_{\rm B} + E_0\Gamma_\alpha^\dagger|0\rangle_{\rm B} \\
&= (\hbar\omega_\alpha + E_0)|n_\alpha = 1\rangle.
\end{aligned} \tag{2.105}$$

We shall now calculate the commutator

$$[H, \Gamma_\alpha^\dagger] = \sum_\alpha \hbar\omega_{\alpha'}[\Gamma_{\alpha'}^\dagger \Gamma_{\alpha'}, \Gamma_\alpha^\dagger]. \tag{2.106}$$

Making use of the relation

$$[AB, C] = A[B, C] + [B, C]A, \tag{2.107}$$

one obtains

$$\begin{aligned}
[H, \Gamma_\alpha^\dagger] - \sum_{\alpha'} \hbar\omega_{\alpha'}\left(\Gamma_{\alpha'}^\dagger[\Gamma_\alpha, \Gamma_\alpha^\dagger][\Gamma_{\alpha'}^\dagger, \Gamma_\alpha^\dagger]\Gamma_{\alpha'}\right) &= \sum_{\alpha'} \hbar\omega_{\alpha'}\Gamma_{\alpha'}^\dagger\delta(\alpha, \alpha') \\
&= \hbar\omega_\alpha\Gamma_\alpha^\dagger.
\end{aligned} \tag{2.108}$$

2.11.2 Single-Particle Self-Energy

The dressing of the single-particle motion due to the coupling to collective vibrations (see Fig. 2.5) can be described in terms of the self-energy mass operator

$$\begin{aligned}
\Sigma_j(\omega) = &\sum_{j',\alpha} |\langle n_\alpha = 1, j'|\delta U|j\rangle|^2 \\
&\times \left(\frac{1}{\hbar\omega - (\varepsilon_{j'} + \hbar\omega_\alpha)} + \frac{1}{\hbar\omega + (\varepsilon_{j'} - \hbar\omega_\alpha)}\right),
\end{aligned} \tag{2.109}$$

where the energies are measured from the Fermi energy. Thus,

$$\frac{1}{\hbar\omega - (\varepsilon_{j'} + \hbar\omega_\alpha)} \quad \text{and} \quad \frac{1}{\hbar\omega + (\varepsilon_{j'} - \hbar\omega_\alpha)}$$

are the energy denominators associated with the graphs (a) and (b) of Fig. 2.5, respectively. The numerators are $\pm|\langle n_\alpha = 1, j'|\delta U|j\rangle|^2$, the upper sign corresponding to the process (a) while the lower sign to process (b). The minus sign is connected with the Pauli principle, which graphically is associated with the crossing of fermion lines shown in graph (b) (see Fig. 2.11), corresponding to a contraction where and odd number of fermions is involved (see e.g. (2.85) and graphs (2) and (4) of Fig. 2.1).

One can rewrite (2.109) as

$$\Sigma_j(\omega) = \sum_{j',\alpha} |\langle n_\alpha = 1, j'|\delta U|j\rangle|^2 \left(\frac{1}{(\hbar\omega - \hbar\omega_\alpha) - \varepsilon_{j'}} + \frac{1}{(\hbar\omega - \hbar\omega_\alpha) + \varepsilon_{j'}} \right)$$

$$= \sum_{j',\alpha} |\langle n_\alpha = 1, j'|\delta U|j\rangle|^2 \frac{2(\hbar\omega - \hbar\omega_\alpha)}{(\hbar\omega - \hbar\omega_\alpha)^2 - \varepsilon_{j'}^2}. \tag{2.110}$$

Two situations are of particular interest (see Sect. 2.4 and App. 8.9.2):
(a) $\hbar\omega - \hbar\omega_\alpha \approx \varepsilon_{j'}$, in which case

$$\Sigma_j(\omega) \approx \sum_{j'\alpha} \frac{|\langle n_\alpha = 1, j'|\delta U|j\rangle|^2}{(\hbar\omega - \hbar\omega_\alpha) - \varepsilon_{j'}}, \tag{2.111}$$

and
(b) $\hbar\omega \approx 0$, where

$$\Sigma_j(\omega) \approx - \sum_{j'\alpha} \frac{2\hbar\omega_\alpha |\langle n_\alpha = 1, j'|\delta U|j\rangle|^2}{(\hbar\omega_\alpha)^2 - \varepsilon_{j'}^2}. \tag{2.112}$$

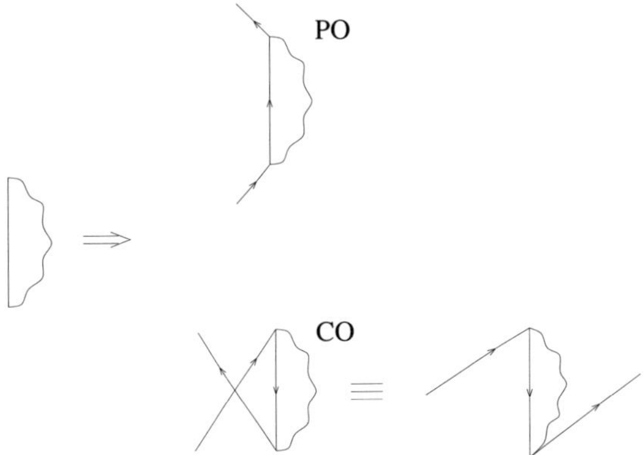

Fig. 2.11. Graphical representation of the polarization (PO) and of the correlation (CO) second contribution to the mass operator (see e.g. Sect. 3.8.5.3 of [30])

3. Electronic Structure

Having introduced the subject in the previous Chapter we shall, in what follows, develop it step by step, explaining in more detail the approximations and working out in detail the equations needed to carry out first principles calculations.

In the present and in the two following Chapters we shall deal with the motion of the electrons, considering the ions as fixed. We start by discussing the validity of such an approximation.

3.1 Born–Oppenheimer Approximation

The total Hamiltonian of a system containing atomic nuclei (ions) and electrons can be written as,

$$
H_{\text{tot}} = -\frac{\hbar^2}{2} \sum_n \frac{\nabla_n^2}{M_n} - \frac{\hbar^2}{2m_e} \sum_i \nabla_i^2 + V_{\text{ion,ion}} + V_{\text{e,e}} + V_{\text{ion,e}} \tag{3.1}
$$

where

$$
V_{\text{ion,ion}} \left(\{ \boldsymbol{R}_n \} \right) = \frac{1}{2} \sum_{m,n \neq m} \frac{Z_n Z_m e^2}{|\boldsymbol{R}_n - \boldsymbol{R}_m|}, \tag{3.2}
$$

is the ion–ion interaction,

$$
V_{\text{e,e}}(\{ \boldsymbol{r}_i \}) = \frac{1}{2} \sum_{j,i \neq j} \frac{e^2}{|\boldsymbol{r}_i - \boldsymbol{r}_j|}, \tag{3.3}
$$

is the electron–electron interaction, and

$$
V_{\text{ion,e}} \left(\{ \boldsymbol{r}_i \}, \{ \boldsymbol{R}_n \} \right) = -\sum_n \sum_i \frac{Z_n e^2}{|\boldsymbol{R}_n - \boldsymbol{r}_i|}, \tag{3.4}
$$

is the electron–ion interaction. Here, the following notation has been used: $\{ \boldsymbol{r}_i \} \equiv (\boldsymbol{r}_1, \boldsymbol{r}_2 \ldots \boldsymbol{r}_N)$ are the coordinates of the N electrons while $\{ \boldsymbol{R}_n \} \equiv (\boldsymbol{R}_1, \boldsymbol{R}_2 \ldots \boldsymbol{R}_{N_{\text{at}}})$, as already defined in Sect. 2.2, are the coordinates of the N_{at} ions.

We carry out the following expansion (in principle exact for any given $\{ \boldsymbol{R}_n \}$) for the wavefunction describing the system,

$$\Psi_Q^{\text{tot}}\left(\{\boldsymbol{r}_i\},\{\boldsymbol{R}_n\}\right) = \sum_{m'} \Phi_{m'}\left(Q,\{\boldsymbol{R}_n\}\right)\Psi_{m'}\left(\{\boldsymbol{r}_i\},\{\boldsymbol{R}_n\}\right), \qquad (3.5)$$

where the electron wavefunctions $\Psi_{m'}$ form a complete set of eigenfunctions corresponding to fixed ion positions, while Q indicates the set of quantum numbers that specify the total state of the system. Inserting Ψ_Q^{tot} in the Schrödinger equation associated with the Hamiltonian defined in (3.1) and multiplying the result from the left by $\Psi_m^*(\{\boldsymbol{r}_i\},\{\boldsymbol{R}_n\})$, one obtains,

$$\Psi_m^* \cdot \sum_{m'} \left[-\frac{\hbar^2}{2}\sum_n \frac{1}{M_n}\nabla_n^2(\Phi_{m'}\Psi_{m'}) - \frac{\hbar^2}{2m_{\text{e}}}\sum_i \nabla_i^2(\Phi_{m'}\Psi_{m'}) \right]$$

$$+\Psi_m^* \cdot \sum_{m'}(V_{\text{ion,e}} + V_{\text{ion,ion}} + V_{\text{e,e}})\Phi_{m'}\Psi_{m'} = \Psi_m^* \cdot E_Q \sum_{m'}\Phi_{m'}\Psi_{m'}, \quad (3.6)$$

where E_Q is the total energy of the system. For simplicity, the arguments of the wavefunctions have been omitted. Making use of the fact that

$$\nabla_i^2(\Phi_{m'}\Psi_{m'}) = \Phi_{m'}\nabla_i^2\Psi_{m'}, \qquad (3.7)$$

and

$$\nabla_n^2(\Phi_{m'}\Psi_{m'}) = \Psi_{m'}\nabla_n^2\Phi_{m'} + 2\boldsymbol{\nabla}_n\Psi_{m'}\cdot\boldsymbol{\nabla}_n\Phi_{m'} + \Phi_{m'}\nabla_n^2\Psi_{m'}, \quad (3.8)$$

one can integrate over the $3N$ electron coordinates, using the orthogonality of the wavefunctions Ψ_m, to obtain

$$\sum_n \frac{-\hbar^2}{2M_n}\left[\nabla_n^2\Phi_m + \sum_{m'}2\langle\Psi_m|\boldsymbol{\nabla}_n|\Psi_{m'}\rangle\cdot\boldsymbol{\nabla}_n\Phi_{m'} + \sum_{m'}\langle\Psi_m|\nabla_n^2|\Psi_{m'}\rangle\Phi_{m'}\right]$$

$$+\sum_{m'}\langle\Psi_m|\frac{-\hbar^2}{2m_{\text{e}}}\sum_i\nabla_i^2 + V_{\text{ion,e}} + V_{\text{e,e}}|\Psi_{m'}\rangle\Phi_{m'} + V_{\text{ion,ion}}\Phi_m = E_Q\Phi_m. \, (3.9)$$

Here Ψ_m is the solution of the electron Hamiltonian for the ions kept fixed at the positions $\{\boldsymbol{R}_n\}$,

$$\left(-\frac{\hbar^2}{2m_{\text{e}}}\sum_i\nabla_i^2 + V_{\text{ion,e}} + V_{\text{e,e}}\right)\Psi_m = \mathcal{E}_m\,\Psi_m, \qquad (3.10)$$

where $\mathcal{E}_m \equiv \mathcal{E}_m(\{\boldsymbol{R}_n\})$ is the total electronic energy of the system in its state labeled by the set of electronic quantum numbers m. Equation (3.9) can be written in a more compact form as (see [46, 47]):

$$\left(\sum_n \frac{-\hbar^2\nabla_n^2}{2M_n} + V_{\text{ion,ion}} + \mathcal{E}_m\right)\Phi_m + \sum_{m'}C(m,m')\Phi_{m'} = E_Q\Phi_m, \, (3.11)$$

where

$$C(m,m') = \sum_n \frac{-\hbar^2}{2M_n}\left[2\langle\Psi_m|\boldsymbol{\nabla}_n|\Psi_{m'}\rangle\cdot\boldsymbol{\nabla}_n + \langle\Psi_m|\nabla_n^2|\Psi_{m'}\rangle\right], \qquad (3.12)$$

is the coupling operator between different ionic configurations. Formally, (3.11) is an infinite set of coupled equations for the ionic wave functions Φ_m with associated eigenvalues E_Q. The coupling term is the matrix element of the ionic operators ∇_n between the electronic wavefunctions.

The *Born–Oppenheimer approximation* amounts to neglecting the off-diagonal matrix elements $C(m, m')$, $m \neq m'$. Within this approximation, one first solves the electronic motion described by (3.10) for fixed ion coordinates, and determines $\mathcal{E}_m$. One then works out the ionic wavefunctions by solving the equation governing the motion of the ions,

$$\left(\sum_n \frac{-\hbar^2 \nabla_n^2}{M_n} + \mathcal{E}_m \left(\{ \boldsymbol{R}_n \} \right) + V_{\text{ion,ion}} \right) \Phi_m = E_Q \, \Phi_m. \tag{3.13}$$

In this equation, the potential energy contains, besides the direct Coulomb ion–ion interaction, $V_{\text{ion,ion}}$, the term $\mathcal{E}_m$, that is, the electronic contribution to the total energy as a function of the ionic coordinates. This term is known as the adiabatic electronic contribution, and can be considered as the "glue" which keeps the atoms together. It is an involved function of the ionic coordinates $\{ \boldsymbol{R}_n \}$. At variance with the ion–ion term $V_{\text{ion,ion}}$, it cannot, as a rule, be expressed as a sum of two-body contributions. Rigorously, the diagonal terms $C(m, m)$ should be included in (3.13). In practice these terms are usually neglected since $m \ll M$.

The Born–Oppenheimer (BO) ansatz corresponds to a decoupling between the ionic and electronic motions which can be understood as follows. The electron mass is much smaller than the ionic mass. Hence, the time scale for the electronic motion is much faster than that for the ionic movement. One can therefore make the ansatz that the electrons adapt themselves instantaneously to the actual position of the ions (the electrons can be considered always in the ground state corresponding to the given ionic configuration). Translating this into the energy domain, the levels corresponding to different electronic eigenstates are often separated by energy gaps which are large with respect to the corresponding separation between ionic eigenstates. Thus, within the BO scheme, the ions do not induce transitions between different electronic states. Of course, these are general arguments which may not be valid in a number of physical systems. One should verify *a posteriori* the smallness of the matrix elements neglected in the Born–Oppenheimer approximation. We will discuss in Chaps. 8 and 9 the effect of these matrix elements on the properties of the system, and provide examples of the limitations of the approximation.

By solving (3.10) for a given ionic configuration $\{ \boldsymbol{R}_n \}$, one obtains the electronic contribution to the total energy of the system. While this electronic equation must necessarily be treated within quantum theory (e.g., in mean field approximation like Hartree–Fock, Density Functional Theory, Quantum Monte Carlo techniques, etc.), the ionic motion described by (3.13) can often be solved at profit within a classical scheme. This means that the total energy of the system can be, as a rule, accurately expressed as $\mathcal{E} + V_{\text{ion,ion}}$. Moreover, if the ionic motion is well approximated by small oscillations around the equilibrium positions, the classical solution of (3.13) coincides with the quantum solution. In fact, it will be shown in Chap. 6 that, in such a case, the problem is equivalent to that of a collection of independent harmonic oscillators.

3.2 Density Functional Theory: Generalities

As we have already discussed in Chap. 2, Hartree–Fock theory allows for a considerable simplification of (3.10). The price to be paid for this simplification is that one looses electron correlation effects, in particular the effective mass contribution to the motion of the electrons and the screening of the bare Coulomb interaction. These can be recovered within the framework of the Density Functional Theory (DFT). In what follows, we summarize here the basic equations of DFT, and how they lead to the so-called Kohn–Sham self-consistent integro-differential equations.

The fundamental variable in DFT is the total electronic density, $\varrho(\boldsymbol{r})$. If $\Psi(\{\boldsymbol{r}_i\})$ is the exact many-body wavefunction of the N-electron system for fixed ions (see (3.10)), one can write

$$\varrho(\boldsymbol{r}) = \int \mathrm{d}^3 r_1...\mathrm{d}^3 r_N \ \Psi^*(\boldsymbol{r}_1...\boldsymbol{r}_N) \left(\sum_i \delta(\boldsymbol{r} - \boldsymbol{r}_i)\right) \Psi(\boldsymbol{r}_1...\boldsymbol{r}_N)$$

$$= N \int \mathrm{d}^3 r_2...\mathrm{d}^3 r_N \ |\Psi(\boldsymbol{r}, \boldsymbol{r}_2...\boldsymbol{r}_N)|^2. \tag{3.14}$$

As Hohenberg and Kohn (HK) have demonstrated [48], there is a one-to-one correspondence between the ground-state electronic density, $\varrho_0(\boldsymbol{r})$, and the external potential acting on the system (e.g. $V_{\text{ion,e}}$)

$$\varrho_0(\boldsymbol{r}) \Leftrightarrow V_{\text{ext}}(\boldsymbol{r}). \tag{3.15}$$

HK also demonstrated that it is possible, given an external potential $V_{\text{ext}}(\boldsymbol{r})$, to build a *functional* of the density which has a minimum in correspondence to the exact electronic density of the ground state. The minimum value of this functional is the ground state energy of the system

$$E_0 = \min_{\varrho} E_{\{V_{\text{ext}}\}}[\varrho], \tag{3.16}$$

where ϱ must be varied under the condition $\int \mathrm{d}^3 r \ \varrho(\boldsymbol{r}) = N$.

We define as functional any scalar F (in the above case the energy E) which depends on the global set of values of a function $f(x)$. This dependence is usually indicated by writing $F[f(x)]$, or simply $F[f]$. In (3.16) the functional $E[\varrho]$ has also a parametrical dependence on another function, namely the external potential.

We briefly recall the formalism of functional derivatives. They are essentially an extension of the concept of the partial derivative of a function F of n variables $F(f_1, f_2...f_n)$, to the case of $F[f(x)]$ defined above. For the partial derivative one has

$$\frac{\partial F}{\partial f_i} = \lim_{\varepsilon \to 0} \frac{F(f_1 \ldots f_i + \varepsilon \ldots f_n) - F(f_1 \ldots f_n)}{\varepsilon}.$$

Consequently, the natural extension to the case in which the discrete index i becomes a continuous variable x (and the set $\{f_i\}$ becomes a function $f(x)$), leads to the following definition of functional derivative:

$$\frac{\delta F[f(x)]}{\delta f(y)} = \lim_{\varepsilon \to 0} \frac{F[f(x) + \varepsilon\delta(x - y)] - F[f(x)]}{\varepsilon}.$$

As an example, let us assume $F = \int \mathrm{d}x \, f(x) \, g(x)$. Then

$$\frac{\delta F}{\delta f} = g(x).$$

Another mathematical tool which will be needed in what follows is the concept of Lagrange multiplier for constrained minimizations.

Using the fact that functional derivatives obey similar rules as the standard derivatives, one searches for the minimum of the energy functional $E[\varrho]$ by equating to zero its functional derivative. The solution of

$$\frac{\delta}{\delta \varrho} E_{\{V_{\mathrm{ext}}\}}[\varrho] = 0, \tag{3.17}$$

gives both the ground state density $\varrho_0(\boldsymbol{r})$ and the corresponding total energy E_0. Following HK and Kohn and Sham (KS) [31], one can write the energy functional in the form,

$$E_{\{V_{\mathrm{ext}}\}}[\varrho] = T_0[\varrho] + \frac{e^2}{2} \int \int \frac{\varrho(\boldsymbol{r})\rho(\boldsymbol{r}')}{|\boldsymbol{r} - \boldsymbol{r}'|} + \int \varrho(\boldsymbol{r})V_{\mathrm{ext}}(\boldsymbol{r}) + E_{\mathrm{xc}}[\varrho]. \tag{3.18}$$

Here $T_0[\varrho]$ is the kinetic energy of an auxiliary system of non-interacting particles which has the same density $\varrho(\boldsymbol{r})$ of the real system. $E_{\mathrm{xc}}[\varrho]$ is the part of the functional containing the exchange and correlation effects, plus the difference between the exact kinetic energy $T[\varrho]$ and that of the auxiliary system $T_0[\varrho]$. Equation (3.18) is actually a definition of $E_{\mathrm{xc}}[\varrho]$ which is called exchange-correlation energy. Since the auxiliary system is non-interacting, its density can be written exactly as a sum of contributions associated to independent single-particle orbitals,

$$\varrho(\boldsymbol{r}) = \sum_{i=1}^{N} |\varphi_i(\boldsymbol{r})|^2. \tag{3.19}$$

where φ_i are known as the Kohn–Sham orbitals (see [49, 50]). We will refer to them simply as KS wavefunctions.

The DFT energy functional can also be written as

$$E_{\{V_{\text{ext}}\}}[\{\varphi_i\}] = -\sum_{i=1}^{N} \int \mathrm{d}^3r \; \varphi_i^*(\boldsymbol{r}) \frac{\hbar^2}{2m_{\text{e}}} \nabla^2 \varphi_i(\boldsymbol{r})$$

$$+ \frac{e^2}{2} \int \int \mathrm{d}^3r \; \mathrm{d}^3r' \; \frac{\left(\sum_{k=1}^{N} |\varphi_k(\boldsymbol{r})|^2\right) \left(\sum_{j=1}^{N} |\varphi_j(\boldsymbol{r}')|^2\right)}{|\boldsymbol{r} - \boldsymbol{r}'|},$$

$$+ \int \mathrm{d}^3r \; \left(\sum_{i=1}^{N} |\varphi_i(\boldsymbol{r})|^2\right) V_{\text{ext}}(\boldsymbol{r}) + E_{\text{xc}}\left[\sum_{i=1}^{N} |\varphi_i(\boldsymbol{r})|^2\right], \quad (3.20)$$

where again, the square brackets indicate a functional dependence. One must hence perform a search for the minimum of $E_{\{V_{\text{ext}}\}}[\{\varphi_i\}]$ with the constraint

$$\int \mathrm{d}^3r \; \varphi_k^*(\boldsymbol{r})\varphi_l(\boldsymbol{r}) - \delta_{kl} = 0, \qquad (3.21)$$

which ensures the orthonormality of the Kohn–Sham orbitals. To search for the minimum, we thus write,

$$\frac{\delta}{\delta \varphi_j^*}\left[E - \sum_{k,l=1}^{N} \lambda_{kl} \left(\int \mathrm{d}^3r \; \varphi_k^*(\boldsymbol{r})\varphi_l(\boldsymbol{r}) - \delta_{kl}\right)\right] = 0, \qquad (3.22)$$

where the Lagrange multipliers λ_{kl} appear explicitly.

By substituting (3.20) into (3.22) and performing the functional derivatives, one obtains

$$-\frac{\hbar^2}{2m_{\text{e}}} \nabla^2 \varphi_j(\boldsymbol{r}) + \left[e^2 \int \frac{\left(\sum_{k=1}^{N} |\varphi_k(\boldsymbol{r}')|^2\right)}{|\boldsymbol{r} - \boldsymbol{r}'|} \mathrm{d}^3r'\right] \varphi_j(\boldsymbol{r}) + V_{\text{ext}}(\boldsymbol{r})\varphi_j(\boldsymbol{r})$$

$$+ \frac{\delta}{\delta \varphi_j^*}\left\{E_{\text{xc}}\left[\sum_{k=1}^{N} |\varphi_k(\boldsymbol{r})|^2\right]\right\} = \sum_{k=1}^{N} \lambda_k \; \varphi_k(\boldsymbol{r}), \qquad (3.23)$$

where the second term in the left-hand side is the Hartree potential $v_{\text{H}}(\boldsymbol{r})$ (already discussed) acting on $\varphi_j(\boldsymbol{r})$, and the fourth term is,

$$\frac{\delta}{\delta \varphi_j^*}\left[E_{\text{xc}}[(\sum_{k=1}^{N} |\varphi_k|^2)]\right] = \frac{\delta \varrho}{\delta \varphi_j^*} \cdot \frac{\delta E_{\text{xc}}}{\delta \varrho} = \varphi_j \cdot \frac{\delta E_{\text{xc}}}{\delta \varrho}. \qquad (3.24)$$

The term $\delta E_{\text{xc}}/\delta \varrho$ is called the exchange-correlation potential $v_{\text{xc}}(\boldsymbol{r})$, and acts *locally* on the wavefunctions (even if, in principle, it contains the *exact* exchange effects according to the HK theorem).

By performing a suitable unitary transformation to the set of wavefunctions φ_j, the resulting system of equations is obtained,

$$H^{\text{KS}}\varphi_j(\boldsymbol{r}) = \lambda_j \varphi_j(\boldsymbol{r}) \qquad \forall j, \qquad (3.25)$$

where

$$H^{\text{KS}} = -\frac{\hbar^2}{2m_{\text{e}}} \nabla^2 + v_{\text{H}}(\boldsymbol{r}) + V_{\text{ext}}(\boldsymbol{r}) + v_{\text{xc}}(\boldsymbol{r}). \qquad (3.26)$$

H^{KS} is known as the Kohn–Sham Hamiltonian. We must note that both $v_{\text{H}}(\boldsymbol{r})$ and $v_{\text{xc}}(\boldsymbol{r})$ depend on the density, hence on the whole set of wavefunctions $\varphi_j(\boldsymbol{r})$. The system must then be solved self-consistently. Once (3.25) has been solved, it is possible to calculate the total energy of the system, E_{tot}, which is given by

$$E_{\{V_{\text{ext}}\}}[\varrho_0] = \sum_{j=1}^{N} \int \varphi_j^*(\boldsymbol{r}) \frac{-\hbar^2}{2m_e} \nabla^2 \varphi_j(\boldsymbol{r}) + \frac{1}{2} \int v_{\text{H}}(\boldsymbol{r})\varrho_0(\boldsymbol{r}) \mathrm{d}^3 r$$

$$+ \int V_{\text{ext}}(\boldsymbol{r})\varrho_0(\boldsymbol{r})\mathrm{d}^3 r + E_{\text{xc}}[\varrho_0]. \tag{3.27}$$

Here N is the number of *occupied* electronic states, and ϱ_0 is the density corresponding to the set of φ_j which solve (3.25), i.e. $\varrho_0 = \sum_{j=1}^{N} \varphi_j^* \varphi_j$. An equivalent expression for the total energy, which can be derived easily from the comparison of (3.27) with the ground state expectation value of (3.26), reads

$$E_{\text{tot}} = \sum_{j=1}^{N} \lambda_j + E_{\text{xc}}[\varrho_0] - \int \mathrm{d}^3 r \; \varrho_0(\boldsymbol{r}) \left[\frac{1}{2}v_{\text{H}}(\boldsymbol{r}) + v_{\text{xc}}(\boldsymbol{r})\right]. \tag{3.28}$$

As it can be seen from (3.28), one has to subtract one half of the Hartree energy from the sum of the eigenvalues, as in Hartree–Fock theory (see footnote nr. 4 in Chap. 2); in addition, one also has to take into account the difference between the exchange-correlation energy and the expectation value of the exchange-correlation potential.

Either choice for the total energy expressions given in (3.27) or (3.28), can be more or less convenient depending on the basis set used. With a plane-wave basis set, the choice given in (3.27) is usually favorable because the kinetic energy term is obtained very easily.

3.3 The Local Density Approximation

Up to this point, no approximations have been introduced. Unfortunately, the exact form for the functional $E_{\text{xc}}[\varrho]$ is not known. One must hence resort to some approximated functional, as it is usually done within the so-called Local Density Approximation (LDA) where the exchange-correlation energy E_{xc} is calculated assuming that in the vicinity of the point $\boldsymbol{r}$ the properties of the (inhomogeneous) electron gas of density $\varrho(\boldsymbol{r})$ can be approximated by those of an infinite homogeneous electron gas of that density. Accordingly, one writes

$$E_{\text{xc}}[\varrho] = \int \mathrm{d}^3 r \; \varepsilon_{\text{xc}}^{\text{hom}}(\varrho(\boldsymbol{r})) \, \varrho(\boldsymbol{r}). \tag{3.29}$$

where $\varepsilon_{\text{xc}}^{\text{hom}}[\varrho]$ is the exchange-correlation energy per particle of the homogeneous electron gas calculated at the local density. Consequently, the exchange-correlation potential $v_{\text{xc}}(\boldsymbol{r})$ becomes

$$v_{\mathrm{xc}}(\boldsymbol{r}) = \frac{\delta}{\delta\varrho(\boldsymbol{r})} E_{\mathrm{xc}}[\varrho] = \varepsilon_{\mathrm{xc}}^{\mathrm{hom}}(\varrho(\boldsymbol{r})) + \varrho(\boldsymbol{r}) \left.\frac{\partial\varepsilon_{\mathrm{xc}}^{\mathrm{hom}}}{\partial\varrho}\right|_{\varrho(\boldsymbol{r})}. \tag{3.30}$$

In the case of the homogeneous electron gas, essentially exact expressions for $\varepsilon_{\mathrm{xc}}^{\mathrm{hom}}(\varrho)$ can be obtained either through many-body diagrammatic calculations or through numerical techniques (e.g., quantum Monte Carlo, see below). Here we discuss how the LDA is implemented in practical calculations.

The exchange and correlation energy per particle, $\varepsilon_{\mathrm{xc}}^{\mathrm{hom}}(\varrho(\boldsymbol{r}))$, can be written as the sum

$$\varepsilon_{\mathrm{xc}}^{\mathrm{hom}}(\varrho(\boldsymbol{r})) = \varepsilon_{\mathrm{x}}(\varrho(\boldsymbol{r})) + \varepsilon_{\mathrm{c}}(\varrho(\boldsymbol{r})), \tag{3.31}$$

of the exchange and correlation energies, respectively. The exchange part, $\varepsilon_{\mathrm{x}}[\varrho]$, is taken from the exact Hartree–Fock result of the homogeneous electron gas,

$$\varepsilon_{\mathrm{x}}[\varrho(\boldsymbol{r})] = -e^2 \frac{3}{4\pi} (3\pi^2\varrho(\boldsymbol{r}))^{1/3}. \tag{3.32}$$

Using the relations introduced after (2.61), one can rewrite (3.32) as,

$$\varepsilon_{\mathrm{x}}(r_s) = -2\mathrm{Ry}\frac{3}{4\pi}\left(\frac{9\pi}{4}\right)^{1/3}\frac{1}{r_{\mathrm{s}}}. \tag{3.33}$$

The correlation part, $\varepsilon_{\mathrm{c}}[\varrho] \equiv \varepsilon_{\mathrm{c}}(r_{\mathrm{s}})$, can be parametrized either using the simple expression given in (2.65), or using more precise results (valid over a broader range of densities) obtained from quantum Monte Carlo calculations for the uniform electron gas [51] which leads to

$$\varepsilon_{\mathrm{c}}(r_{\mathrm{s}}) = \frac{2\mathrm{Ry}\alpha}{1 + \beta\sqrt{r_{\mathrm{s}}} + \gamma r_{\mathrm{s}}}, \quad \text{for } r_{\mathrm{s}} \geq 1, \tag{3.34}$$

with $\alpha = -0.1423$, $\beta = 1.0529$ and $\gamma - 0.3334$, and

$$\varepsilon_{\mathrm{c}}(r_{\mathrm{s}}) = 2\mathrm{Ry}(\delta + \lambda \ln r_{\mathrm{s}} + \nu r_{\mathrm{s}} + \sigma r_{\mathrm{s}} \ln r_{\mathrm{s}}), \quad \text{for } r_{\mathrm{s}} < 1, \tag{3.35}$$

with $\delta = -0.048$, $\lambda = 0.0311$, $\nu = -0.0116$ and $\sigma = 0.002$.

The exchange-correlation potential, $v_{\mathrm{xc}}(r_{\mathrm{s}})$ can be extracted using (3.30). In particular, for the exchange part (3.32) one finds the result already shown in (2.61),

$$v_{\mathrm{x}}^{\mathrm{LDA}}(\boldsymbol{r}) = -e^2 \frac{1}{\pi}\left(3\pi^2\varrho(\boldsymbol{r})\right)^{1/3} \approx -0.9848\, e^2\, \varrho^{1/3}.$$

Using r_{s} as basic variable, the expression which corresponds to (3.30) is

$$v_{\mathrm{xc}}(r_{\mathrm{s}}) = \varepsilon_{\mathrm{xc}}(r_{\mathrm{s}}) - \frac{1}{3}r_{\mathrm{s}}\frac{\partial\varepsilon_{\mathrm{xc}}}{\partial r_{\mathrm{s}}}. \tag{3.36}$$

Inserting (3.33) in the above equation, we obtain

$$v_{\mathrm{x}}(r_{\mathrm{s}}) = -2Ry\frac{1}{\pi}\left(\frac{9\pi}{4}\right)^{1/3}\frac{1}{r_{\mathrm{s}}} = -\frac{16.62}{r_{\mathrm{s}}}\,\mathrm{eV}. \tag{3.37}$$

It should be noted that the functional dependence of $v_{\mathrm{x}}^{\mathrm{LDA}}(\boldsymbol{r})$ is the same as that found by Slater in the early 1950's, by averaging the exchange potential over all the occupied states, although the prefactor differs. In fact,

$$v_{\mathrm{x}}^{\mathrm{Slater}}(\boldsymbol{r}) = -\frac{3}{2\pi}e^2\left(3\pi^2\varrho(\boldsymbol{r})\right)^{1/3} = \frac{3}{2}v_{\mathrm{x}}^{\mathrm{LDA}}(\boldsymbol{r}) \tag{3.38}$$

For illustration, the resulting correlation potential (as given by (3.34), (3.35) and (3.36)) and the corresponding exchange term (3.37) are plotted versus r_{s} in Fig. 3.1.

3.3.1 Discussion of Kohn–Sham Method

The main advantages of the Kohn–Sham based methods outlined in the previous section (and denoted usually as first-principles or *ab initio* methods) arise from the fact that they do not require any phenomenological input, i.e., there is *no* adjustable parameter in the calculation. As a consequence,

1. calculations can be performed for different systems using the same ingredients: there is no need to check the transferability of the model, at variance with the case of semi-empirical schemes;
2. there is no restriction based on the system symmetry: low-symmetry systems can be studied, in principle, with no additional difficulty.

During the last decade, first-principles methods have become useful in condensed matter physics also for applications to complex and realistic systems such as surfaces, defects, amorphous systems, and aggregates or clusters

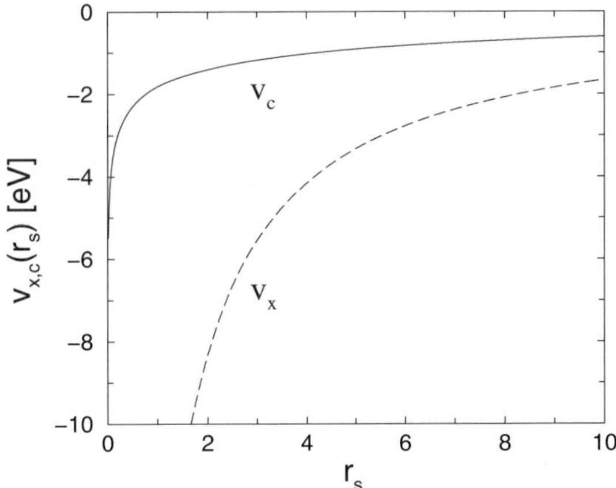

Fig. 3.1. The LDA exchange and correlation potentials, v_{x} (*dashed line*) and v_{c} (*continuous line*) versus r_{s}. The exchange potential is given by (3.37) and the correlation part by (3.34), (3.35) and (3.36). The latter correspond to the parametrization of [51]

(see e.g., [32] and refs. therein). Two factors contributed in making this possible: first, the advances in computer power (hardware progress); second, recent developments in theoretical approaches. The latter, in particular, benefit from three novel methods: (a) the Density Functional Theory (DFT) (Sect. 3.2), (b) the theory of Norm-Conserving Pseudo-Potentials (NCPP), developed in the 1980's [52, 53] (see Sect. 3.3.2), and (c) the so-called Car–Parrinello Method (CP), introduced in 1985 [54] (see Chap. 7).

The CP method is now widely used by condensed-matter researchers. The basis set is usually that of plane-waves, although applications of localized basis functions have also become important. These are more natural in the treatment of finite systems, and have a long tradition in molecular and chemical physics aside from also in nuclear physics [45, 55]. Although the optimization of Fourier-transform-based algorithms nowadays makes the use of non-localized plane-wave basis suitable also for finite systems, we find it more convenient to choose a spherical basis to start our description of possible implementations of the Kohn–Sham equations. This will be done in Sect. 3.4. In Sect. 3.3.2 we briefly summarize the theory of *ab initio*, norm conserving pseudopotentials.

3.3.2 Pseudopotentials

An introduction to the theory of atomic pseudopotentials (PP) is given in [56] (see also Appendix 3.6.1). One starts from the observation that core electrons play a minor role in the chemical bonding between atoms to form molecules and solids, the key role being played by valence electrons, as one can deduce from the periodicity of the Mendeleev table. Thus one can assume that core electrons remain frozen in their atomic configuration ("frozen core" approximation). However, the requirement of orthogonality to core states also influences the valence wavefunctions outside the core region. Hence valence electrons are to some extent affected by the presence of core electrons.

The use of a pseudopotential allows one to eliminate the core electrons, approximately taking into account their effects on the valence states in terms of an effective potential. The orthogonality to core states has the same effects as a repulsive potential acting on valence wavefunctions only. However, a question arises: can one introduce such an effective potential fully within an *ab initio* scheme, i.e. without adjustable parameters? The answer is positive, but one must accept the complication that pseudopotentials become *nonlocal*, hence requiring the use of angular projection operators when they are used in molecular or solid-state calculations. This is explained in Appendix 3.6.1.

A good pseudopotential should be *transferable*. That is, a pseudo-atom should behave as the original many-electrons atom in different structural and chemical situations. Because of well-known relations connecting logarithmic derivatives, scattering amplitudes, and matching conditions (see e.g. [57]), which can be summarized by the expression

$$-\frac{1}{2}r^2|\varphi_\varepsilon(\boldsymbol{r})|^2\frac{\mathrm{d}}{\mathrm{d}\varepsilon}\left\{\frac{\mathrm{d}}{\mathrm{d}r}\ln\varphi_\varepsilon(\boldsymbol{r})\right\}\Big|_{r=r_{\mathrm{c}},\varepsilon=\varepsilon'} = \int_0^{r_{\mathrm{c}}}|\varphi_{\varepsilon'}(\boldsymbol{r})|^2 r^2\mathrm{d}^3 r \quad (3.39)$$

(where ε labels the energy and r_{c} is the core radius), good transferability can be achieved if the pseudo- and full-core potentials have similar logarithmic derivatives in the energy range which is of interest for chemical bonding, i.e. typically in the range 10-20 eV around the isolated atom eigenvalues, corresponding in the solid to the resulting band widths and in the molecule to the energy separation between 'hybridized' levels.

Smooth and optimally transferable pseudopotentials have been construct- ed that exactly reproduce the valence wavefunctions outside a given core region for a fixed, reference atomic configuration [52, 53]. In this configura- tion, the logarithmic derivative of the pseudowavefunction is identical to the corresponding full-core one,

$$\left(\frac{R'_\ell}{R_\ell}\right)^{\mathrm{PS}}_{\epsilon=\epsilon_\ell,r_{\mathrm{c}}} = \left(\frac{R'_\ell}{R_\ell}\right)^{\mathrm{FC}}_{\epsilon=\epsilon_\ell,r_{\mathrm{c}}}. \quad (3.40)$$

Here, ϵ_ℓ is the atomic bound-state energy[1] and r_{c} is the core radius. Moreover, away from the reference energy ϵ_ℓ, logarithmic derivatives follow the all- electron ones up to first order in the energy difference $\epsilon - \epsilon_\ell$, i.e.

$$\frac{\mathrm{d}}{\mathrm{d}\epsilon}\Big|_{\epsilon=\epsilon_\ell}\left(\frac{R'_\ell(\epsilon,r)}{R_\ell(\epsilon,r)}\right)^{\mathrm{PS}} = \frac{\mathrm{d}}{\mathrm{d}\epsilon}\Big|_{\epsilon=\epsilon_\ell}\left(\frac{R'_\ell(\epsilon,r)}{R_\ell(\epsilon,r)}\right)^{\mathrm{FC}}. \quad (3.41)$$

Hence, discrepancies are only of second order in the energy difference. More recently, other authors (see, e.g., [58]) introduced modified schemes to pro- duce even softer pseudopotentials[2] without loosing much in transferability as compared to those of [52, 53].

3.4 Kohn–Sham Equations in a Spherical Basis

As discussed above, the Kohn–Sham equations describe the ground state of an N-electron system within a single particle scheme, yielding N independent wavefunctions $\varphi_j(\boldsymbol{r})$ obeying the Schrödinger-like equation of the type of (3.25),

$$\left[-\frac{\hbar^2}{2m_{\mathrm{e}}}\nabla^2 + v_{\mathrm{eff}}(\boldsymbol{r})\right]\varphi_j(\boldsymbol{r}) = \lambda_j\,\varphi_j(\boldsymbol{r}) \quad (3.42)$$

---

[1] The main quantum number n has been omitted for simplicity (see Appendix 3.6.1).

[2] The term softer pseudopotential indicates that fewer basis wavefunctions are required to determine the Kohn–Sham orbitals as compared to other pseudopo- tentials that require a larger number of basis sets to achieve similar accuracy.

where $v_{\text{eff}}(\boldsymbol{r})$ is the effective potential felt by the electrons which can be written as in (3.26), namely,

$$v_{\text{eff}}(\boldsymbol{r}) = V_{\text{ext}}(\boldsymbol{r}) + v_{\text{H}}(\boldsymbol{r}) + v_{\text{xc}}(\boldsymbol{r}). \tag{3.43}$$

In order to solve (3.42), we can expand the KS orbitals $\varphi_i(\boldsymbol{r})$ in spherical basis functions $\Phi_j(\boldsymbol{r})$ [59]. We make a slight change of notation compared to the previous Sections, and the index i will refer here to the Kohn–Sham orbitals. The index j will label the basis functions which are solutions of a radially symmetric problem. We also replace the vector notation for the electron coordinate, because in the spherical basis we shall separate the radial coordinate from the polar angles. Consequently, we write the expansion of a generic Kohn–Sham orbital as

$$\varphi_i(\boldsymbol{r}) = \sum_{j=1}^{j_{\max}} c_j^{(i)} \, \Phi_j(\boldsymbol{r}), \tag{3.44}$$

where the coefficients $c_j^{(i)}$ can be obtained by diagonalization of the Hermitian matrix associated to (3.42) written on the Φ_j basis. The cutoff $j_{\max}$ must be selected so that the results of the calculation are stable. The basis functions Φ_j are the solutions of the single-particle Schrödinger equation with a suitable spherical potential $v_{\text{Basis}}(r)$ which depends only on the radial coordinate r,

$$\left[-\frac{\hbar^2}{2m} \nabla^2 + v_{\text{Basis}}(r) \right] \Phi_j(\boldsymbol{r}) = \epsilon_j \, \Phi_j(\boldsymbol{r}). \tag{3.45}$$

The wavefunctions $\Phi_j(\boldsymbol{r})$ can be written in terms of the spherical harmonics, $Y_{\ell,m}(\theta, \phi)$, as

$$\Phi_j(\boldsymbol{r}) = R_{n_j, \ell_j}(r) \, Y_{\ell_j, m_j}(\theta, \phi) \tag{3.46}$$

where $R_{n_j, \ell_j}(r)$ are the solutions of the radial equation corresponding to (3.45). The angles θ and ϕ are the standard polar and azimuthal angles, and the coordinate axis (x, y, z) are chosen so that they correspond to convenient symmetry axis of the molecule (see Appendix 3.6.2).

Following standard notation, n, ℓ and m indicate the principal quantum number, the orbital angular momentum and its z-projection, respectively. These are the quantum numbers associated to a particular solution Φ_j. In other words, the index j labels the set of quantum numbers n_j, ℓ_j and m_j. The spherical harmonics can be written in terms of the generalized Legendre polynomials $\Theta_{\ell,|m|}(\cos\theta)$ according to (see e.g. [57])

$$Y_{\ell,m}(\theta, \phi) = (-1)^{(m+|m|)/2} \, \Theta_{\ell,|m|}(\cos\theta) \, e^{im\phi},$$

where

$$\Theta_{\ell,|m|}(\cos\theta) = \left[\frac{(2\ell + 1)}{4\pi} \frac{(\ell - |m|)!}{(\ell + |m|)!} \right]^{1/2} P_\ell(\cos\theta),$$

and $P_\ell(\cos\theta)$ are the standard Legendre polynomials. Note that the spherical harmonics are defined for any value of m in the range $-\ell \le m \le \ell$, and $Y_{\ell,m}$ obeys the orthonormality condition

$$\int_0^\pi d\theta \sin\theta \int_0^{2\pi} d\phi \, Y_{\ell,m}^*(\theta,\phi) \, Y_{\ell',m'}(\theta,\phi) = \delta_{\ell,\ell'} \, \delta_{m,m'}.$$

3.4.1 The Expansion of V_{ext} in Spherical Harmonics

In order to deal with (3.42) within a spherical basis it is convenient to expand the effective potential itself in spherical harmonics. Since the symmetry of the molecule is determined by the ionic positions, the corresponding expansion of the external potential plays a fundamental role. In what follows we discuss, separately, the local and non-local parts of V_{ext} (the reason for the appearance of the non-local part is discussed in Sect. 3.3.2 and Appendix 3.6.1).

The local part of V_{ext}. For the local part of the external ionic potential we can write

$$v_{\text{loc}}(\boldsymbol{r}) = \sum_{L,M} v_{L,M}^{\text{loc}}(r) \, Y_{L,M}(\theta,\phi). \tag{3.47}$$

Similar expressions hold for the electron–electron potentials v_{H} and v_{xc} and for the electronic density ϱ, with the corresponding multipole functions $v_{L,M}^{(\text{H})}(r)$, $v_{L,M}^{(\text{xc})}(r)$ and $\varrho_{L,M}(r)$, respectively.

Given a set of N_{at} ionic positions $\{\boldsymbol{R}_n\}$, the total local potential at position $\boldsymbol{r}$ is given by

$$v_{\text{loc}}(\boldsymbol{r}) = \sum_{n=1}^{N_{\text{at}}} v_{\text{loc}}^{(n)}(|\boldsymbol{r} - \boldsymbol{R}_n|), \tag{3.48}$$

where the ionic potential is spherically symmetric around the ion. Let us now discuss how (3.48) can be transformed to the form shown in (3.47).

To this end, we first consider the contribution of the n-th ion, $v_{\text{loc}}^{(n)}(|\boldsymbol{r} - \boldsymbol{R}_n|)$, to the total potential defined in (3.48). Following Löwdin [60], we choose a rotated coordinate system (x', y', z'), centered at $\boldsymbol{r} = 0$, whose z'-axis is pointing along $\boldsymbol{R}_n$. The orientation of the x'- and y'-axis does not concern us now and will be specified below. In this coordinate system, we look for an expansion of the form

$$v_{\text{loc}}^{(n)}(|\boldsymbol{r} - \boldsymbol{R}_n|) = \sum_{L,M'} v_{L,M'}^{(\text{loc})}(r, R_n) \, Y_{L,M'}(\theta',\phi'), \tag{3.49}$$

where $r = |\boldsymbol{r}|$, $R_n = |\boldsymbol{R}_n|$, and the angles are defined with respect to the rotated coordinate system. According to this convention we have $|\boldsymbol{r} - \boldsymbol{R}_n|^2 = r^2 + R_n^2 - 2rR_n \cos\theta'$, and $v_{L,M'}^{(\text{loc})}(r, R_n)$ becomes

$$v_{L,M'}^{(\text{loc})}(r, R_n) = 2\pi \, \delta_{M',0} \sqrt{\frac{2L+1}{4\pi}} \int_{-1}^1 dx \, v_{\text{loc}}^{(n)}(|\boldsymbol{r} - \boldsymbol{R}_n|) \, P_L(x), \tag{3.50}$$

where $x = \cos\theta'$. Note that the Kronecker symbol $\delta_{M',0}$ reflects the arbitrariness in the orientation of the x'- and y'-axis. For simplicity, we assume that the x'- and y'-axis result from rotations of the natural axis with the Euler angles, α, β and γ, given respectively by $\alpha = \phi_n$, $\beta = \theta_n$ and $\gamma = 0$ (ϕ_n and θ_n corresponding to the azimuthal and polar angles of $\boldsymbol{R}_n$). This is consistent with our choice of z' pointing along $\boldsymbol{R}_n$. It follows that $Y_{L,M'}(\theta',\phi')$ can be expressed in terms of the natural axis angles according to (see, e.g., [57])

$$Y_{L,M'}(\theta',\phi') = \sum_{M=-L}^{L} D_{M,M'}^{(L)}(\phi_n,\theta_n,0)\, Y_{L,M}(\theta,\phi). \tag{3.51}$$

In the present case, only terms with $M' = 0$ need to be taken into account and the transformation matrix D can be written simply as

$$D_{M,0}^{(L)}(\phi_n,\theta_n,0) = \sqrt{\frac{4\pi}{2L+1}}\, Y_{L,M}^{*}(\theta_n,\phi_n). \tag{3.52}$$

Thus, (3.49) becomes

$$v_{\mathrm{loc}}^{(n)}(|\boldsymbol{r}-\boldsymbol{R}_n|) = \sum_{L,M} v_{L,M}^{(\mathrm{loc})}(r,R_n,\theta_n,\phi_n)\, Y_{L,M}(\theta,\phi), \tag{3.53}$$

where

$$v_{L,M}^{(\mathrm{loc})}(r,R_n,\theta_n,\phi_n) = 2\pi\, Y_{L,M}^{*}(\theta_n,\phi_n)\, \int_{-1}^{1} \mathrm{d}x\; v_{\mathrm{loc}}^{(n)}(f(x))\, P_L(x),$$

where the argument of $v_{\mathrm{loc}}^{(n)}$ is $f(x) = \sqrt{r^2 + R_n^2 - 2rR_n x}$. This result can now be used in (3.48) to obtain the total local ionic potential at position $\boldsymbol{r}$ and, according to (3.47), we finally have

$$v_{L,M}^{(\mathrm{loc})}(r) = \sum_{n=1}^{N_{\mathrm{at}}} v_{L,M}^{(\mathrm{loc})}(r,R_n,\theta_n,\phi_n).$$

The non-local part of $\boldsymbol{V}_{\mathrm{ext}}$. The non-local part of the pseudopotential reflects the different effects that core electrons have on valence electron wavefunctions of different angular momentum. Let us consider the n-th ion at position $\boldsymbol{R}_n$. With respect to the center of the ion, a non-local pseudopotential takes the form

$$v_{\mathrm{nloc}}^{(n)}(r'') = \sum_{\ell_0=0}^{\infty} v_{\mathrm{nloc}}^{(n)}(r'',\ell_0)\, |\ell_0\rangle\langle\ell_0|, \tag{3.54}$$

where $r'' = |\boldsymbol{r}''|$ is the distance from the n-th ion and $|\ell_0\rangle\langle\ell_0|$ is the projection operator on a state with angular momentum ℓ_0. Thus, different components of the valence electron wavefunction see different pseudopotentials. In order to deal with (3.54), it is convenient to expand the basis functions given in (3.46) in spherical harmonics around the center of the ion.

To do this, we split the calculation into two parts. We start by writing the spherical harmonics $Y_{\ell_j,m_j}(\theta,\phi)$ in the rotated system defined above, that is,

$$Y_{\ell_j,m_j}(\theta,\phi) = \sum_{m'_j=-\ell_j}^{\ell_j} D^{(\ell_j)*}_{m_j,m'_j}(\phi_i,\theta_i,0)\, Y_{\ell_j,m'_j}(\theta',\phi').$$

Then $\Phi_j(\boldsymbol{r})$ becomes

$$\Phi_j(\boldsymbol{r}) = \sum_{m'_j=-\ell_j}^{\ell_j} D^{(\ell_j)*}_{m_j,m'_j}(\phi_n,\theta_n,0)\, \Phi'_j(\boldsymbol{r}'), \tag{3.55}$$

where $|\boldsymbol{r}'| = |\boldsymbol{r}|$, and

$$\Phi'_j(\boldsymbol{r}') = R_{n_j,\ell_j}(r')\, Y_{\ell_j,m'_j}(\theta',\phi').$$

As a next step, remembering that the radial vector from the center of the ion $\boldsymbol{r}'' = \boldsymbol{r}' - \boldsymbol{R}_n$, and defining θ'' and ϕ'' as its corresponding polar and azimuthal angles, we look for an expansion of the form

$$\Phi'_j(\boldsymbol{r}') = \sum_{\ell,m} \alpha_{\ell,m}(r'',R_n,n_j,\ell_j,m'_j)\, Y_{\ell,m}(\theta'',\phi'') \tag{3.56}$$

where

$$\alpha_{\ell,m} = \int_0^\pi d\theta''\, \sin\theta'' \int_0^{2\pi} d\phi''\, R_{n_j,\ell_j}(r')\, Y_{\ell_j,m'_j}(\theta',\phi')\, Y^*_{\ell,m}(\theta'',\phi'').$$

Since $\phi' = \phi''$, the integration over ϕ'' is immediate and yields a factor $2\pi\delta_{m,m'_j}$. The integration over θ'' can be performed by noting that, according to the geometric relation $\boldsymbol{r}'' = \boldsymbol{r}' - \boldsymbol{R}_n$, the following results hold: $r''^2 = r'^2 + R_n^2 - 2r'R_n\cos\theta'$ and $r'^2 = r''^2 + R_n^2 - 2r''R_n\cos\gamma$, where $\gamma = \pi - \theta''$. Thus, we have

$$r' = (r''^2 + R_n^2 + 2r''R_n\cos\theta'')^{1/2},$$

and

$$\cos\theta' = \frac{r'^2 + R_n^2 - r''^2}{2r'R_n}.$$

Using these relations, we finally obtain

$$\alpha_{\ell,m} = 2\pi\,\delta_{m,m'_j} \int_{-1}^1 dx''\, R_{n_j,\ell_j}(r')\, \Theta_{\ell_j,|m|}(x')\, \Theta_{\ell,|m|}(x''), \tag{3.57}$$

where $x'' = \cos\theta''$ and $x' = \cos\theta'$. This equation is evaluated numerically in an appropriate range of distances r'', by taking into account the fact that the non-local pseudopotential has in general a rather short range as measured from the center of the ion (see Appendix 3.6.1).

3.4.2 The Expansion of $\varrho(r)$, v_{H} and v_{xc}

According to the expansion of $\varphi_i(\boldsymbol{r})$ in the spherical basis, (3.44), the total electronic density can be written as

$$\varrho(\boldsymbol{r}) = \sum_{i=1}^{N}\sum_{j=1}^{M} c_j^{(i)}\, \Phi_j(\boldsymbol{r}) \sum_{j'=1}^{M} c_{j'}^{(i)*}\, \Phi_{j'}^*(\boldsymbol{r}) = \sum_{j,j'} C_{j,j'}\, \Phi_j(\boldsymbol{r})\, \Phi_{j'}^*(\boldsymbol{r}), \quad (3.58)$$

where

$$C_{j,j'} = \sum_{i=1}^{N} c_j^{(i)}\, c_{j'}^{(i)*} = C_{j',j}^*.$$

As mentioned above, the electronic density admits a multipole expansion similar to (3.47), where now the density multipoles $\varrho_{L,M}$ are given by

$$\varrho_{L,M}(r) = \sum_{j,j'} C_{j,j'}\, R_{n_j,\ell_j}(r)\, R_{n_{j'},\ell_{j'}}(r) \int \mathrm{d}\Omega\, Y_{\ell_j,m_j}\, Y_{L,M}^*\, Y_{\ell_{j'},m_{j'}}^*. \quad (3.59)$$

The angular integration in the above equation,

$$I_\Omega = (-)^M \int \mathrm{d}\Omega\, Y_{\ell_j,m_j}(\theta,\phi)\, Y_{L,-M}(\theta,\phi)\, Y_{\ell_{j'},m_{j'}}^*(\theta,\phi),$$

can be expressed in terms of $3j$ coefficients or Wigner symbols as (see, e.g., [57])

$$(-)^{M+m_{j'}} \sqrt{\frac{(2\ell_j+1)(2L+1)(2\ell_{j'}+1)}{4\pi}} \begin{pmatrix} \ell_j & L & \ell_{j'} \\ 0 & 0 & 0 \end{pmatrix} \begin{pmatrix} \ell_j & L & \ell_{j'} \\ m_j & -M & -m_{j'} \end{pmatrix}.$$

Using the results $C_{j,j'} = C_{j',j}^*$ and $Y_{L,M}^* = (-)^M Y_{L,-M}$, it is easy to verify from (3.58) and (3.59) that $\varrho_{L,M}^*(r) = (-)^M \varrho_{L,-M}(r)$, which shows that $\varrho(\boldsymbol{r})$ is a real quantity. In the particular case in which the coefficients $c_j^{(i)}$ in (3.44) are real, then $C_{j,j'}$ and therefore $\rho_{L,M}$ are also real and one has, $\varrho_{L,-M}(r) = (-)^M \varrho_{L,M}(r)$. The same holds for the multipoles $v_{L,M}^{(\mathrm{loc})}(r)$, $v_{L,M}^{(\mathrm{H})}(r)$ and $v_{L,M}^{(\mathrm{xc})}(r)$.

The direct Coulomb interaction, or Hartree term, is given by

$$v_{\mathrm{H}}(\boldsymbol{r}) = e^2 \int \mathrm{d}^3 r'\, \frac{\varrho(\boldsymbol{r}')}{|\boldsymbol{r}-\boldsymbol{r}'|}. \quad (3.60)$$

In order to obtain the corresponding multipole components $v_{L,M}^{(\mathrm{H})}(r)$, one inserts in (3.60) the well known relation (see e.g. [55])

$$\frac{1}{|\boldsymbol{r}-\boldsymbol{r}'|} = \sum_{\ell=0}^{\infty}\sum_{m=-\ell}^{\ell} \frac{4\pi}{2\ell+1}\, \frac{r_<^\ell}{r_>^{\ell+1}}\, Y_{\ell,m}(\theta,\phi)\, Y_{\ell,m}^*(\theta',\phi'),$$

where $r_< = \min(r,r')$ and $r_> = \max(r,r')$. This, together with the multipole expansion of $\varrho(\boldsymbol{r}')$ and the orthonormality property of the spherical harmonics, yields

$$v_{L,M}^{(H)}(r) = \frac{4\pi e^2}{2L+1} \int_0^\infty dr'\ r'^2\ \frac{r_<^L}{r_>^{L+1}}\ \varrho_{L,M}(r'),$$

which can be explicitly written as

$$v_{L,M}^{(H)}(r) = \frac{4\pi e^2}{2L+1}\ \frac{1}{r}\left[\int_0^r dr'\ r'^2 \left(\frac{r'}{r}\right)^L \varrho_{L,M}(r')\right.$$
$$\left. + \int_r^\infty dr'\ r'^2 \left(\frac{r}{r'}\right)^{L+1} \varrho_{L,M}(r')\right]. \tag{3.61}$$

Concerning the exchange-correlation potential, $v_{\rm xc}(\boldsymbol{r})$, the calculation of the corresponding multipoles, $v_{L,M}^{({\rm xc})}(r)$, needs to be performed fully numerically, according to the standard relation,

$$v_{L,M}^{({\rm xc})}(r) = \int_0^\pi d\theta \sin\theta \int_0^{2\pi} d\phi\ v_{\rm xc}(\boldsymbol{r})\ Y_{L,M}^*(\theta,\phi).$$

Regarding the computational aspect of the present problem, the calculation of $v_{L,M}^{({\rm xc})}(r)$ in fact constitutes the most time consuming part of the present approach.

Finally, the multipole expansion of the local part of $v_{\rm eff}$ in (3.43) is obtained as the sum of $v_{L,M}^{({\rm loc})}(r)$, $v_{L,M}^{(H)}(r)$ and $v_{L,M}^{({\rm xc})}(r)$. We denote this sum $v_{L,M}^{({\rm loc,eff})}$ and we use it in (3.77), when calculating the matrix elements of the local part of $v_{\rm eff}$. In Appendix 3.6.2, these matrix elements, as well as those of the kinetic energy and of the non-local part of $v_{\rm eff}$, are explicitly evaluated.

3.5 Kohn–Sham Equations in a Cylindrical Basis

In cylindrical coordinates (ρ, z, ϕ), the LDA equations can be solved by expanding the KS orbitals φ on a basis of cylindrical wavefunctions

$$\varphi_i(\rho, z, \phi) = \sum_j c_j^{(i)}\ \phi_j(\rho, z, \phi), \tag{3.62}$$

(see [18] for more details) where the index j stands for the set of quantum numbers $j \equiv (n, \ell, m)$ and $\phi_j(\rho, z, \phi)$ are the basis functions, solutions of $-(\hbar^2/2m_{\rm e})\nabla^2\phi_j = \varepsilon_j\ \phi_j$, within a cylindrical box of radius ρ_0 and length $2z_0$, subject to the boundary conditions $\phi_j(\rho_0, z, \phi) = 0$ and $\phi_j(\rho, \pm z_0, \phi) = 0$, where $0 \le \rho \le \rho_0$, $-z_0 \le z \le z_0$ and $0 \le \phi \le 2\pi$. This case corresponds to a basis potential $v_{\rm Basis} = 0$ inside the box, and is the cylindrical equivalent of plane waves inside a cube.

The basis functions can be decomposed according to

$$\phi_j(\rho, z, \phi) = R_{|m|,n}(\rho)\ Z_\ell(z)\ \Phi_m(\phi),$$

where

$$\Phi'' = -m^2\, \Phi\,,$$

$$Z'' = -k_\ell^2\, Z\,,$$

and

$$R''(x) + \frac{1}{x}R'(x) + \left(1 - \frac{m^2}{x^2}\right)R(x) = 0\,,$$

with $x = D_{|m|,n}\rho$. This yields [55]

$$\Phi_m(\phi) = \frac{1}{\sqrt{2\pi}}\mathrm{e}^{\mathrm{i}m\phi}\,,$$

where m is a (positive or negative) integer number,

$$Z_\ell(z) = \frac{1}{\sqrt{z_0}}\sin(k_\ell\, z)\,, \qquad \text{for even } \ell \geq 2\,,$$

and

$$Z_\ell(z) = \frac{1}{\sqrt{z_0}}\cos(k_\ell\, z)\,, \qquad \text{for odd } \ell \geq 1\,,$$

with $k_\ell = \pi\ell/2z_0$. For the function R one has,

$$R_{|m|,n}(\rho) = \frac{\sqrt{2}}{\rho_0 J_{|m|+1}(D_{|m|,n}\rho_0)}\; J_{|m|}(D_{|m|,n}\rho)\,,$$

where $J_{|m|}$ is a Bessel function of integer order, $D_{|m|,n} = x_{|m|,n}/\rho_0$, with $n \geq 1$ and $J_{|m|}(x_{|m|,n}) = 0$. For $|m| \gg 1$ and $n \gg 1$, one has $x_{|m|,n} \approx n\pi + (|m| - 1/2)\pi/2$. The eigenvalues ε_j are then given by

$$\varepsilon_j = \frac{\hbar^2}{2m_\mathrm{e}}\left[\left(\frac{x_{|m|,n}}{\rho_0}\right)^2 + \left(\frac{\pi\ell}{2z_0}\right)^2\right].$$

In what follows, we briefly discuss how the matrix elements between the basis functions ϕ_j, and associated with the different terms in the KS equations, are calculated.

(1.) Kinetic energy: The kinetic energy term is diagonal, yielding

$$\int \mathrm{d}^3r\; \phi_{j'}^*(-)\frac{\hbar^2}{2m_\mathrm{e}}\nabla^2\; \phi_j = \varepsilon_j\; \delta_{n,n'}\; \delta_{\ell,\ell'}\; \delta_{m,m'}\,.$$

(2.) Ionic pseudopotentials: For the local part, we have

$$V_{j',j}^{(\mathrm{loc})}(n) = \int \mathrm{d}^3r\; \phi_{j'}^*\; v_{\mathrm{loc}}^{(\mathrm{n})}(|\boldsymbol{r} - \boldsymbol{R}_n|)\; \phi_j\,,$$

where $\boldsymbol{R}_n = (\rho_n, z_n, \phi_n)$ is the position vector of n-th ion. The integrations are performed numerically using the relation

$$|\boldsymbol{r} - \boldsymbol{R}_n|^2 = \rho^2 + \rho_n^2 - 2\rho\rho_n \cos(\phi - \phi_n) + (z - z_n)^2.$$

The non-local part of V_{ext} depends on the value of the angular momentum of the electron wavefunction. In order to compute its matrix elements, it is convenient (as we did for the case of spherical coordinates) to project the basis function ϕ_j onto the ion at $\boldsymbol{R}_n$, rather than expressing the projection operator in the original coordinates (ρ, z, ϕ). This can be accomplished carrying out the expansion,

$$\phi_j^{(n)}(\rho, z, \phi) = \sum_{\ell_0, m_0} \Phi_{\ell_0 m_0, j}^{(n)}(r) \, Y_{\ell_0, m_0}(\theta_s, \phi_s),$$

where Y_{ℓ_0, m_0} are the spherical harmonics, and θ_s and ϕ_s the corresponding spherical angles measured with respect to the atomic position $\boldsymbol{R}_n$. In terms of the projected wavefunctions $\Phi_{\ell_0 m_0, j}^{(n)}(r)$, the non-local matrix elements become

$$V_{j',j}^{(\text{nloc})}(n) = \sum_{\ell_0, m_0} \int_0^\infty \mathrm{d}r \; r^2 \; \Phi_{\ell_0 m_0, j'}^{*(n)}(r) \; v_{\text{nloc}}^{(n)}(r, \ell_0) \; \Phi_{\ell_0 m_0, j}^{(n)}(r).$$

(3.) Hartree term: To obtain v_{H}, it is convenient to use the expansion [55]

$$\frac{1}{|\boldsymbol{r} - \boldsymbol{r}'|} = \frac{2}{\pi} \sum_{m=-\infty}^{\infty} e^{\mathrm{i}m(\phi - \phi')} \int_0^\infty \mathrm{d}k \; \cos[k(z - z')] \, I_{|m|}(k\rho_<) \, K_{|m|}(k\rho_>),$$

where $I_{|m|}$ and $K_{|m|}$ are modified Bessel functions of integer order, together with the cylindrical expansion

$$\varrho(\boldsymbol{r}) = \sum_{L,M} \varrho_{L,M}(\rho) \, Z_L(z) \, \Phi_M(\phi),$$

with Z and Φ as defined above. In the case of Z_L, it holds $k_L = \pi L/2z_0$ with $L > 0$. The matrix elements of v_{H} thus become

$$v_{j',j}^{(H)} = \frac{2e^2}{\pi} \frac{1}{(2\pi z_0)^{3/2}} \sum_L (-)^{(L+\ell'+\ell-1)/2}$$

$$\times \int_0^{\rho_0} \mathrm{d}\rho \; \rho \; R_{|m'|,n'}(\rho) \, R_{|m|,n}(\rho) \int_0^{\rho_0} \mathrm{d}\rho' \; \rho' \; \varrho_{L,m'-m}(\rho')$$

$$\times \int_0^\infty \mathrm{d}k \; I_{|m'-m|}(k\rho_<) \, K_{|m'-m|}(k\rho_>) \frac{kk_L}{k^2 - k_L^2} \; \sin(2kz_0) \, [\Delta_{\ell',\ell}^{(-)}(k) - \Delta_{\ell',\ell}^{(+)}(k)],$$

where

$$\Delta_{\ell',\ell}^{(-)}(k) = \frac{1}{k^2 - (k_{\ell'} \pm k_\ell)^2}.$$

(4.) Exchange and correlation potentials: The matrix elements of v_{xc} are obtained by direct numerical integration.

3.6 Appendices

3.6.1 Basics on Pseudopotentials

To solve the Kohn–Sham equations (3.25) and (3.26), one needs to perform a number of simplifications, in particular regarding the external potential, $V_{\text{ext}}(\boldsymbol{r})$, due to the interaction of electrons with the atomic nuclei. The question is whether one can consider only the 'valence' electrons explicitly, by incorporating the most tightly bound electrons, i.e. those forming the so called 'core' of the atom, into an effective electron–ion potential.

This simplification is justified by the fact that only the valence electrons play a role in the formation of an aggregate or molecule, extending their wavefunctions far beyond the size of the atom and giving rise to the 'chemical bonds'. The core electrons, being so closely bound to the nuclei, play a very minor role in the formation of such interatomic bonds. In some circumstances one may be interested in the effects of their 'polarization', but these are in general small and will be neglected in the present discussion.

If one wishes to consider the valence electrons only, the question remains of how the effects of the core electrons can be effectively implemented. This leads us to the concept of 'pseudopotentials'. These are functions constructed in order to work accurately in different 'chemical environments'. In other words, they are 'transferable' from one type of cluster to another. The price to be paid to obtain this simplification, is that pseudopotentials are usually angular momentum dependent, as discussed below.

To derive the pseudopotentials, one needs to perform an all-electron calculation of the atom of interest. Such a calculation presents a number of difficulties, including the consideration of relativistic effects in the case of atoms with relatively large atomic number (see e.g. [52, 53]). As a rule, the calculations are implemented by employing the KS scheme within the LDA. Since the atom has spherical symmetry, the KS equation reduces simply to the radial equation for the wavefunctions $R_{n,\ell}(r)$, where n (≥ 1) is the main quantum number and ℓ (≥ 0) the angular momentum quantum number. The radial equation reads,

$$\left(-\frac{\hbar^2}{2m_{\text{e}}}\frac{\mathrm{d}^2}{\mathrm{d}r^2} + \frac{\hbar^2}{2m_{\text{e}}}\frac{\ell(\ell+1)}{r^2} + v_{\text{eff}}(r) \right) rR_{n,\ell}(r) = \varepsilon_{n,\ell}\, rR_{n,\ell}(r), \quad (3.63)$$

where (see also (3.43))

$$v_{\text{eff}}(r) = V_{\text{ext}}(r) + v_{\text{H}}(r) + v_{\text{xc}}(r), \tag{3.64}$$

with

$$V_{\text{ext}}(r) = -\frac{Z_{\text{at}}e^2}{r}, \tag{3.65}$$

Z_{at} being the atomic number. The full electronic wavefunction is given by

$$\Phi_{n,\ell,m}(\boldsymbol{r}) = R_{n,\ell}(r)\, Y_{\ell,m}(\theta,\phi), \tag{3.66}$$

where $Y_{\ell,m}(\theta,\phi)$ are the spherical harmonics and $-\ell \le m \le \ell$ (see e.g. [57]).

To construct a pseudopotential, $V_\ell^{\mathrm{PS}}(r)$, we need to invert (3.63) once the $R_{n,\ell}(r)$ are known. This amounts to dividing by $R_{n,\ell}(r)$, an operation which runs into difficulties due to the $(n-1)$ nodes of $R_{n,\ell}(r)$, yielding a division by zero for $n > 1$. The idea is then to construct a 'pseudo-wavefunction', $R_\ell^{\mathrm{PS}}(r)$, which does not have such nodes. Note that for simplicity, we have omitted the index n from the latter since it is a nodeless wavefunction.

The condition imposed to the $R_\ell^{\mathrm{PS}}(r)$ is that it coincides with the all-electron wavefunction for distances $r > r_{\mathrm{c}}(\ell)$, i.e.

$$R_\ell^{\mathrm{PS}}(r) = R_{n,\ell}(r) \quad \text{for} \quad r > r_{\mathrm{c}}(\ell), \tag{3.67}$$

where $r_{\mathrm{c}}(\ell)$ is an appropriate matching distance, larger than the value at which $R_{n,\ell}(r)$ has its last node (see Fig. 3.2).

Using the normalization condition of R, together with (3.67), one obtains that 'inside' the core, i.e. for $r < r_{\mathrm{c}}(\ell)$,

$$\int_0^{r_{\mathrm{c}}} \mathrm{d}r \; r^2 \, (R_\ell^{\mathrm{PS}}(r))^2 = \int_0^{r_{\mathrm{c}}} \mathrm{d}r \; r^2 \, R_{n,\ell}^2(r). \tag{3.68}$$

This leads automatically to the matching of the logarithmic derivative (3.40) at $r = r_{\mathrm{c}}(\ell)$, up to first order in energy differences $\varepsilon - \varepsilon_{n,\ell}$ (see (3.41)). Thus, together with the obvious requirement

$$\varepsilon_\ell^{\mathrm{PS}} = \varepsilon_{n,\ell} \tag{3.69}$$

one can invert (3.63) to obtain,

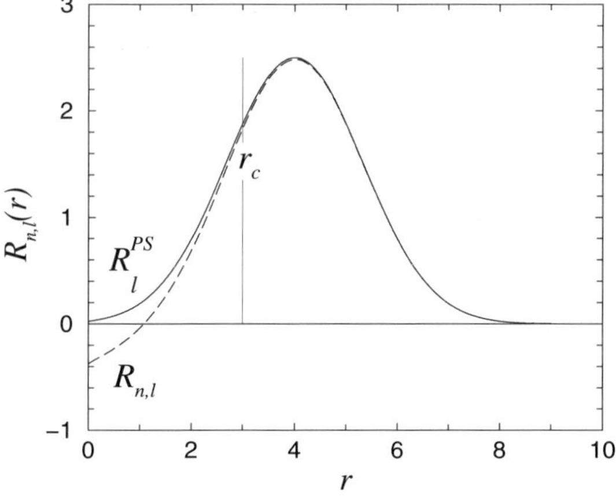

Fig. 3.2. Schematic illustration of the radial behavior of the full-electron-calculation wavefunction, $R_{n,\ell}(r)$ (*dashed line*), and its corresponding pseudo-wavefunction, $R_\ell^{\mathrm{PS}}(r)$ (*continuous line*), as a function of distance r from the center of the atom. The location of the matching point, r_{c}, beyond which (3.67) is valid is indicated by the vertical line. The case shown here corresponds to $\ell = 0$ and $n = 2$

$$V_\ell(r) = \varepsilon_\ell^{\mathrm{PS}} - \frac{\hbar^2}{2m_e}\frac{\ell(\ell+1)}{r^2} + \frac{\hbar^2}{2m_e r R_\ell^{\mathrm{PS}}(r)}\frac{\mathrm{d}^2}{\mathrm{d}r^2}\left(r R_\ell^{\mathrm{PS}}(r)\right). \qquad (3.70)$$

Finally, since the pseudopotential must be used in a self-consistent equation, we have to substract from $V_\ell^{\mathrm{PS}}(r)$ the v_{H} and v_{xc} contributions of the valence electrons, i.e.

$$V_\ell^{\mathrm{PS}}(r) = V_\ell(r) - v_{\mathrm{H}}^{\mathrm{PS}}[\varrho_{\mathrm{val}}^{\mathrm{PS}}] - v_{\mathrm{xc}}^{\mathrm{PS}}[\varrho_{\mathrm{val}}^{\mathrm{PS}}] \qquad (3.71)$$

calculated from their total density

$$\varrho_{\mathrm{val}}^{\mathrm{PS}}(r) = \sum_{\ell \in \mathrm{val}} N_\ell \, |R_\ell^{\mathrm{PS}}|^2. \qquad (3.72)$$

where N_ℓ is the number of valence electrons in the ℓ-state.

Clearly, the pseudopotential $V_\ell^{\mathrm{PS}}(r)$ obeys the general behavior,

$$V_\ell^{\mathrm{PS}}(r) \to -\frac{Z_{\mathrm{ion}}e^2}{r} \quad \text{for} \quad r \to \infty, \qquad (3.73)$$

where Z_{ion} is the ionic charge of the pseudo-atom and it can be decomposed into a 'local' part independent of ℓ, which we denote $v_{\mathrm{loc}}(r)$, and a 'non-local' one, denoted as $v_{\mathrm{nloc}}(r, \ell)$, which depends on ℓ. The local part obeys (3.73), while $v_{\mathrm{nloc}}(r, \ell)$ is of short range.

As an example let us consider the case of Carbon. Its valence electron configuration is $2s^2 2p^2$ and $Z_{\mathrm{ion}} = 4$. Therefore, for these four electrons two pseudopotentials $V_\ell^{\mathrm{PS}}(r)$ are needed, i.e., $V_0^{\mathrm{PS}}(r)$ and $V_1^{\mathrm{PS}}(r)$. Higher energy states with $\ell > 1$ would require additional pseudopotentials. In this case, one can choose $v_{\mathrm{loc}}(r)$ such that $v_{\mathrm{nloc}}(r, \ell)$ vanishes for $r > 0.5$ Å (see Fig. 3.3).

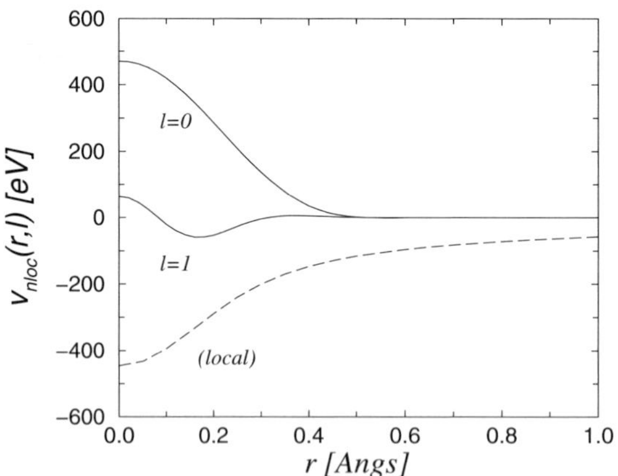

Fig. 3.3. Pseudopotentials for Carbon, obtained from the parametrization reported in [53], for the local part $v_{\mathrm{loc}}(r)$ (*dashed line*) and the non-local part $v_{\mathrm{nloc}}(r, \ell)$ (*continuous lines*) for $\ell = 0$ and $\ell = 1$. We note that for $r > 0.5$ Å, $v_{\mathrm{loc}}(r)$ coincides with the asymptotic Coulomb tail given in (3.73)

3.6.2 Matrix Elements in a Spherical Basis

To obtain the KS orbitals $\varphi(\boldsymbol{r})$, we diagonalize the hermitian matrix associated with (3.42), whose matrix elements are calculated within the spherical basis functions $\Phi_j(\boldsymbol{r})$ as

$$H_{j',j} = \int d^3r \; \Phi_{j'}^*(\boldsymbol{r}) \left[-\frac{\hbar^2}{2m_{\mathrm{e}}} \nabla^2 + v_{\mathrm{eff}}(\boldsymbol{r}) \right] \Phi_j(\boldsymbol{r}). \tag{3.74}$$

The lowest eigenvalues of the $M \times M$ matrix H, λ_i, provide the occupied KS single-particle energies of (3.42), and the corresponding eigenvectors the set of coefficients $\{c_j^{(i)}\}$ introduced in (3.44). Let us consider separately the different terms which enter into (3.74).

The matrix elements of the kinetic energy term, $K_{j,j'}$, can be calculated according to (3.45) as

$$K_{j',j} = \varepsilon_j \; \delta_{j,j'} - \int d^3r \; \Phi_{j'}^*(\boldsymbol{r}) \; v_{\mathrm{Basis}}(r) \; \Phi_j(\boldsymbol{r}) \,. \tag{3.75}$$

According to (3.46) and the orthonormality property of the $Y_{\ell,m}$, $K_{j',j}$ reduces to,

$$\begin{aligned}
K_{j',j} = {}& \varepsilon_j \delta_{j,j'} \\
& - \delta_{\ell_j,\ell_{j'}} \delta_{m_j,m_{j'}} \int_0^\infty dr \; r^2 \; R_{n_{j'},\ell_{j'}}(r) \; v_{\mathrm{Basis}}(r) \; R_{n_j,\ell_j}(r) \,.
\end{aligned} \tag{3.76}$$

The local part of the effective potential, $v_{\mathrm{eff}}^{(\mathrm{loc})}(\boldsymbol{r})$, admits a multipole expansion of the form

$$v_{\mathrm{eff}}^{(\mathrm{loc})}(\boldsymbol{r}) = \sum_{L,M} v_{L,M}^{(\mathrm{loc,eff})}(r) \; Y_{L,M}(\theta,\phi) \,. \tag{3.77}$$

The corresponding matrix elements, $V_{j',j}^{(\mathrm{loc})}$, are then given by

$$V_{j',j}^{(\mathrm{loc})} = \sum_{L,M} \int_0^\infty dr \; r^2 \; R_{n_{j'},\ell_{j'}}(r) \; v_{L,M}^{(\mathrm{loc,eff})}(r) \; R_{n_j,\ell_j}(r) \cdot I_\Omega, \tag{3.78}$$

where

$$I_\Omega = \int d\Omega \; Y_{\ell_{j'},m_{j'}}^*(\theta,\phi) \; Y_{L,M}(\theta,\phi) \; Y_{\ell_j,m_j}(\theta,\phi) \,. \tag{3.79}$$

Let us consider first the contribution of the n-th ion to the matrix elements of the non-local part of the pseudopotential, $V_{j',j}^{(\mathrm{nloc})}(n)$. The projection operator in (3.54), $|\ell_0\rangle\langle\ell_0|$, can be written in terms of spherical harmonics referred to the coordinate axis attached to the ion as

$$|\ell_0\rangle\langle\ell_0| = \sum_{m_0=-\ell_0}^{\ell_0} Y_{\ell_0,m_0}(\theta_1'',\phi_1'') \; Y_{\ell_0,m_0}^*(\theta_2'',\phi_2''), \tag{3.80}$$

where the double prime indicates the choice of the coordinate system centered at the ion and the different subindices, 1 and 2, correspond to different angular variables. In this coordinate system, $V_{j',j}^{(\mathrm{nloc})}$ is given by (see (3.54))

$$V_{j',j}^{(\mathrm{nloc})}(n) = \sum_{\ell_0=0}^{\infty} \int_0^{\infty} \mathrm{d}r'' \; r''^2 \; v_{\mathrm{nloc}}^{(n)}(r'',\ell_0) \sum_{m_0=-\ell_0}^{\ell_0} I_{\Omega_1''} \, I_{\Omega_2''}, \qquad (3.81)$$

where

$$I_{\Omega_1''} = \int \mathrm{d}\Omega_1'' \; \Phi_{j'}^*(\boldsymbol{r}) \, Y_{\ell_0,m_0}(\theta_1'',\phi_1''), \qquad (3.82)$$

and

$$I_{\Omega_2''} = \int \mathrm{d}\Omega_2'' \; \Phi_j(\boldsymbol{r}) \, Y_{\ell_0,m_0}^*(\theta_2'',\phi_2''). \qquad (3.83)$$

According to (3.55), (3.56) and (3.57), the basis function $\Phi_j(\boldsymbol{r})$ can now be written as

$$\Phi_j(\boldsymbol{r}) = \sum_{m_j'=-\ell_j}^{\ell_j} D_{m_j,m_j'}^{(\ell_j)*}(\phi_n,\theta_n,0)$$

$$\times \sum_{\ell=0}^{\infty} \alpha_{\ell,m_j'}(r'',R_n,n_j,\ell_j,m_j') \, Y_{\ell,m_j'}(\theta_2'',\phi_2''), \qquad (3.84)$$

and the angular integrations yield, for $|m_0| \le \ell_{j'}$,

$$I_{\Omega_1''} = D_{m_{j'},m_0}^{(\ell_{j'})}(\phi_n,\theta_n,0) \, \alpha_{\ell_0,m_0}(r'',R_n,n_{j'},\ell_{j'},m_0), \qquad (3.85)$$

and

$$I_{\Omega_2''} = D_{m_j,m_0}^{(\ell_j)*}(\phi_n,\theta_n,0) \, \alpha_{\ell_0,m_0}(r'',R_n,n_j,\ell_j,m_0). \qquad (3.86)$$

Thus, (3.81) becomes

$$V_{j',j}^{(\mathrm{nloc})}(n) = \sum_{\ell_0,m_0} D_{m_{j'},m_0}^{(\ell_{j'})}(\phi_n,\theta_n,0)$$

$$\times D_{m_j,m_0}^{(\ell_j)*}(\phi_n,\theta_n,0) \, F_{j,j'}^{(n)}(\ell_0,m_0), \qquad (3.87)$$

where

$$F_{j,j'}^{(n)}(\ell_0,m_0) = \int_0^{\infty} \mathrm{d}r'' \; r''^2 \; \alpha_{\ell_0,m_0}(r'',R_n,n_{j'},\ell_{j'},m_0) \, v_{\mathrm{nloc}}^{(n)}(r'',\ell_0)$$

$$\times \alpha_{\ell_0,m_0}(r'',R_n,n_j,\ell_j,m_0). \qquad (3.88)$$

Equation (3.87) can be further simplified by noting that

$$D_{m_j,m_0}^{(\ell_j)*}(\phi_i,\theta_i,0) = (-)^{m_j+m_0} \, D_{-m_j,-m_0}^{(\ell_j)}(\phi_i,\theta_i,0),$$

and

$$
D^{(\ell_{j'})}_{m_{j'},m_0} \, D^{(\ell_j)}_{-m_j,-m_0} = \sum_{L,M,m} (-)^{M+m} \, (2L+1) \begin{pmatrix} \ell_{j'} & \ell_j & L \\ m_{j'} & -m_j & -M \end{pmatrix}
$$

$$
\times D^{(L)}_{M,m} \begin{pmatrix} \ell_{j'} & \ell_j & L \\ m_0 & -m_0 & -m \end{pmatrix} \tag{3.89}
$$

where the arguments of D have been omitted for simplicity. It follows that $m = 0$, since the sum $m_0 - m_0 - m$ must vanish. Thus, only the function $D^{(L)}_{M,0}$, which was already present in (3.52), is required here again. All together, $V^{(\mathrm{nloc})}_{j,j'}(n)$ now reads

$$
V^{(\mathrm{nloc})}_{j',j}(n) = \sum_{L,M} A^{j,j'}_{L,M}(\theta_n,\phi_n) \sum_{\ell_0,m_0} (-)^{m_0} \begin{pmatrix} \ell_{j'} & \ell_j & L \\ m_0 & -m_0 & 0 \end{pmatrix}
$$

$$
\times F^{(n)}_{j,j'}(\ell_0,m_0), \tag{3.90}
$$

where

$$
A^{j,j'}_{L,M}(\theta_n,\phi_n) = \sqrt{4\pi(2L+1)} \, Y^*_{L,M}(\theta_n,\phi_n) \, (-)^{M+m_j} \begin{pmatrix} \ell_{j'} & \ell_j & L \\ m_{j'} & -m_j & -M \end{pmatrix}.
$$

The generalization to the total matrix elements of the non-local part of the pseudopotential amounts to sum over the index n and is therefore straightforward.

4. Electronic Response to Time-Dependent Perturbations

In Chap. 3 we discussed how to determine the ground state of a system of ions and electrons within the DFT-LDA. We have seen that this ground state corresponds to a minimum of the total energy written as a functional of the electronic density. If a weak external time-dependent perturbation is applied to the electronic system (for instance, a low intensity electromagnetic field of a sufficiently high frequency such that one can neglect the response of the ionic degrees of freedom), the electronic density will start oscillating around the minimum energy configuration. For such weak external fields, one can, to a very good approximation, retain only the linear part of the response. In this approximation, the frequency of the induced oscillations will be determined by the curvature of the energy surface around its minimum. As the density oscillates, the effective potential changes and, in turn, induces a change in the density itself. Therefore, to describe these electron 'vibrations' we must require self-consistency between the variations of the density and those of the effective potential. In this way, a time dependent theory can be obtained starting, e.g., from Hartree, Hartree–Fock or LDA energy functional within the linear response approximation. However, the first two approaches are intrinsically based on a mean-field approximation and neglect correlations beyond those arising from the Pauli principle. On the other hand, the time-dependent extension of the LDA (TDLDA), could also be derived, making again a local approximation, from an exact theorem which is the time-dependent counterpart of the HK one (see e.g. [61]) and which leads to the so-called time-dependent DFT (TDDFT). In the following, we derive explicitly the TDLDA equations from the LDA Hamiltonian without going through the TDDFT formalism. The related framework known as the Random Phase Approximation (RPA)(see e.g. [10]) will be also mentioned.

4.1 Linear Response: RPA and TDLDA

Let us denote by H_0 an unperturbed, time-independent self-consistent-field Hamiltonian of the form

$$H_0 = -\frac{\hbar^2}{2m_{\mathrm{e}}}\nabla^2 + v_{\mathrm{eff}}[\varrho_0], \tag{4.1}$$

where the effective potential $v_{\text{eff}}[\varrho_0]$ includes the Hartree, ionic, and, in the case of the Kohn–Sham equations of DFT, also an exchange-correlation potential (see e.g. (3.26)). H_0 is evaluated at the self-consistent density of the ground state, ϱ_0. The associated set of time-dependent Schrödinger equations is

$$i\hbar \frac{\partial}{\partial t}\varphi_h(t) = H_0\varphi_h(t) = \varepsilon_h\varphi_h(t) \tag{4.2}$$

whose solutions $\varphi_h(t) = \exp(-i\varepsilon_h t/\hbar)\varphi_h(0)$ are the (stationary) one-particle states of the system, where $\varphi_h(0)$ is the solution of (3.25). The label h refers here to occupied states. As a result of a perturbation, electrons can be promoted from occupied states h, which become 'hole' states, to empty (or unoccupied) states p, which we denote as 'particle' states.

We specify the perturbation which is added to the Hamiltonian (4.1) by writing it as,

$$V_{\text{pert}}(\boldsymbol{r},t) = V_{\text{pert}}(\boldsymbol{r})\exp(-i\omega t) + \text{h.c.} \tag{4.3}$$

The perturbed density will be then

$$\varrho(\boldsymbol{r},t) = \varrho_0(\boldsymbol{r}) + \delta\varrho(\boldsymbol{r},\omega)\cdot\exp(-i\omega t) + \text{h.c.}, \tag{4.4}$$

where ϱ_0 is the ground state density. Since the Hamiltonian H_0 is density dependent (through the term v_{eff}) it must acquire, in addition to the external perturbation, a term like

$$\delta H(\boldsymbol{r},t) = \delta v_{\text{eff}}(\boldsymbol{r},\omega)\cdot\exp(-i\omega t) + \text{h.c.}, \tag{4.5}$$

where

$$\delta v_{\text{eff}}(\boldsymbol{r},\omega) = \int d^3r' \frac{\delta v_{\text{eff}}(\boldsymbol{r})}{\delta\varrho(\boldsymbol{r}')}\,\delta\varrho(\boldsymbol{r}',\omega). \tag{4.6}$$

The total Hamiltonian is then

$$H = H_0 + [(V_{\text{pert}} + \delta v_{\text{eff}})\cdot\exp(-i\omega t) + \text{h.c.}]. \tag{4.7}$$

It can be shown that, if one writes the perturbed wavefunctions as

$$\varphi_h'(t) = \exp(-i\varepsilon_h t/\hbar)\,[\varphi_h + \delta\varphi_h^{(-)}\exp(-i\omega t) + \delta\varphi_h^{(+)}\exp(i\omega t)], \tag{4.8}$$

and inserts them into the Schrödinger equation $i\hbar\partial\varphi_h'(t)/\partial t = H\varphi_h'(t)$, using the Hamiltonian (4.7) and neglecting second order terms, the following equation for the self-consistent density variation is obtained (see Appendix 4.4.1, in particular (4.52)),

$$\delta\varrho(\boldsymbol{r},\omega) = \int d^3r' \Pi^0(\boldsymbol{r},\boldsymbol{r}',\omega)\,V_{\text{tot}}(\boldsymbol{r}') \tag{4.9}$$

with

$$V_{\text{tot}}(\boldsymbol{r}') = V_{\text{pert}}(\boldsymbol{r}') + \int d^3r_1 \frac{\delta v_{\text{eff}}(\boldsymbol{r}')}{\delta\varrho(\boldsymbol{r}_1)}\delta\varrho(\boldsymbol{r}_1,\omega). \tag{4.10}$$

Here, Π^0 is the density–density correlation function, representing the probability amplitude of a change of density at $\boldsymbol{r}$ induced by a change in the density at $\boldsymbol{r}'$ caused by $V_{\text{pert}}(\boldsymbol{r}')$. It is defined (for real wavefunctions) as:

$$\Pi^0(\boldsymbol{r},\boldsymbol{r}',\omega) = \sum_{h=1}^{\text{occ}} \sum_{p\in\{\text{unocc}\}} \varphi_h^*(\boldsymbol{r})\varphi_p(\boldsymbol{r})\; \varphi_h(\boldsymbol{r}')\varphi_p^*(\boldsymbol{r}')$$

$$\left(\frac{1}{\hbar\omega - (\varepsilon_p - \varepsilon_h)} - \frac{1}{\hbar\omega + (\varepsilon_p - \varepsilon_h)}\right). \tag{4.11}$$

From this equation one can see that the poles of Π^0 occur at energies $\hbar\omega$ corresponding to unperturbed particle–hole (p–h) transitions, $\varepsilon_p - \varepsilon_h$. These poles could cause instabilities in the numerical calculation of (4.11). To avoid this problem, it is customary to shift the poles from the real axis to the complex plane by adding a small imaginary quantity $+i\eta$ in the denominators of (4.11)[1]. It can be shown that this is completely equivalent to averaging $\Pi^{(0)}$ around $\hbar\omega$ by means of a Lorentzian function having width equal to η.

The density response to an arbitrary external field can be calculated from (4.10) by neglecting the residual interaction term $\delta v_{\text{eff}}/\delta\varrho$. This yields the so-called unperturbed response of the system, where the induced density variation is given by

$$\delta\varrho^0(\boldsymbol{r},\omega) = \int \mathrm{d}^3r'\; \Pi^0(\boldsymbol{r},\boldsymbol{r}',\omega)\, V_{\text{pert}}(\boldsymbol{r}'), \tag{4.12}$$

and is also known as the independent particle response. Accordingly, it will be indicated with the superscript ^{0}. Similarly to (4.12), the self-consistent expression (4.10) can be rewritten as,

$$\delta\varrho(\boldsymbol{r},\omega) = \int \mathrm{d}^3r'\; \Pi(\boldsymbol{r},\boldsymbol{r}',\omega)\, V_{\text{pert}}(\boldsymbol{r}'), \tag{4.13}$$

where $\Pi(\boldsymbol{r},\boldsymbol{r}',\omega)$ is the so-called RPA density–density correlation function. It can be shown, using (4.13) and (4.10), that Π obeys the following Dyson-type equation,

$$\Pi(\boldsymbol{r},\boldsymbol{r}',\omega) = \Pi^0(\boldsymbol{r},\boldsymbol{r}',\omega)$$

$$+ \int \mathrm{d}^3r_1 \int \mathrm{d}^3r_2\; \Pi^0(\boldsymbol{r},\boldsymbol{r}_1,\omega)\, K(\boldsymbol{r}_1,\boldsymbol{r}_2)\, \Pi(\boldsymbol{r}_2,\boldsymbol{r}',\omega), \tag{4.14}$$

which can be written formally[2] as

$$\Pi = (1 - \Pi^0\, K)^{-1}\, \Pi^0. \tag{4.15}$$

---

[1] Care must be taken in choosing the sign of the small imaginary parts: the choice leading to a causal (instead of time-ordered) Π is usually adopted within the framework of linear response theory, while a time-ordered quantity, which allows one to apply Wick's theorem, is useful within the framework of many-body perturbation theory (see the discussion in Sect. 13 of [62]).

[2] There are other equivalent ways to express Π, e.g., $\Pi = \Pi^0(1 - K\Pi^0)^{-1}$ or $\Pi = \Pi^0(\Pi^0 - \Pi^0 K\Pi^0)^{-1}\Pi^0$

Here K is a shorthand notation for the kernel $\delta v_{\text{eff}}/\delta\varrho$ in (4.10), containing the total residual interaction between electrons, i.e.,

$$K(\boldsymbol{r}_1,\boldsymbol{r}_2) \equiv \frac{\delta v_{\text{eff}}(\boldsymbol{r}_1)}{\delta\varrho(\boldsymbol{r}_2)} = \frac{e^2}{|\boldsymbol{r}_1 - \boldsymbol{r}_2|} + f_{\text{xc}}(\boldsymbol{r}_1,\boldsymbol{r}_2), \tag{4.16}$$

where the first term on the right-hand side comes from the density variation of the Hartree potential (and is usually responsible for the largest part of the difference between Π_0 and Π [33]). If one starts from the exact time-dependent extension of the HK theorem, f_{xc} is in general a time- or frequency-dependent quantity, and we should write it as

$$f_{\text{xc}}(\boldsymbol{r}_1,\boldsymbol{r}_2;t_1,t_2) = \frac{\delta^2 E_{\text{xc}}}{\delta\rho(\boldsymbol{r}_1)\delta\rho(\boldsymbol{r}_2)} = \frac{\delta v_{\text{xc}}(\boldsymbol{r}_1,t_1)}{\delta\rho(\boldsymbol{r}_2,t_2)}, \tag{4.17}$$

considering the density variation of the exact DFT exchange-correlation potential, v_{xc} [61]. In LDA, one has:

$$f_{\text{xc}}(\boldsymbol{r}_1,\boldsymbol{r}_2;t_1,t_2) = \delta(\boldsymbol{r}_1 - \boldsymbol{r}_2)\,\delta(t_1 - t_2)\,\frac{\mathrm{d}v_{\text{xc}}}{\mathrm{d}\varrho}, \tag{4.18}$$

and the kernel is hence frequency-independent.

Let us mention that e.g. in nuclear physics, as well as in other contexts, one refers to the above equations as the self-consistent RPA equations [10]. However, in solid state physics RPA refers sometimes to a theory where only the Coulomb part of the residual interaction is considered, i.e. RPA is used as a synonym of "linearized time-dependent Hartree" equations. It is so, e.g., in the theory originally proposed by Bohm and Pines [63], in which only the Coulomb interaction is used to solve the Hartree–Fock equations and to build Π^0 accordingly, and then as the kernel to solve (4.14). Another possible source of misunderstanding about the meaning of RPA among physicists from different fields comes from the use (quite common within the solid state physics community) to name RPA the "independent-quasiparticle" approximation for the irreducible polarizability of the many-body theory [33, 62], i.e. the approximation in which this polarizability is written in terms of quasiparticle orbitals and energies as in (4.11). The present formalism, in which one solves (4.15) with a kernel which includes also the derivative of the exchange-correlation potential (4.16), is generally known as, and so will be called in the following, time-dependent LDA (TDLDA), avoiding in this way any possible misunderstanding.

Finally, it can be shown that the RPA, or TDLDA, propagator Π admits the following spectral representation,

$$\begin{aligned}
\Pi(\boldsymbol{r},\boldsymbol{r}',\omega) &= \sum_n \langle 0|\varrho(\boldsymbol{r})|n\rangle\langle n|\varrho(\boldsymbol{r}')|0\rangle \left(\frac{1}{\hbar\omega - E_n} - \frac{1}{\hbar\omega + E_n}\right) \\
&= \sum_n \delta\varrho_n^*(\boldsymbol{r})\,\delta\varrho_n(\boldsymbol{r}') \left(\frac{1}{\hbar\omega - E_n} - \frac{1}{\hbar\omega + E_n}\right)
\end{aligned} \tag{4.19}$$

where the poles of Π occur at the excited states energies E_n corresponding to the vibrational modes of the electronic density. As discussed above for $\Pi^{(0)}$, these poles are usually shifted to the complex plane, namely the denominators of (4.19) contain an imaginary term $+i\eta$. These modes are usually called plasmons. In the next Section, we discuss how the quantities $\delta\varrho_n(r') = \langle n|\varrho(r')|0\rangle$ introduced in (4.19) are related with the transition probabilities to the excited states n.

4.2 Strength Function and Sum Rules

From the induced density $\delta\varrho(\boldsymbol{r},\omega)$ one can extract physical quantities of interest such as the dynamical linear polarizability, $\alpha(\omega)$ [64], according to

$$\alpha(\omega) = \int d^3r \, \delta\varrho(\boldsymbol{r},\omega) \, V_{\text{pert}}(\boldsymbol{r}). \tag{4.20}$$

Making use of (4.13), $\alpha(\omega)$ can be written as,

$$\alpha(\omega) = \int d^3r \, d^3r' \, V_{\text{pert}}(\boldsymbol{r}) \, \Pi(\boldsymbol{r},\boldsymbol{r}',\omega) \, V_{\text{pert}}(\boldsymbol{r}'). \tag{4.21}$$

The excitation of the system by a short-duration perturbation is conveniently described by the strength function $\mathcal{S}(E)$, defined by

$$\mathcal{S}(E) \equiv \sum_n |\langle n|V_{\text{pert}}|0\rangle|^2 \, \delta(E - E_n), \tag{4.22}$$

where $E = \hbar\omega$ is the excitation energy, and $|0\rangle$ and $|n\rangle$ denote the ground and excited states of the system, respectively. A useful relation between (4.21) and (4.22) can be derived by adding, as already mentioned, a small imaginary term $+i\eta$ to the denominators in (4.19). Making use of the relation

$$\lim_{\eta\to 0} \frac{1}{E - E_n + i\eta} = P\frac{1}{E - E_n} - i\pi\delta(E - E_n)$$

it can be shown, using (4.19) in (4.21), that

$$-\frac{1}{\pi}\text{Im}\,\alpha(E) = \sum_n \left|\int d^3r \, \delta\varrho_n(\boldsymbol{r}) \, V_{\text{pert}}(\boldsymbol{r})\right|^2 \delta(E - E_n) = \mathcal{S}(E) \tag{4.23}$$

where, by definition,

$$\int d^3r \, \delta\varrho_n(\boldsymbol{r}) \, V_{\text{pert}}(\boldsymbol{r}) = \langle n|V_{\text{pert}}|0\rangle. \tag{4.24}$$

Some general properties of the response of the many-particle system to the external field V_{pert} can be analyzed and understood in a transparent way in terms of *sum rules* (see e.g. [6, 10, 65, 66]). The sum rules are defined as the *k-moments* of the strength function,

$$M_k(V_{\mathrm{pert}}) \equiv \int_0^{+\infty} \mathrm{d}E\, E^k \mathcal{S}(E) = \sum_n E_n^k \, |\langle 0|V_{\mathrm{pert}}|n\rangle|^2 \,. \tag{4.25}$$

This quantity can be also expressed in terms of the ground state expectation value of a suitable combination of commutators of the field V_{pert} with the Hamiltonian H_0 [10, 65, 66]. This fact is useful because the expectation value can be calculated using the LDA ground state. This can be seen for instance in the particularly interesting case of the first moment M_1, also called energy weighted sum rule (EWSR), for which the following equality holds

$$M_1(V_{\mathrm{pert}}) = \int_0^\infty \mathrm{d}E\, E\mathcal{S}(E) = \frac{1}{2}\langle 0|\,[V_{\mathrm{pert}}, [H_0, V_{\mathrm{pert}}]]\,|0\rangle$$

$$= \frac{\hbar^2}{2m_e} \int \mathrm{d}^3 r\, |\nabla V_{\mathrm{pert}}(\boldsymbol{r})|^2 \varrho(\boldsymbol{r}). \tag{4.26}$$

This is Thouless theorem [67], generalized to density-dependent, yet local Hamiltonians [6, 10, 65, 66]. In the case of LDA, however, the non-locality associated with the pseudopotentials contained in H_0, may induce discrepancies between the numerically calculated value of M_1 and the r.h.s. of (4.26). These discrepancies are in general expected to be small (see e.g. [68] for a discussion in a specific case, and also [69, 70]).

The energy weighted sum rule (EWSR) M_1 has a simple physical interpretation, in that the quantity $\hbar\nabla V_{\mathrm{pert}}$ indicates the momentum transferred to an electron by an external field in a sudden collision. The total energy transferred is the sum of the associated kinetic energies. It is independent of the interactions acting among the particles (electrons) because the energy is absorbed before the system is disturbed from equilibrium. Consequently, the r.h.s. of (4.26) is model independent, depending only on the mass m_e of the electrons [10]. It gives an upper limit for the energy that a system of $N = \int \mathrm{d}^3 r\, \varrho(\boldsymbol{r})$ electrons can absorb when perturbed by a (short duration) external perturbation V_{pert} [71, 72].

In the case of a spherically symmetric density distribution $\varrho(r)$, it is particularly convenient to use a multipole expansion of the external field. For instance, in the case of the Coulomb field one has a multiple expansion in spherical harmonics of different angular momentum numbers L as discussed in Sect. 3.4.2. For our purposes, it thus turns out useful to consider a single component of the form $V_{\mathrm{pert}}(\boldsymbol{r}) = e\, r^L Y_{L,0}(\Omega)$. In this case, (4.26) simplifies to

$$M_1(V_{\mathrm{pert}}) = \frac{\hbar^2 e^2}{2m_e} \frac{L(2L+1)}{4\pi} \int \mathrm{d}^3 r\, r^{2L-2}\, \varrho(r). \tag{4.27}$$

Other interesting moments are the cases $k = -1, 0, 3$ [6, 66]. In particular, the moment $M_{-1}(V_{\mathrm{pert}})$ is simply proportional to the static polarizability [6],

$$M_{-1}(V_{\mathrm{pert}}) = \int_0^{+\infty} \mathrm{d}E\, \frac{\mathcal{S}(E)}{E} = \frac{1}{2}\alpha(0). \tag{4.28}$$

In the case of molecular systems, the external perturbation is usually the electromagnetic field, and often, since its wavelength is much larger than the size of the system, the dipole approximation applies: the field can be considered constant in space, and the associated potential is linear, that is

$$V_{\text{pert}}(\boldsymbol{r}) \propto \boldsymbol{r}. \tag{4.29}$$

In this case, (4.26) takes a very simple form and can be evaluated analytically. Inserting the β component of the dipole electromagnetic field $V_{\text{pert}} = ex_\beta$ in (4.26) one has

$$M_1(L = 1) = \frac{\hbar^2}{2m_{\text{e}}}\, e^2\, N, \quad \text{for } \beta = x, y, z \tag{4.30}$$

where N is the number of electrons in the system. This is also known as the *Thomas–Reiche–Kuhn* (TRK) or *f-sum* rule. This result is, as already mentioned above, model independent. It also constitutes an important constraint which becomes useful for quantifying the accuracy of the numerical calculations, as well as to measure the collectivity of individual states.

The strength function is related to the cross section for photon absorption by the following equation (Golden Rule)

$$\sigma(E) = \frac{4\pi^2}{\hbar c}\, E\, \mathcal{S}(E) = -\frac{4\pi}{\hbar c}\, E\, \text{Im}\,\alpha(E), \tag{4.31}$$

and the total cross section is given by

$$\int \mathrm{d}E\,\sigma(E) = \frac{2\pi^2 e^2 \hbar}{m_{\text{e}} c}\, N. \tag{4.32}$$

If a single vibrational state exhausts a large fraction of the total cross section, (4.32) tells us that also a large fraction of the total number of electrons take part in the vibration. Therefore the state is said to be a collective excitation. Hence an experimental measure of the cross section can be used to determine the collective nature of a state.

Another useful quantity is the fraction of the oscillator strength absorbed by a specific state n, defined as

$$f_n = \frac{2m_{\text{e}}}{\hbar^2}\, |\langle 0|x_\beta|n\rangle|^2\, E_n. \tag{4.33}$$

From the definition of EWSR, (4.25), and using (4.30), one has

$$\sum_n f_n = \frac{2m_{\text{e}}}{\hbar^2 e^2} M_1(L = 1) = N, \quad \forall\, \beta. \tag{4.34}$$

When the dipole strength is concentrated into a single peak, its centroid can be well approximated by

$$E_1 = \sqrt{\frac{M_1}{M_{-1}}} = \sqrt{\frac{\hbar^2 e^2 N}{m_{\text{e}}\alpha(0)}}. \tag{4.35}$$

This relation can be used to predict the dipole resonance peak energy from the measured static dipole polarizability $\alpha(0)$ and viceversa [6, 10].

4.3 Transition Density and Plasmons

In the formulation we have just developed, all quantities are expressed in coordinate space. If one solves the RPA equation (4.10) and obtains the propagator Π, one can know the energy of the plasmons and their associated transition probabilities (as described in the last section). In addition to that, from the diagonal part $\Pi(\boldsymbol{r}, \boldsymbol{r}, \omega)$, calculated in the neighborhood of a plasmon energy E_n, one can extract the quantities $\delta\varrho_n(\boldsymbol{r})$. These quantities are called *transition densities*. They have been already used in (4.23) and (4.24). It can be shown, using (4.13) that

$$-\frac{1}{\pi}\mathrm{Im}\ \delta\varrho(\boldsymbol{r}, \omega) = \sum_n \delta\varrho_n(\boldsymbol{r})\ \langle n|V_{\mathrm{pert}}|0\rangle\ \delta(\hbar\omega - E_n), \tag{4.36}$$

that is, one could have access to the same quantities looking at the induced density calculated close to the plasmon energy, E_n.

Although we have understood from this discussion that the transition density has a clear physical meaning, namely it corresponds to the change in density associated with the transition to the excited plasmon state, in the present formulation neither the transition density nor any other quantity contains explicitly the information on the particle–hole content of the plasmon. Often, one is interested in extracting the wavefunction of the plasmon expressed on a basis of particle–hole excitations. For this aim, a different formulation of the TDLDA equations is needed.

In Appendix 4.4.2 it is shown that the TDLDA equation can be transformed to the form [10],

$$(\varepsilon_p - \varepsilon_h - \hbar\omega)\ X_{ph} = -\sum_{p'h'}(A_{ph,p'h'}X_{p'h'} + B_{ph,p'h'}Y_{p'h'}),$$

$$(\varepsilon_p - \varepsilon_h + \hbar\omega)\ Y_{ph} = -\sum_{p'h'}(A^*_{ph,p'h'}Y_{p'h'} + B^*_{ph,p'h'}X_{p'h'}), \tag{4.37}$$

where A and B are shorthand notations for the two-body matrix elements of the residual interaction,

$$A_{ph,p'h'} = \int\int \mathrm{d}^3r_1\ \mathrm{d}^3r_2\ \varphi^*_{h'}(\boldsymbol{r}_2)\varphi_{p'}(\boldsymbol{r}_2)\ \frac{\delta v_{\mathrm{eff}}(\boldsymbol{r}_1)}{\delta\varrho(\boldsymbol{r}_2)}\ \varphi^*_p(\boldsymbol{r}_1)\varphi_h(\boldsymbol{r}_1),$$

$$B_{ph,p'h'} = \int\int \mathrm{d}^3r_1\ \mathrm{d}^3r_2\ \varphi_{h'}(\boldsymbol{r}_2)\varphi^*_{p'}(\boldsymbol{r}_2)\ \frac{\delta v_{\mathrm{eff}}(\boldsymbol{r}_1)}{\delta\varrho(\boldsymbol{r}_2)}\ \varphi^*_p(\boldsymbol{r}_1)\varphi_h(\boldsymbol{r}_1). \tag{4.38}$$

If the difference $\varepsilon_p - \varepsilon_h$ is added to the diagonal part of the matrix A, the set of equations (4.37) can be written as

$$\begin{pmatrix} A & B \\ -B^* & -A^* \end{pmatrix}\begin{pmatrix} X \\ Y \end{pmatrix} = \hbar\omega\begin{pmatrix} X \\ Y \end{pmatrix}. \tag{4.39}$$

Note that the matrix A is Hermitian, while B is symmetric. Once a basis in the p–h space is chosen (constructed from the particle and hole states having appropriate quantum numbers that are allowed by the selection rules), (4.39) can be explicitly written in the form of an eigenvalue equation. The eigenvalues E_n are the plasmon energies and can be obtained by diagonalization, whereas if they are searched for by looking at the poles of (4.19) repeated solutions of (4.15) at many different frequencies, are needed. The matrix solution of the TDLDA may become computationally prohibitive for systems with many electronic states, since the size of the numerical calculation scales with the square of the number of p–h configurations. This is not the case for the formulation in the r-space, where the scale parameter is the square of the radial mesh, connected with the size of the system.

Moreover, in the matrix formulation of the TDLDA the eigenvectors are provided in the form of the amplitudes $X_{ph}^{(n)}$ and $Y_{ph}^{(n)}$. From the definitions (4.55) and (4.56) given in Appendix 4.4.2,

$$X_{\mathrm{ph}} = \int \mathrm{d}^3 r \; \varphi_p^*(\boldsymbol{r})\delta\varphi_h^{(-)}(\boldsymbol{r}),$$

and

$$Y_{\mathrm{ph}}^* = \int \mathrm{d}^3 r \; \varphi_p^*(\boldsymbol{r})\delta\varphi_h^{(+)}(\boldsymbol{r}),$$

one can see that X corresponds to the probability amplitude that the variation of the hole wavefunction h is associated to a transition to the particle state p. On the other hand, Y corresponds to the probability amplitude that a particle p already present in the (correlated) ground state makes a transition to the hole state h.

Equation (4.39) can be directly derived by assuming that the system has electronic vibrational excitations which are superpositions of p–h states. In the language of second quantization (see Appendix 2.11.1), these excitations $|n\rangle$ are created by operators Γ_n acting on the "plasmon vacuum",

$$\Gamma_n^\dagger = \sum_{ph} \left[X_{ph}^{(n)} a_p^\dagger \, a_h - Y_{ph}^{(n)} a_h^\dagger \, a_p \right],$$

$$\Gamma_n^\dagger |0\rangle = |0\rangle \qquad \Gamma_n|0\rangle = 0. \tag{4.40}$$

By linearizing the equation of motion $H\Gamma_n|0\rangle = E_n\Gamma_n|0\rangle$, (4.39) is obtained (see e.g. Appendix 2.11.1, [72], and refs. therein). It will prove useful to list some general properties of the solution of (4.39):

1. If (4.39) has a solution at energy E_n, it also admits a solution at negative energy $-E_n$ with associated amplitudes $Y_{ph}^{*(n)}, X_{ph}^{*(n)}$.
2. The eigenvalues are real if the ground state is a true minimum of the energy surface; if one (or more) eigenvalues happen to be pure imaginary, this is a signature of the fact that the ground state one has started from, is indeed a saddle point.
3. The solution are orthonormal with respect to the scalar product defined by the matrix

$$N = \begin{pmatrix} 1 & 0 \\ 0 & -1 \end{pmatrix},$$

that is, the normalization of an eigenvector is $\sum_{ph} X_{ph}^2 - Y_{ph}^2 = 1$.

Once the $X_{ph}^{(n)}$, $Y_{ph}^{(n)}$ amplitudes have been obtained, the strength of the state $|n\rangle$ can be expressed as

$$\mathcal{S}(E_n) = \left| \sum_{ph} (X_{ph}^{(n)} - Y_{ph}^{(n)}) \langle p|V_{\text{pert}}|h\rangle \right|^2, \tag{4.41}$$

and its contribution to the EWSR calculated as $E_n \mathcal{S}(E_n)$. Making use of the $X_{ph}^{(n)}$ and $Y_{ph}^{(n)}$ amplitudes, the transition density associated to the plasmon state $|n\rangle$ can be obtained from the relation

$$\delta \varrho_n(\boldsymbol{r}) = \sum_{ph} (X_{ph}^{(n)} - Y_{ph}^{(n)}) \, \varphi_p(\boldsymbol{r}) \varphi_h^*(\boldsymbol{r}). \tag{4.42}$$

This relation plays a basic role in dealing with the electron–plasmon coupling which will be discussed at length in Chap. 8.

We conclude this section by mentioning an additional result which may be useful in practical implementations of TDLDA. Starting from (4.49), (4.55) and (4.56), with $V_{\text{pert}} = 0$, one obtains the following expression

$$X_{ph}^{(n)} = \frac{V_{p,h}}{E_n - (\varepsilon_p - \varepsilon_h)}, \tag{4.43}$$

$$Y_{ph}^{(n)} = -\frac{V_{h,p}}{E_n + (\varepsilon_p - \varepsilon_h)}. \tag{4.44}$$

If the particle and hole energies are known, and one resorts to some simplified model for the residual interaction, like for example an effective two-body separable interaction between particle-hole states of the type,

$$K(\boldsymbol{r}, \boldsymbol{r}') = \kappa \, F(\boldsymbol{r}) \, F(\boldsymbol{r}') \tag{4.45}$$

in the place of (4.16), then (4.43) and (4.44) allow estimating the $X_{ph}^{(n)}$, $Y_{ph}^{(n)}$ amplitudes from, e.g., the experimental plasmon energy E_n (see Appendix 4.4.3). This approach will be used in Chapt. 5 in discussing the response of C_{60} molecules to a dipole field. In this case, it may turn out useful to assume for the single-particle field $F(\boldsymbol{r})$ the simple form

$$F(\boldsymbol{r}) = z. \tag{4.46}$$

4.4 Appendices

4.4.1 Linear Response

In this Appendix, we derive the different expressions used in Sect. 4.1. Let us consider a time-dependent external field $V_{\text{pert}}(\boldsymbol{r}) \exp(-i\omega t) + \text{h.c.}$ acting on the system. The total Hamiltonian H becomes time-dependent in the form (4.7),

$$H = H_0 + [(V_{\text{pert}}(\boldsymbol{r}) + \delta v_{\text{eff}}(\boldsymbol{r}, \omega)) \exp(-i\omega t) + \text{h.c.}],$$

where $H_0 \varphi_h = \varepsilon_h \varphi_h$ and δv_{eff} has been defined in (4.6). The Schrödinger equation we wish to solve is,

$$i\hbar \frac{\partial \varphi_h'(t)}{\partial t} = H \varphi_h'(t),$$

where the perturbed wavefunction $\varphi_h'(t)$ is assumed to be of the form (4.8),

$$\varphi_h'(t) = \exp(-i\varepsilon_h t/\hbar) \left[\varphi_h + \delta\varphi_h^{(-)} \exp(-i\omega t) + \delta\varphi_h^{(+)} \exp(i\omega t) \right].$$

Inserting this expression for $\varphi_h'(t)$ into the Schrödinger equation, and neglecting second-order terms, we arrive at the two following relations,

$$(H_0 - \varepsilon_h - \hbar\omega)\delta\varphi_h^{(-)}(\boldsymbol{r}, \omega) = -(V_{\text{pert}}(\boldsymbol{r}) + \delta v_{\text{eff}}(\boldsymbol{r}, \omega))\varphi_h(\boldsymbol{r})$$

$$(H_0 - \varepsilon_h + \hbar\omega)\delta\varphi_h^{(+)}(\boldsymbol{r}, \omega) = -(V_{\text{pert}}(\boldsymbol{r}) + \delta v_{\text{eff}}(\boldsymbol{r}, \omega))^* \varphi_h(\boldsymbol{r}). \quad (4.47)$$

Next, we expand the unknown functions $\delta\varphi_h^{(\mp)}$ using the unperturbed basis functions $\varphi_p(\boldsymbol{r})$,

$$\delta\varphi_h^{(\mp)}(\boldsymbol{r}, \omega) = \sum_{p \in \{\text{unocc}\}} c_p^{(\mp)}(\omega)\varphi_p(\boldsymbol{r}), \quad (4.48)$$

where the index p represents the unoccupied states, i.e., the states lying above the Fermi energy. Multiplying (4.47) from the left by $\varphi_p^*(\boldsymbol{r})$ and integrating over $\boldsymbol{r}$, one finds,

$$\delta\varphi_h^{(-)}(\boldsymbol{r}, \omega) = - \sum_{p \in \{\text{unocc}\}} \frac{V_{p,h}}{\varepsilon_p - \varepsilon_h - \hbar\omega} \varphi_p(\boldsymbol{r}),$$

$$\delta\varphi_h^{(+)}(\boldsymbol{r}, \omega) = - \sum_{p \in \{\text{unocc}\}} \frac{V_{h,p}^*}{\varepsilon_p - \varepsilon_h + \hbar\omega} \varphi_p(\boldsymbol{r}), \quad (4.49)$$

where $V_{p,h}$ is defined as

$$V_{p,h} = \int d^3 r' \, \varphi_p^*(\boldsymbol{r}') \left(V_{\text{pert}}(\boldsymbol{r}') + \delta v_{\text{eff}}(\boldsymbol{r}', \omega)\right) \varphi_h(\boldsymbol{r}').$$

The time-dependent density of the system, $\varrho(\boldsymbol{r}, t)$, can be written to first order in V_{pert} as,

$$\varrho(\boldsymbol{r}, t) = \sum_{h=1}^{\text{occ}} |\varphi_h'(t)|^2$$

$$= \sum_{h=1}^{\text{occ}} (\varphi_h^* + \delta\varphi_h^{*(-)} e^{i\omega t} + \delta\varphi_h^{*(+)} e^{-i\omega t})(\varphi_h + \delta\varphi_h^{(-)} e^{-i\omega t} + \delta\varphi_h^{(+)} e^{i\omega t})$$

$$= \sum_{h=1}^{\text{occ}} |\varphi_h(\boldsymbol{r})|^2 + \sum_{h=1}^{\text{occ}} \sum_{p \in \{\text{unocc}\}}$$

$$\times \left(-\frac{\varphi_h^*(\boldsymbol{r})\varphi_p(\boldsymbol{r})V_{p,h}}{\varepsilon_p - \varepsilon_h - \hbar\omega} - \frac{\varphi_h(\boldsymbol{r})\varphi_p^*(\boldsymbol{r})V_{h,p}}{\varepsilon_p - \varepsilon_h + \hbar\omega} \right) \, \mathrm{e}^{-\mathrm{i}\omega t} + \text{h.c.}, \qquad (4.50)$$

which has the form

$$\varrho(\boldsymbol{r},t) = \varrho_0(\boldsymbol{r}) + \delta\varrho(\boldsymbol{r},\omega) \cdot \exp(-\mathrm{i}\omega t) + \delta\varrho^*(\boldsymbol{r},\omega) \cdot \exp(\mathrm{i}\omega t). \qquad (4.51)$$

Therefore, by comparing the last two equations and writing explicitly $V_{p,h}$, one finds

$$\delta\varrho(\boldsymbol{r},\omega) = \int \mathrm{d}^3 r' \, \left(V_{\text{pert}}(\boldsymbol{r}') + \delta v_{\text{eff}}(\boldsymbol{r}',\omega) \right) \times$$

$$\sum_{h=1}^{\text{occ}} \sum_{p \in \{\text{unocc}\}} \left[\frac{\varphi_h^*(\boldsymbol{r})\varphi_p(\boldsymbol{r})\varphi_p^*(\boldsymbol{r}')\varphi_h(\boldsymbol{r}')}{\hbar\omega - (\varepsilon_p - \varepsilon_h)} - \frac{\varphi_h(\boldsymbol{r})\varphi_p^*(\boldsymbol{r})\varphi_p(\boldsymbol{r}')\varphi_h^*(\boldsymbol{r}')}{\hbar\omega + \varepsilon_p - \varepsilon_h} \right] \qquad (4.52)$$

This has the same structure of (4.9). The density–density correlation function $\Pi^{(0)}$ takes the form (4.11) in the case that the wavefunctions φ are real. If one neglects the residual interaction (i.e. one sets $\delta v_{\text{eff}} = 0$), the expression (4.12) is obtained.

4.4.2 TDLDA or RPA in the Configuration Space

In this Appendix we derive, starting from the equations of Sect. 4.1, the TDLDA in configuration space, that is, using a basis made up of particle-hole excitations. This basis is taken to be discrete, athough the particle states may be above the ionization threshold of the molecule. These states are usually calculated by setting the system in a large cell (see also Chap. 7). The TDLDA equations in a p–h basis take a matrix form and our purpose is to derive the matrix elements involved.

Let us start by writing the induced density in the same form as in the second line of (4.50) (see also (4.51)), namely

$$\delta\varrho(\boldsymbol{r},\omega) = \sum_{h=1}^{\text{occ}} \varphi_h^* \, \delta\varphi_h^{(-)} + \delta\varphi_h^{*(+)} \, \varphi_h, \qquad (4.53)$$

where $\delta\varphi_h^{(\mp)}$ is the solution of (4.47). The external perturbation V_{pert} provides the system with an initial boost and starts its (self-consistent) motion. The system is expected to maintain this (normal mode) motion when the perturbation is turned off. Consequently, the TDLDA equation of motion does not contain V_{pert}. This was already evident by looking at (4.14). In this Appendix, we consider, for the sake of clarity, the limit $V_{\text{pert}} \to 0$ from the beginning.

We multiply the first of the equations (4.47) from the right by φ_p^* and integrate over $\boldsymbol{r}$, obtaining

$$(\varepsilon_p - \varepsilon_h - \hbar\omega) \int d^3r \; \varphi_p^* \, \delta\varphi_h^{(-)} = -\sum_{h'} \int\int d^3r_1 \, d^3r_2 \; \varphi_p^*(\boldsymbol{r}_1) \frac{\delta v_{\text{eff}}(\boldsymbol{r}_1)}{\delta\varrho(\boldsymbol{r}_2)}$$

$$\times \left[\varphi_{h'}^*(\boldsymbol{r}_2) \, \delta\varphi_{h'}^{(-)}(\boldsymbol{r}_2) + \varphi_{h'}(\boldsymbol{r}_2) \, \delta\varphi_{h'}^{*(+)}(\boldsymbol{r}_2) \right] \varphi_h(\boldsymbol{r}_1), \tag{4.54}$$

where the explicit expression of δv_{eff} (4.6), together with (4.53), has been used. We write now the expansion (4.48) in the form,

$$\delta\varphi_h^{(-)}(\boldsymbol{r}, \omega) = \sum_{p'} X_{p'h}(\omega) \, \varphi_{p'}(\boldsymbol{r}) \tag{4.55}$$

and similarly,

$$\delta\varphi_h^{*(+)}(\boldsymbol{r}, \omega) = \sum_{p'} Y_{p'h}(\omega) \, \varphi_{p'}^*(\boldsymbol{r}). \tag{4.56}$$

Then (4.54) becomes

$$(\varepsilon_p - \varepsilon_h - \hbar\omega) X_{ph}$$
$$= -\sum_{p'h'} \int\int d^3r_1 \, d^3r_2 \; \varphi_{h'}^*(\boldsymbol{r}_2)\varphi_{p'}(\boldsymbol{r}_2) \frac{\delta v_{\text{eff}}(\boldsymbol{r}_1)}{\delta\varrho(\boldsymbol{r}_2)} \varphi_p^*(\boldsymbol{r}_1)\varphi_h(\boldsymbol{r}_1) \cdot X_{p'h'}$$
$$- \sum_{p'h'} \int\int d^3r_1 \, d^3r_2 \; \varphi_{h'}(\boldsymbol{r}_2)\varphi_{p'}^*(\boldsymbol{r}_2) \frac{\delta v_{\text{eff}}(\boldsymbol{r}_1)}{\delta\varrho(\boldsymbol{r}_2)} \varphi_p^*(\boldsymbol{r}_1)\varphi_h(\boldsymbol{r}_1) \cdot Y_{p'h'} \tag{4.57}$$

In this way, the first of the equations (4.37) is obtained. Similar steps performed by starting from the second of the equations (4.47), lead to the second of equations (4.37).

4.4.3 Plasmon Wavefunction Components for a Separable Interaction

Assuming the residual interaction to be separable of the form (4.45), that is,

$$K(\boldsymbol{r}, \boldsymbol{r}') \equiv \frac{\delta v_{\text{eff}}(\boldsymbol{r})}{\delta\varrho(\boldsymbol{r}')} = \kappa \, F(\boldsymbol{r}) \, F(\boldsymbol{r}') \tag{4.58}$$

we can write (4.57), and the corresponding equation for Y_{ph}, as

$$(\varepsilon_p - \varepsilon_h - \hbar\omega) X_{ph} = -\kappa \, F_{p,h} \sum_{p'h'} \left(X_{p'h'} F_{h'p'}^* + Y_{p'h'} F_{p'h'}^* \right)$$

$$(\varepsilon_p - \varepsilon_h + \hbar\omega) Y_{ph} = -\kappa \, F_{h,p}^* \sum_{p'h'} \left(Y_{p'h'} F_{p'h'} + X_{p'h'} F_{h'p'} \right) \tag{4.59}$$

where the matrix elements $F_{p,h}$ are given by

$$F_{p,h} = \int d^3r' \; \varphi_p^*(\boldsymbol{r}') \, F(\boldsymbol{r}') \, \varphi_h(\boldsymbol{r}').$$

Note that the sums in the r.h.s. of (4.59), being independent of the indices (p, h), are constants which can be determined by the normalization condition

$\sum_{ph}(X_{ph}^2 - Y_{ph}^2) = 1$. Thus, if the plasmon energy E_n is known, one can calculate the corresponding coefficients $X_{ph}^{(n)}, Y_{ph}^{(n)}$ using (4.59) with $\hbar\omega = E_n$.

4.4.4 Spherical TDLDA

Let us consider a spherical basis to solve (4.10). The single-particle wavefunctions for particles, $\varphi_p(\boldsymbol{r})$, and holes, $\varphi_h(\boldsymbol{r})$, are expanded in the basis of spherical wavefunctions (see Sect. 3.4),

$$\varphi_p(\boldsymbol{r}) = \sum_k c_p^{(k)}\varphi_k(\boldsymbol{r}) = \sum_{a_k m_k} c_p^{(a_k m_k)} R_{a_k}(r) Y_{l_k m_k}(\Omega) \tag{4.60}$$

and

$$\varphi_h(\boldsymbol{r}) = \sum_i c_h^{(i)}\varphi_i(\boldsymbol{r}) = \sum_{a_i m_i} c_h^{(a_i m_i)} R_{a_i}(r) Y_{l_i m_i}(\Omega) \tag{4.61}$$

where the label $a \equiv (n, l)$, stands for the number of nodes and the orbital angular momentum, and $\Omega \equiv \{\theta, \phi\}$. Making use of the above relations one can write

$$\varphi_h^*(\boldsymbol{r})\varphi_p(\boldsymbol{r}) = \sum_{\lambda\mu} \Phi_{ph}^{\lambda\mu}(r) Y_{\lambda\mu}(\Omega), \tag{4.62}$$

where

$$\Phi_{ph}^{\lambda\mu}(r) = \sum_{a_k m_k a_i m_i} (-1)^{m_i} \sqrt{\frac{(2l_i + 1)(2l_k + 1)(2\lambda + 1)}{4\pi}}$$

$$\times \begin{pmatrix} l_i & l_k & \lambda \\ 0 & 0 & 0 \end{pmatrix} \begin{pmatrix} l_i & l_k & \lambda \\ -m_i & m_k & \mu \end{pmatrix} c_h^{*a_i m_i} c_p^{a_k m_k} R_{a_i}(r) R_{a_k}(r). \tag{4.63}$$

The unperturbed response function (polarization propagator) becomes

$$\Pi^0(\boldsymbol{r}, \boldsymbol{r}', \omega) = \sum_{\lambda\mu\lambda'\mu'} Y_{\lambda\mu}(\Omega)\, \Pi^0_{\lambda\mu,\lambda'\mu'}(r, r', \omega)\, Y^*_{\lambda'\mu'}(\Omega'), \tag{4.64}$$

with

$$\Pi^0_{\lambda\mu,\lambda'\mu'}(r, r', \omega) = \sum_{ph} \Phi_{ph}^{\lambda\mu}(r)\, G(\varepsilon_p - \varepsilon_h, \hbar\omega)\, \Phi_{ph}^{*\lambda'\mu'}(r'), \tag{4.65}$$

and $G(\varepsilon_p - \varepsilon_h, \hbar\omega) = [\hbar\omega - (\varepsilon_p - \varepsilon_h) + i\eta]^{-1} - [\hbar\omega + (\varepsilon_p - \varepsilon_h) + i\eta]^{-1}$. Assuming now for purposes of illustration that the external perturbation has the simple form $V_{\text{pert}} = er^L Y_{LM}(\Omega)$, (4.12) yields

$$\delta\varrho^0(\boldsymbol{r}, \omega) = e \int d^3r'\, r'^L Y_{LM}(\Omega') \Pi^0(\boldsymbol{r}, \boldsymbol{r}', \omega)$$

$$= e \sum_{\lambda\mu} Y_{\lambda\mu}(\Omega) \int_0^\infty dr' r'^{2+L} \Pi^0_{\lambda\mu,LM}(r, r', \omega). \tag{4.66}$$

In particular, for a dipolar field applied along the z-axis, i.e. in the case of $L = 1$ and $M = 0$, one has

$$\delta\varrho^0(\boldsymbol{r}, \omega) = e \sum_{\lambda\mu} Y_{\lambda\mu}(\Omega) \int \mathrm{d}r' r'^3 \Pi^0_{\lambda\mu,10}(r, r', \omega).$$

(4.67)

In the case in which the system is spherical, the propagator takes the form

$$\Pi^0_{\lambda\mu,\lambda'\mu'} = \delta_{\lambda\lambda'}\delta_{\mu\mu'}\Pi^0_{\lambda\mu,\lambda\mu}.$$

Consequently, only the component $\lambda = \lambda' = 1$, $\mu = \mu' = 0$ contributes in (4.64). However, if the internal field created by the ions is non-spherical, a number of different (λ, μ) values can contribute. In particular $\lambda = 0$ and $\lambda = L$, but also higher multipoles. The same arguments apply to the corrrelated induced density $\delta\varrho$.

In order to study the perturbed response of the system we need to specify the residual interaction kernel. According to (4.16),

$$K(\boldsymbol{r}_1, \boldsymbol{r}_2) = \frac{e^2}{|\boldsymbol{r}_1 - \boldsymbol{r}_2|} + \delta(\boldsymbol{r}_1 - \boldsymbol{r}_2) \frac{\mathrm{d}v_{\mathrm{xc}}}{\mathrm{d}\varrho},$$

(4.68)

which can be expanded in spherical harmonics (see Sect. 3.4.2) using the relations,

$$\frac{1}{|\boldsymbol{r}_1 - \boldsymbol{r}_2|} = \sum_{\ell=0}^{\infty} \sum_{m=-\ell}^{\ell} \frac{4\pi}{2\ell + 1} \frac{r_<^\ell}{r_>^{\ell+1}} Y_{\ell,m}(\Omega) Y_{\ell,m}^*(\Omega'),$$

where $r_<$ $(r_>)$ is the smaller (larger) between r_1 and r_2,

$$\delta(\boldsymbol{r}_1 - \boldsymbol{r}_2) = \frac{1}{r^2}\delta(r_1 - r_2) \sum_{\ell,m} Y_{\ell,m}(\Omega) Y_{\ell,m}^*(\Omega'),$$

and

$$\frac{\mathrm{d}v_{\mathrm{xc}}[\varrho(\boldsymbol{r})]}{\mathrm{d}\varrho} = \sum_{\ell,m} \left(\frac{\mathrm{d}v_{\mathrm{xc}}(r)}{\mathrm{d}\varrho}\right)_{\ell,m} Y_{\ell,m}(\Omega).$$

We are now in the position to expand (4.10) in multipoles, i.e.

$$\delta\varrho_{LM}(r_1) = \delta\varrho^0_{LM}(r_1) + \sum_{\lambda\mu} \int_0^\infty \mathrm{d}r_4 r_4^2 \, B^{\lambda\mu}_{LM}(r_1, r_4) \, \delta\varrho_{\lambda\mu}(r_4)$$

(4.69)

where

$$\delta\rho^0_{LM}(r_1) = e \sum_{ph\lambda\mu} \Phi^{\lambda\mu}_{ph}(r_1) \, G(\varepsilon_p - \varepsilon_h, \hbar\omega) \int_0^\infty \mathrm{d}r_2 \, r_2^{2+L} \Phi^{*LM}_{ph}(r_2)$$

(4.70)

and

$$B_{LM}^{\lambda\mu}(r_1, r_4) = \sum_{ph} \Phi_{ph}^{\lambda\mu}(r_1)\, G(\varepsilon_p - \varepsilon_h, \hbar\omega) \left[\left(\frac{dv_{\rm xc}(r_4)}{d\varrho} \right)_{L,M} \Phi_{ph}^{*LM}(r_4) \right.$$

$$+ \frac{4\pi}{2L+1} \frac{e^2}{r_4} \left(\int_0^{r_4} dr_3\, r_3^2 \left(\frac{r_3}{r_4} \right)^L \Phi_{ph}^{*LM}(r_3) \right.$$

$$\left. \left. + \int_{r_4}^{\infty} dr_3\, r_3^2 \left(\frac{r_4}{r_3} \right)^{L+1} \Phi_{ph}^{*LM}(r_3) \right) \right]. \qquad (4.71)$$

Once the different elements are calculated, one has to solve the matrix relation

$$\begin{pmatrix} \delta\rho_{L_1 M_1} \\ \delta\rho_{L_2 M_2} \\ \vdots \end{pmatrix} = \begin{pmatrix} \delta\rho^0_{L_1 M_1} \\ \delta\rho^0_{L_2 M_2} \\ \vdots \end{pmatrix} + \begin{pmatrix} B_{L_1 M_1}^{L_1 M_1} & B_{L_1 M_1}^{L_2 M_2} & \cdots \\ B_{L_2 M_2}^{L_1 M_1} & B_{L_2 M_2}^{L_2 M_2} & \cdots \\ \vdots & \vdots & \ddots \end{pmatrix} \begin{pmatrix} \delta\rho_{L_1 M_1} \\ \delta\rho_{L_2 M_2} \\ \vdots \end{pmatrix} \quad (4.72)$$

to obtain the induced density.

4.4.5 Cylindrical TDLDA

In the case of elongated systems it is useful to consider cylindrical coordinates and to switch to a matrix form via projection onto the cylindrical basis $\varphi_{n,\ell,m}(\rho, z, \phi)$ introduced in (3.62). In this case the operator $\Pi(\boldsymbol{x}, \boldsymbol{x}', \omega)$ depends on 3 continuous variables, ρ, ρ', ω, and on four discrete quantum numbers, l, m, l', m'. The cylindrical multipoles of the polarization propagator, $\Pi_{l_1 m_1 l_2 m_2}(\rho_1, \rho_2, \omega)$, are defined as

$$\Pi_{l_1 m_1 l_2 m_2} = \int d\tilde\Omega_1 d\tilde\Omega_2\, \tilde{Y}_{l_1 m_1}^*(\tilde\Omega_1)\, \Pi(\boldsymbol{x}_1, \boldsymbol{x}_2, \omega)\, \tilde{Y}_{l_2 m_2}(\tilde\Omega_2), \qquad (4.73)$$

where $\boldsymbol{x} = (\rho, \phi, z) = (\rho, \tilde\Omega)$ are the cylindrical coordinates, and $\tilde{Y}_{lm}(\tilde\Omega)$ the (ϕ, z) basis wavefunctions (cylindrical harmonics, see [18, 73] for details). The free polarization propagator $\Pi^0_{l_1 m_1 l_2 m_2}(\rho_1, \rho_2, \omega)$ is defined in a similar way. Equation (4.14) becomes,

$$\Pi^0_{l_1 m_1 l_2 m_2}(\rho_1, \rho_2, \omega) = \left[\, \delta_{l_1 l_4} \delta_{m_1 m_4} \delta(\rho_1 - \rho_4) - \sum_{l_3 m_3 l_4 m_4} \int d\rho_3\, d\rho_4\, \rho_3\, \rho_4 \right.$$

$$\times\, \Pi^0_{l_1 m_1 l_3 m_3}(\rho_1, \rho_3, \omega)\, K_{l_3 m_3 l_4 m_4}(\rho_3, \rho_4, \omega) \, \right] \Pi_{l_4 m_4 l_2 m_2}(\rho_4, \rho_2, \omega) \quad (4.74)$$

where the multipoles of the kernel $K(\boldsymbol{x}_1, \boldsymbol{x}_2)$ are given by

$$K_{l_1 m_1 l_2 m_2}(\rho_1, \rho_2) = \int d\tilde\Omega_1 d\tilde\Omega_2\, \tilde{Y}_{l_1 m_1}^*(\tilde\Omega_1)\, K(\boldsymbol{x}_1, \boldsymbol{x}_2)\, \tilde{Y}_{l_2 m_2}(\tilde\Omega_2). \quad (4.75)$$

The explicit form of the cylindrical multipoles of the free polarization propagator $\Pi^0_{l_1 m_1 l_2 m_2}(\rho_1, \rho_2, \omega)$ and of the kernel $K_{l_1 m_1 l_2 m_2}(\rho_1, \rho_2)$ are given elsewhere [73].

Once (4.74) has been solved for $\Pi_{l_1 m_1 l_2 m_2}(\rho_1, \rho_2, \omega)$, one can calculate the dynamical linear polarizability as a function of the cylindrical multipoles of the polarization propagator (4.73) and of the external field, here denoted as $V_{l,m}^{\text{pert}}$, as

$$\alpha(\omega) = \sum_{l_1 m_1 l_2 m_2} \int_0^{\rho_0} \mathrm{d}\rho_1 \, \mathrm{d}\rho_2 \, \rho_1 \rho_2 \; V_{l_1 m_1}^{\text{pert}}(\rho_1) \; \Pi_{l_1 m_1 l_2 m_2}(\rho_1, \rho_2, \omega)$$

$$V_{l_2 m_2}^{\text{pert}}(\rho_2), \qquad (4.76)$$

from which the physical information can be extracted.

5. Applications to Carbon Structures and Metal Clusters

In this Chapter we describe the results obtained by applying the theoretical approaches discussed in Chaps. 3 and 4 to the study of carbon structures such the C_{60} fullerene, carbon nanotubes and linear carbon chains, and sodium metal clusters such as Na_8.

5.1 C_{60} Fullerene

We start with the paradigm of the whole fullerene family, the C_{60} buckyball (see Fig. 1.3). We will first discuss its Kohn–Sham electronic structure, and the associated symmetry aspects related to its geometrical configuration. In this dicussion, the spherical coordinates will be used as an application of the theory described in Sect. 3.4. The electromagnetic response of the buckyball is discussed afterwards, in particular the structure of the low-lying plasmons peaks of C_{60}, using the theoretical tools developed in Appendix 4.4.4.

5.1.1 Kohn–Sham States for C_{60}

Exploiting the high isotropy of the C_{60} molecule which makes it almost spherical, we can expand $V_{\text{ext}}(\boldsymbol{r})$ in terms of spherical harmonics (see Sect. 3.4) to obtain,

$$V_{\text{ext}}(\boldsymbol{r}) = \sum_{L,M} \sqrt{\frac{4\pi}{2L+1}} \, S_{L,M} \, V_L(r) \, Y_{L,M}(\hat{r}) \,, \tag{5.1}$$

where $S_{L,M} = \sum_{i=1}^{N_{\text{at}}} Y_{L,M}^*(\hat{R}_i)$, while the functions $V_L(r)$ can be obtained using a Löwdin transformation [60]. Because of the icosahedral symmetry of C_{60}, the coefficients $S_{L,M} = (-)^M S_{L,-M}$ are different from zero for selected values of the angular momentum $L = 0, 6, 10, 12, 16, 18, 20, 22$, etc., and for M equal to a integer multiple of 5 (see Table 5.1 and Appendix 5.5.1).

To solve the associated Kohn–Sham equation (3.42), we expand the wavefunctions $\phi_j(\boldsymbol{r})$ in spherical basis functions according to (3.44). The latter are solutions of the radial Schrödinger equation where the basis potential $v_{\text{Basis}}(r)$ is taken as the spherical component ($L = 0$) of $v_{\text{loc}}(\boldsymbol{r})$.

Table 5.1. The first 36 spherical multipole coefficients $S_{L,M}$ for C$_{60}$. The (0,0) value has not been included in the list, and it is given by $S_{0,0} = 60/\sqrt{4\pi} \approx 16.9257$. These results were obtained by taking the nearest neighbor distances between carbon ions to be 1.453 Å on pentagons and 1.369 Å between pentagons, yielding a radius of C$_{60}$ of $R = 3.526$ Å. Note that $S_{L,M} = (-)^M S_{L,-M}$

L	M	$S_{L,M}$	L	M	$S_{L,M}$	L	M	$S_{L,M}$
6	-5	1.1267	12	10	-3.5030	18	5	-11.4228
6	0	1.4123	16	-15	-5.7891	18	10	9.1501
6	5	-1.1267	16	-10	7.4652	18	15	-18.0509
10	-10	-6.2107	16	-5	5.8094	20	-20	-4.7526
10	-5	11.3725	16	0	-7.8676	20	-15	2.6717
10	0	-7.1379	16	5	-5.8094	20	-10	-8.6471
10	5	-11.3725	16	10	7.4652	20	-5	5.3272
10	10	-6.2107	16	15	5.7891	20	0	-1.7930
12	-10	-3.5030	18	-15	18.0509	20	5	-5.3272
12	-5	-2.1762	18	-10	9.1501	20	10	-8.6471
12	0	-4.2113	18	-5	11.4228	20	15	-2.6717
12	5	2.1762	18	0	20.2541	20	20	-4.7526

The results for the occupied energy levels in the buckyball are displayed in Fig. 5.1. The single-particle energy gap between the highest-occupied and

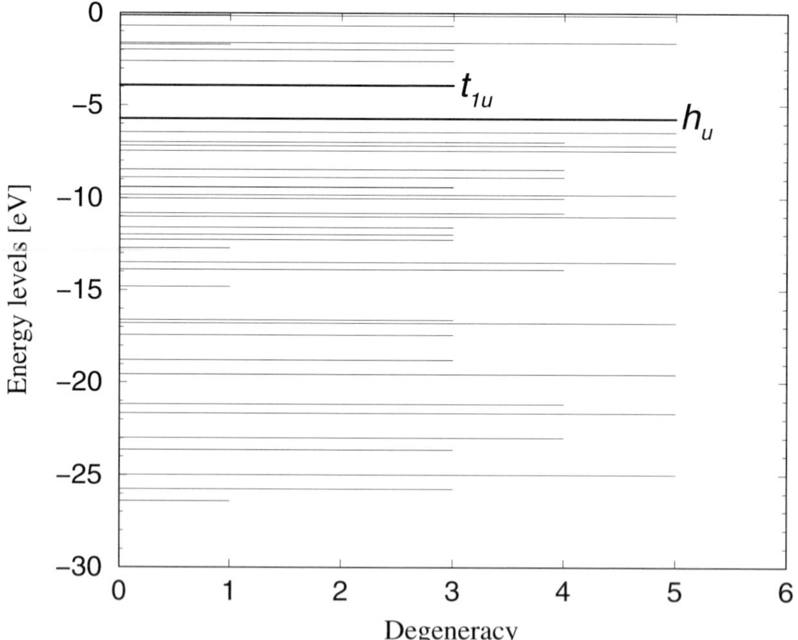

Fig. 5.1. Energy levels of C$_{60}$ calculated in LDA (after [59]). The HOMO (h_u) and LUMO (t_{1u}) levels are indicated

the lowest-unoccupied molecular orbital states (HOMO–LUMO gap) is predicted to be 1.9 eV [59] (see Fig. 5.1). The value of 1.86 ± 0.1 eV measured in [27], cannot in principle be directly compared with this LDA gap because the experimental value also includes self-energy and electron–hole interaction effects. However, these effects are found to cancel each other to a large extent. Theory thus provides an overall account of the experimental findings. Remembering that in the spherical basis, the KS eigenstates can be characterized in terms of different values of the quantum numbers n, ℓ and m carried by the basis eigenfunctions, it is found that the HOMO and LUMO states have predominantly $n_i = 1$ character[1] ($\approx 70\%$) and have no components with $n_i = 0$. Consequently, one can identify these states with $\pi-$like orbitals which are mainly directed perpendicularly to the surface of the hollow sphere. Using the icosahedral symmetry group, those states are found to have t_{1u} and h_u symmetry (see Appendix 5.5.1).

In the spherical basis they are written (using the notation $|n\ell m\rangle$, see (3.44)) as[2]

$$
\begin{aligned}
|\text{HOMO}\rangle \approx &-0.8(|151\rangle - |15\bar{1}\rangle) + 0.6(|152\rangle + |15\bar{2}\rangle)\\
&-0.7(|153\rangle - |15\bar{3}\rangle) + 0.5(|154\rangle + |15\bar{4}\rangle)\\
&+0.7(|155\rangle + |15\bar{5}\rangle)
\end{aligned}
\tag{5.2}
$$

and

$$
\begin{aligned}
|\text{LUMO}\rangle \approx &\, 0.7|150\rangle - 0.5(|151\rangle + |15\bar{1}\rangle) + 0.7(|154\rangle - |15\bar{4}\rangle)\\
&-0.3(|155\rangle - |15\bar{5}\rangle)
\end{aligned}
\tag{5.3}
$$

where, for illustration, only coefficients $|c_j^{(H,L)}| > 0.2$ (see (3.44)) have been displayed.

Important contributions to the HOMO–LUMO gap arise from the very large $S_{10,M}$ ($M = 0, \pm 5, \pm 10$) term in (5.1). This is due to the fact that the angular momentum of the levels defining the gap is mainly $\ell = 5$. The energy range spanned by the occupied levels (band width) is predicted to be ≈ 21 eV. This number is controlled, to a large extent, by the non-local part of the pseudopotential, as well as by the terms of the local part with L up to 18 (note in particular the large value of $S_{18,M}$ in Table 5.1). This is because the highest bound states arise from spherical states with ℓ up to 9. Both the local and non-local $S_{18,M}$ terms are also important for the existence of the HOMO–LUMO gap, since without their inclusion the HOMO would result only partially occupied.

5.1.2 The Optical Absorption of C$_{60}$

Making use of the single-particle basis discussed previously and displayed in Fig. 5.1, the dipole response of C$_{60}$ has been worked out in the TDLDA

---

[1] Here n indicates the number of nodes of the radial wavefunction.

[2] The bar on top of the number indicates the negative sign.

using the theory discussed in Chap. 4 (see also Appendices 4.4.4 and 5.5.1 for details). The present approach yields a static polarizability of the system of about 88.6 Å^3, which can be compared to measurements of $\alpha(0) \approx 80$ Å^3 for a single molecule [74]. One can also compare the TDLDA result with experimental findings obtained from the cubic solid phase (fullerites), using the Clausius–Mossotti relation linking the dielectric constant of the solid, $\varepsilon(0)$, and $\alpha(0)$,

$$\alpha(0) = \frac{3v}{4\pi} \frac{\varepsilon(0) - 1}{\varepsilon(0) + 2} \tag{5.4}$$

where $v = a_{\mathrm{fcc}}^3/4$ is the volume of the unit cell. Using the experimental value of the fcc–lattice constant of fullerite ($a_{\mathrm{fcc}} \approx 14.2$ Å) and the experimental result $\varepsilon(0) \approx 4.5$ [75], one can estimate the molecular polarizability to be about 92 Å^3.

The results for both the unperturbed particle–hole and RPA strength functions $S(\omega)$ of C$_{60}$ are displayed in Fig. 5.2. The theoretical results exhaust 95% and 90% of the TRK energy weighted sum rule (4.30), respectively. The particle–hole components associated with the most important peaks displayed by the free dipole-strength function $S^{(0)}(\omega)$ are reported in Table 5.2. After the residual interaction is switched on, these configurations become heavily mixed. However, it is still possible to identify the different peaks below ≈ 6 eV with $\pi - \pi$ transitions ($n = 1 \to n = 1$ orbitals), while for increasing energies transitions connecting σ orbitals ($n = 0$) with π and δ orbitals ($n = 1$ and 2, respectively) dominate the response.

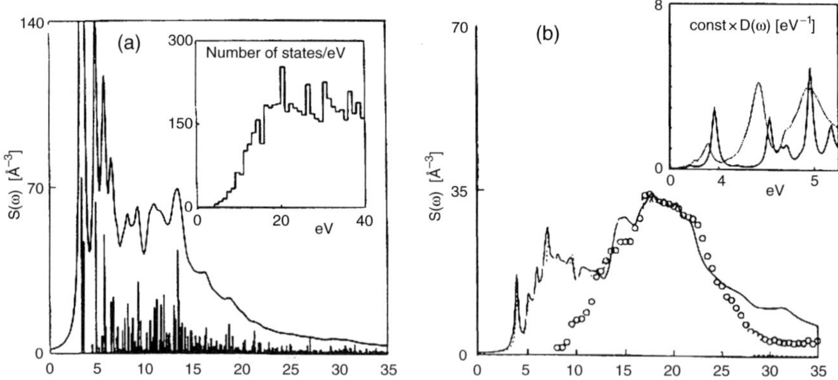

Fig. 5.2. The electromagnetic response of C$_{60}$ (after [59]). (a) The unperturbed response $S^{(0)}(E)$ (inset: density of states corresponding to particle–hole transition energies). The *continuous line* represents the calculated spectrum [59] convoluted with a Lorentzian function having width $\Gamma = 0.06E$. (b) RPA response (*full line*) compared to the experimental points (*open circles*) [76] (to be noted that the cut-off at ≈ 10 eV is due to the experimental set up). Inset: detailed spectrum at low energies [59] compared to experimental results (*broken curve*) [79]

Table 5.2. Selected particle–hole transitions involved in the dipole response of C$_{60}$ (see also Fig. 5.2). From left to right the table displays: transition energy (ΔE), hole (E_h) and particle (E_p) energies (in eV), percentage of the TRK sum rule exhausted, quantum numbers of the most important particle–hole configurations coupled by the dipole field with quantum numbers n_h,ℓ_h and n_p,ℓ_p. The numbers in brackets represent the percentage of spherical components in these wavefunctions

ΔE[eV]	$E_h \to E_p$	EWSR (%)	n_h (%) ; ℓ_h (%)	n_p (%) ; ℓ_p (%)
3.29	-7.40 ; -4.11	3.35	1 (80) ; 4 (88)	1 (68) ; 5 (82)
3.50	-6.03 ; -2.52	2.12	1 (71) ; 5 (97)	1 (47) ; 6 (97)
4.74	-6.03 ; -1.28	2.88	1 (71) ; 5 (97)	1 (46) ; 6 (71)
5.60	-7.27 ; -1.67	2.25	1 (79) ; 4 (83)	1 (71) ; 5 (74)
6.25	-6.03 ; 0.22	0.54	1 (71) ; 5 (97)	2 (42) ; 4 (19)
6.48	-8.93 ; -2.45	1.08	0 (89) ; 8 (53)	0 (18) ; 9 (15)
7.67	-9.33 ; -1.66	0.68	1 (70) ; 3 (74)	2 (49) ; 2 (60)
9.11	-11.1 ; -1.99	1.36	0 (87) ; 8 (89)	0 (48) ; 9 (46)
10.3	-10.1 ; 0.22	0.59	0 (79) ; 9 (73)	0 (19) ; 10 (12)
13.6	-15.3 ; -1.67	0.51	0 (88) ; 6 (81)	1 (71) ; 5 (74)
16.1	-8.93 ; 7.17	0.35	0 (89) ; 8 (52)	3 (62) ; 7 (65)
16.2	-17.9 ; -1.67	0.49	0 (91) ; 6 (90)	1 (71) ; 5 (74)
17.3	-15.3 ; 1.97	0.25	0 (90) ; 7 (91)	1 (48) ; 6 (53)
18.2	-13.3 ; 4.96	0.37	0 (94) ; 7 (70)	0 (29) ; 8 (21)

Theory provides an overall account of the experimental findings [76] (see Fig. 5.2), also for the width of the plasmon resonance (see e.g. [77]) located around 17 eV and amounting to about 12 eV. Consequently, Landau damping, that is the decay of the collective modes into single (unperturbed) particle–hole excitations (examples of which are collected in Table 5.2), can be viewed as the main relaxation mechanism of the above described collective plasmon mode [59, 78].

The experimental data at low excitation energy corresponds to the optical/UV absorption spectra of C$_{60}$ in hexane solution, at room temperature [79]. Nine transitions below 6.4 eV have been observed displaying the following energies (oscillator strength): 3.04 eV (0.005), 3.30 eV (-), 3.78 eV (0.37), 4.06 and 4.35 eV (0.10), 4.84 eV (2.27), 5.46 eV (0.22), 5.88 eV (3.09) and 6.36 eV (-). Theory predicts 3.4 eV (0.048), 3.91 eV (0.59), 4.38 eV (0.041), 5.05 eV (0.50) 5.30 eV (0.14), 5.42 eV (0.21), 5.70 eV (0.084), 5.92 eV (0.95) 6.35 eV (0.48) [59]. Although there is an overall correspondence between the calculations and the experimental findings (see also [80]), theory predicts energies which are blue shifted as compared with the observations. In particular, while the observed summed oscillator strength below 6.4 eV amounts to 6.1% of the TRK sum rule, the TDLDA prediction is 3.1%. To be noted that the value associated with the unperturbed response function in the same energy interval is 33.4%, in keeping with the fact that correlations among these basis states is repulsive.

Similar blue shifts have been found by essentially all TDLDA calculations of the photoabsorption spectrum carried out in atomic clusters, in particular in metal clusters (see e.g. [5, 6] and refs. therein). As will be discussed below (see Chap. 8), this is because TDLDA calculations neglect the coupling between different plasmons states. By taking these couplings into account, i.e. the coupling between states containing one and two plasmons, it should be possible to correct the energy centroid of the plasmon peaks. Also, to provide these states with a width which essentially accounts for the experimental findings. Similar results have been found in the study of the photoabsorption spectrum of other finite many-body systems like e.g. atomic nuclei [72].

To gain further insight into the structure of the low-lying plasmon peaks in the optical response of C_{60}, we can apply the formalism discussed in Sect. 4.3. Assuming the residual interaction to be separable as in (4.45), and $F(\boldsymbol{r})$ given by (4.46), we can estimate the $X^{(n)}$ and $Y^{(n)}$ components of the n-th plasmon peak at energy E_n, as indicated in Appendix 4.4.3. That is,

$$X_{ph}^{(n)} = \kappa' \, \frac{F_{p,h}}{E_n - E_{ph}}, \quad Y_{ph}^{(n)} = -\kappa' \, \frac{F_{h,p}}{E_n + E_{ph}}, \tag{5.5}$$

where κ' is determined by the normalization condition and $E_{\mathrm{ph}} = \varepsilon_p - \varepsilon_h$ is the associated particle–hole energy difference. The results are shown in Fig. 5.3, where the quantities $|X_{ph}^{(n)}|$ are plotted as a function of E_{ph}. We note that $|Y_{ph}^{(n)}|$ being typically one order of magnitude smaller than $|X_{ph}^{(n)}|$. One can see that a number of unperturbed particle–hole excitations contribute to the different experimental peaks. That is, the corresponding wavefunctions are linear combinations of particle–hole excitations.

In summary, the optical/UV part of the dipole response of C_{60} is dominated by collective excitations (plasmons), linear combinations of many excitations between particle–hole transitions among π-orbitals whose contributions add up coherently in the photoabsorption process, while the Mie

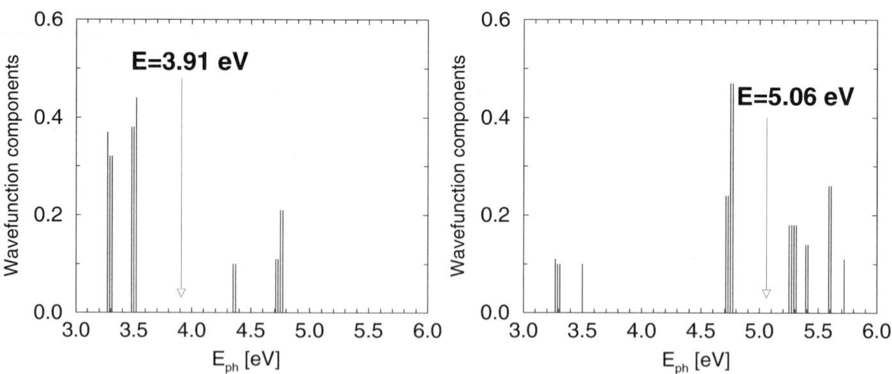

Fig. 5.3. Particle–hole components $|X_{ph}^{(n)}|$ (*vertical lines*) of the RPA wavefunctions (5.5), corresponding to the first two peaks (whose locations are indicated by the *vertical arrows*) in the optical response of C_{60} shown in the inset of Fig. 5.2(b). Degenerate particle–hole transitions has been shifted in energy by a small amount to visualize their components

resonance observed at ≈ 20 eV is built by a coherent superposition of not only $\sigma - \sigma$ transitions but also, and to a large extent, from particle–hole excitations connecting σ- with π, δ-orbitals, as well as final states with a higher number of nodes. The inclusion of these transitions is essential to fulfill the energy weighted sum rule. The width of the Mie resonance is due to the decay of the plasmon into single-particle configurations (Landau damping).

Studies of this type have been extended to other systems such as C_{60} covered by a layer of Li atoms ($Li_{12}C_{60}$ [81]), and smaller fullerenes, e.g. C_{28} [82], and $C_{28}H_n$ [83].

5.2 Carbon Nanotubes

Carbon nanotubes (see [13] and refs. therein) are cylindrical molecules with a diameter as little as one nanometer and a length up to many micrometers. They can essentially be thought as a single layer of graphite that has been wrapped into a cylinder. In other words, a tube is made up of sp^2 (trivalent) carbon atoms that form an hexagonal network. There are several methods for making nanotubes [84], but the carbon arc method seems to be the most efficient one. The growth mechanism of the nanotubes is a complex subject in itself, being the result of the competition between two sets of carbon atoms which are present near the cathode surface: the carbon ions anisotropically accelerated across the arc gap, and the thermally evaporated carbon from the cathode with isotropic velocity distribution [85]. In other words, the introduction of a preferential axis in the reaction zone due to the accelerated carbons results in elongated structures.

Analysis shows that such a preferential axis always exist in all the methods used to generate nanotubes. For instance, in the catalytic growth of single-wall nanotubes (SWNT), the catalytic particles provide the asymmetry in the three dimensions. The dual distribution of carbon species seems to agree with the dual distribution of products, namely nanotubes and nanoparticles (which are polyhedral).

The primary symmetry classification of a carbon nanotube is as either being "chiral" or "achiral". Chiral nanotubes are those characterized by a pair of (two component) chiral vectors n, m with $n \neq m$ and $n, m \neq 0$. They have a spiral symmetry; i.e. their mirror image cannot be superimposed to the original one. An achiral carbon nanotube, instead, is one whose mirror image has an identical structure to that of the original system. Armchair and zigzag carbon nanotubes are the only two cases of achiral nanotubes, and are associated with $m = 0$ (armchair) or $m = n$ (zigzag). The indices (n, m) are particularly important for the electronic structure of the nanotubes: tubes for which $n - m = 3i$, where i is an integer, are metallic, all others are semiconducting [13].

In analogy to a C_{60} molecule, one can build an "ideal" armchair, single-wall nanotube by bisecting a C_{60} molecule at the equator and joining the resulting hemispheres with a cylindrical tube made out of an arbitrary number of belts, each built out of five benzoid rings. Each belt thus adds ten carbon atoms to the molecule. Fullerene C_{70} is obtained by adding one belt to C_{60}, while e.g. C_{90}, C_{110}, C_{150} and C_{190} are obtained by adding three, five, nine and thirteen belts, respectively (see Fig. 5.4). Armchair or zigzag tubules are obtained when the C_{60} molecule is bisected normal to a five-fold or a three-fold axis, respectively.

We discuss in what follows the electronic structure, electromagnetic response and static polarizability of short armchair nanotubes C_{110}, C_{150} and C_{190} by employing the LDA and TDLDA methods discussed above (see also [86]).

5.2.1 Electronic Structure

The LDA results for the electronic structure of C_{110}, C_{150} and C_{190} around the HOMO–LUMO region are shown in Fig. 5.5. The HOMO–LUMO gap undergoes strong variations within the three structures considered. Metallic, i.e. zero-gap systems, are also possible e.g. in the case of infinitely long zigzag ($n = m$) nanotubes (see [13]). Recent scanning tunneling microscope studies of tubes several microns in length have verified these expectations [87, 88]. On the other hand, short nanotubes could display a finite gap, due to the additional confinement of the valence electrons along the nanotube axis. Recent measurements on short tubes (a few Å) have provided evidence for such an effect [89].

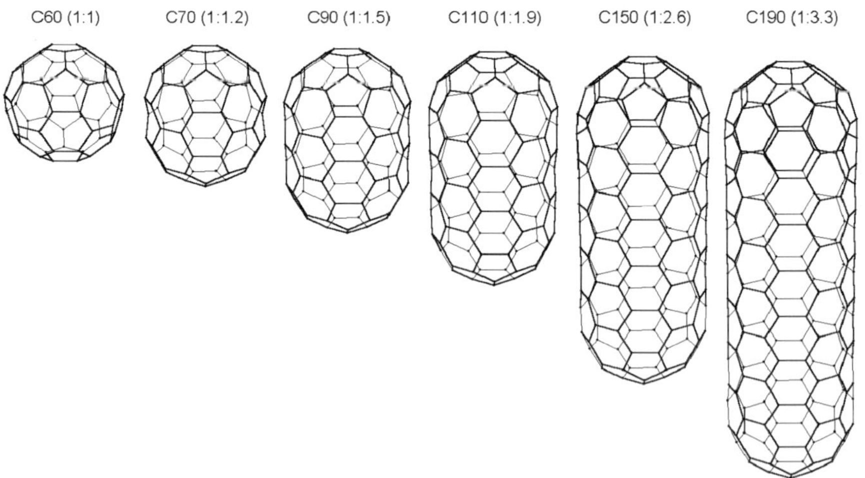

Fig. 5.4. Molecular models of icosahedral C_{60} and derived fullerene nanotubes C_{70}, C_{90}, C_{110}, C_{150} and C_{190}. The ratios $R_\perp/R_\parallel$ between the shortest and the longest radii $R_\perp$ and $R_\parallel$, respectively, are also indicated in parenthesis

5.2.2 Electromagnetic Response and Static Polarizabilities

The results of TDLDA calculations for the longitudinal dipole electromagnetic response for these elongated systems in the UV/optical range below 10 eV are shown in Fig. 5.6 for the case of C_{190}, where the TDLDA response is also compared to the unperturbed response.

In Table 5.3 we collect the energies of the lowest energy peaks, along with the fraction of the exhausted EWSR M_1, and the associated oscillator strength, for the three considered structures. In the same table we report also the corresponding values for the unperturbed response and the total fraction of EWSR and oscillator strength found below 7 eV. Beyond this energy spectral features give room to the incipient Mie resonance. We note that for increasing nanotube length, i.e. in going from C_{110} to C_{190}, the lowest energy absorption peak is redshifted by an amount up to ~ 0.3 eV, concomitant with an "accumulation" of strength at low energies, testifying also an increasing collective character of these states. This softening of the long wavelength dipole response of finite systems is closely connected with the electron spill out and the coupling of the dipole collective vibration and the quadrupole deformation of the system. Such an effect has also been observed in the case of deformed atomic nuclei (see e.g. [72, 90] and refs. therein).

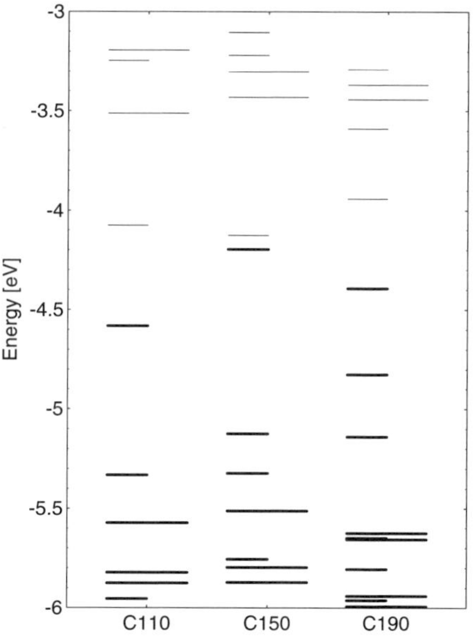

Fig. 5.5. LDA results, around the HOMO–LUMO, for the electronic energy levels of the incipient fullerene nanotubes C_{110}, C_{150}, C_{190}. The *thick (thin) lines* denote the occupied (empty) states, while their lengths are proportional to the corresponding degeneracies, either 1 or 2 in this case

Table 5.3. Calculated values for the lowest energy peaks displayed by the response functions of C_{110}, C_{150} and C_{190}. The latter is shown in Fig. 5.6. Cols. 2-4: LDA values for the peak energy, fraction of EWSR and oscillator strength. Cols. 5-7: the corresponding TDLDA values. Cols. 8,9: TDLDA percentage of the EWSR and oscillator strength below 7 eV

N	E_{LDA}	$M_1[\%]$	f	E_{TDLDA}	$M_1[\%]$	f	$M_1[\%]$	f
110	1.3	1.4	6.2	1.7	0.11	0.4	3.0	13.4
150	1.1	2.0	12.0	1.6	0.22	1.3	3.9	23.7
190	0.9	2.1	16.0	1.4	0.33	2.5	4.5	34.2

The TDLDA static dipole polarizabilities for C_{110}, C_{150} and C_{190} have also been calculated. The longitudinal components α^{zz} are collected in Table 5.4 and displayed in Fig. 5.7. These results seem to confirm an incipient metallic character of these elongated molecules, as already pointed out in a study of shorter tubes [86].

Table 5.4. Calculated static longitudinal dipole polarizabilities (Å^3) for C_{110}, C_{150} and C_{190}. Col. 2: LDA values. Col. 3: TDLDA values

N	α^{zz}_{LDA}	α^{zz}_{TDLDA}
110	1059	228
150	2065	424
190	3780	698

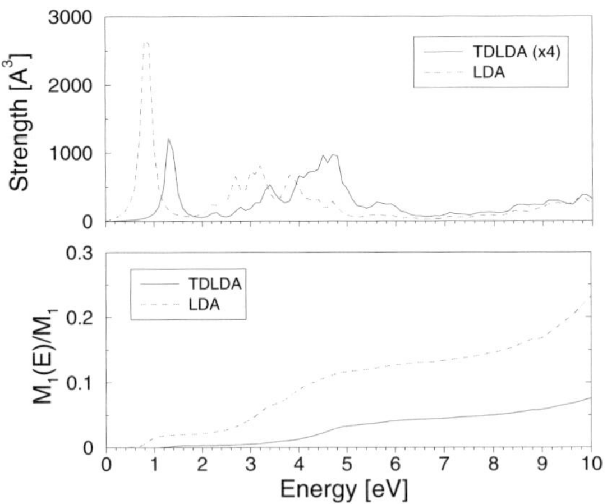

Fig. 5.6. Longitudinal dipole response for C_{190}. Upper panel: strength $\mathcal{S}(E)$ as a function of energy, in the UV/optical region up to 10 eV, both for single particle (LDA and full response (TDLDA). The TDLDA strength has been multiplied by a factor 4 for clarity. Lower panel: accumulated fraction of the dipole EWSR (M_1) as a function of energy

5.3 Linear Carbon Chains

Let us now consider linear carbon chains. To illustrate the main aspects of these systems we will discuss their electronic spectrum (Sect. 5.3.1), and electromagnetic response (Sect. 5.3.2). The LDA and TDLDA calculations reported here were obtained making use of cylindrical coordinates (see Sect. 3.5 and Appendix 4.4.5).

5.3.1 Electronic Structure

There is a great deal of literature regarding theoretical calculations for linear carbon chains (see e.g. [18, 20, 91, 92, 93, 94, 95, 96, 97, 98, 99, 100]). We first discuss two typical examples for odd and even N chains, where we compare the electronic structures of C_5 and C_6 (see Fig. 5.8).

Due to the axial symmetry of the chains, the azimuthal quantum number m is a good quantum number. States having $m = 0$ are denoted as σ-states (admitting up to two electrons), while states having $m = \pm 1$ are denoted as π-states and are doubly-degenerate (admitting up to four electrons). These states are displayed by the short and long lines for the σ and the π states, respectively. No δ ($m = 2$) states, or states with $m \geq 2$ are found among occupied states nor among the lowest unoccupied states. In addition, the states are either symmetric $(+)$ or antisymmetric $(-)$ under the transformation $z \to -z$, and are denoted as $\sigma^\pm$ or $\pi^\pm$. Thus, their parity $(\pm)$ is a

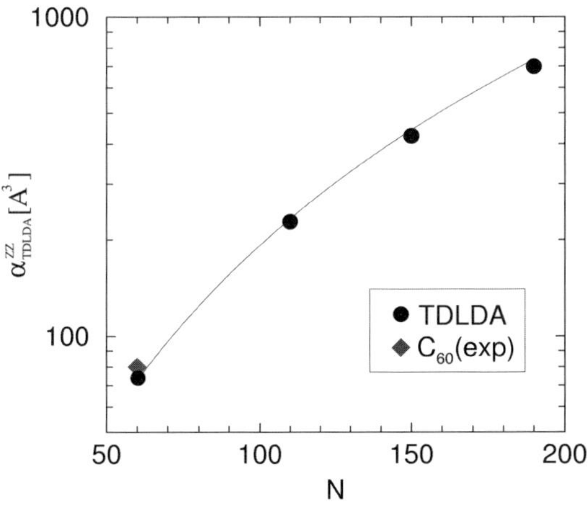

Fig. 5.7. Static longitudinal dipole polarizabilities calculated in TDLDA for C_{110}, C_{150} and C_{190}. The calculated and measured values for C_{60} are also reported [74]. The continuous line is a fit as described in [86] (see also [18])

good quantum number. For any symmetric/antisymmetric pair of states, the symmetric ones lies always below the antisymmetric states.

There are $(N+1)$ occupied σ states corresponding to the $(N-1)$ carbon–carbon bonds, plus two dangling bonds (i.e. two unsaturated bonds) at the tips of the chain. The first $(N-1)$ σ states turn out to be the lowest in energy and well separated from the remaining levels (located below -12 eV in Fig. 5.8). The two σ dangling states always occur at higher energy values (around -8 eV in Fig. 5.8).

Regarding the π states, there are $n_\pi = (2N-2)/4$ fully occupied levels for odd N carbon chains, while for even N chains there are $N/2-1$ fully occupied levels plus a remaining half-occupied π state. In the case of C_5, there are two fully occupied π states and for C_6 there are two fully occupied and one half-occupied π state.

In Fig. 5.8 also the lower (unperturbed) longitudinal particle–hole dipole transitions are displayed, which can be excited by an external field proportional to the spherical harmonics $Y_{1,0}$ (see Sec. 5.3.2). In this case, the selection rules

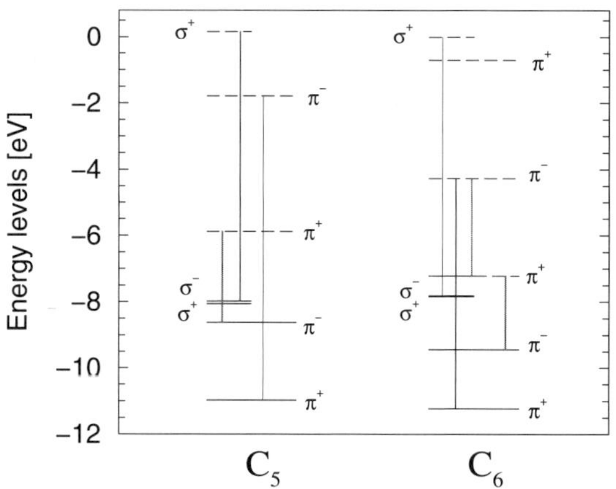

Fig. 5.8. Energy spectrum of carbon chains C_5 (left side) and C_6 (right side) calculated within the LDA near the HOMO–LUMO region. *Full/dashed lines* are used for occupied/empty states. *Long/short lines* for π/σ states (doubly/singly degenerate; see text). The even chain C_6 has an open-shell structure with the HOMO being a half-filled π^+ state at -7.2 eV. The *vertical lines* show (unperturbed) longitudinal dipole transitions. To be noted is that each line connecting double degenerate π states corresponds, actually, to *two* particle–hole transitions, fulfilling the dipole selection rules, (5.6), with the exception of the transitions involving the HOMO state in C_6, where the partial occupation forbids one of the two HOMO $\to$ LUMO transitions and allows a HOMO-1 $\to$ HOMO transition. Hence we are showing a total of five transitions on each panel. The chains have been relaxed using the Car–Parrinello method (see Chap. 7), yielding bond lengths around 1.3 Å. Thus, for simplicity, the constant bond length of 1.3 Å has been used in the calculations. This value is also close to the experimental one and very similar to the values obtained in various theoretical works (from 1.29 Å to 1.31 Å, see e.g. [99])

$$\Delta m = 0, \qquad \pi_i \times \pi_f = -1, \tag{5.6}$$

apply. Here, $\pi_{i,f}$ denotes the parity of the initial or final state, and must not be confused with the same notation used to label single-particle π states with $m = \pm 1$. Because of the partial occupancy of the HOMO state, found in the case of even N chains, there are two low lying single-particle transitions with approximately the same energy: one is due to transitions between the HOMO and the first unoccupied π state, and the other corresponding to transitions from the first lower-lying π state and the HOMO.

In Fig. 5.9 we show the probability densities for the highest occupied σ and π states in C_9 on the x-z plane. At the tips of the chain, one can observe the localization of the charge density associated with σ states. For the π states the corresponding density is delocalized along the chain[3]. In Fig. 5.10 a 3-D plot of the total charge density of C_9 is shown. The sp-bonding (or cumulenic) structure of the chain is clearly seen. Ten peaks are evident, corresponding to the eight bonds among the nine atoms of the chain, located between the peaks, plus the two dangling bonds at the tips of the chain.

The LDA electronic structure of linear carbon chains C_N has been calculated for odd N up to $N = 29$, and for even N up to $N = 20$ and additionally for $N = 30$, and the results are shown in Fig. 5.11 [18]. Experiments have established the existence of such systems for odd $N \leq 21$ and for even $N \leq 10$ (see [101, 102]). The evolution of the electronic structure as a function of N clearly indicates the formation of an occupied σ "band" in the range (-22,-15) eV. The π "bands" are found above a "gap" which, for long chains tends to a value of ≈ 4 eV. The two remaining occupied σ states (dangling bonds) are always located within the π "bands" in the neighborhood of the HOMO.

The HOMO characteristics alternates between even and odd N chains. For odd N chains, the HOMO is always a σ^- state when $N \leq 9$, then it becomes a π state with parity $(-)^{n_\pi+1}$, where n_π is the number of occupied π states. For even N, the HOMO is always a half-filled π state with parity $(-)^{N/2+1}$. Regarding the HOMO energy values, they alternate between even and odd N chains, such that the values for even N are always higher than for the neighboring odd N ones. Both values tend to about -6 eV for large N. The energy of the LUMO coincides with that of the HOMO for even N (open-shell) chains, while for odd N it is always a π state with parity opposite to that of the HOMO state. The HOMO–LUMO gap for odd N chains decreases as a function of N and reaches the value of 0.6 eV for C_{29}.

The features discussed above have direct consequences on the dipole electromagnetic response of the system, which will be addressed in the next Section. Longitudinal dipole single-particle transitions obey the selection rule $\Delta m = 0$ (see (5.6) and Fig. 5.8). Hence, they are allowed only between $\sigma - \sigma$ and $\pi - \pi$ states, the energies of the particle–hole excitations associated with π states being smaller than those associated with σ states. Furthermore, the

---

[3] In the case of C_9 there is an accidental degeneracy of the higher occupied π state with the two σ-tip states.

lower-lying σ band is completely filled, while the anti-bonding σ band is completely empty, thus allowing inter-band transitions only. On the other hand, the lower-lying π band is semi-occupied, lying across the Fermi energy, thus allowing both intra-band and inter-band transitions. As a consequence, $\sigma - \sigma$ and $\pi - \pi$ transitions are expected to contribute to different frequency regions of the response of the system. In particular, the (UV/optical) plasmon resonance is expected to be associated with $\pi - \pi$ transitions. Concerning the transverse dipole response, associated to the operator $Y_{1,\pm 1}$, the selection rules $\Delta m = \pm 1$, $\pi_i \times \pi_f = -1$ (see (5.6)) allow only $\sigma - \pi$ transitions. Because the occupied σ states lie below -15 eV (with the exception of the two tip states, see also Fig. 5.11), the lower transitions to unoccupied states have energies of the order of $\simeq 10$ eV or more and the associated response is expected to show no features below this energy. These qualitative considerations will be confirmed by the quantitative results described in the next Section.

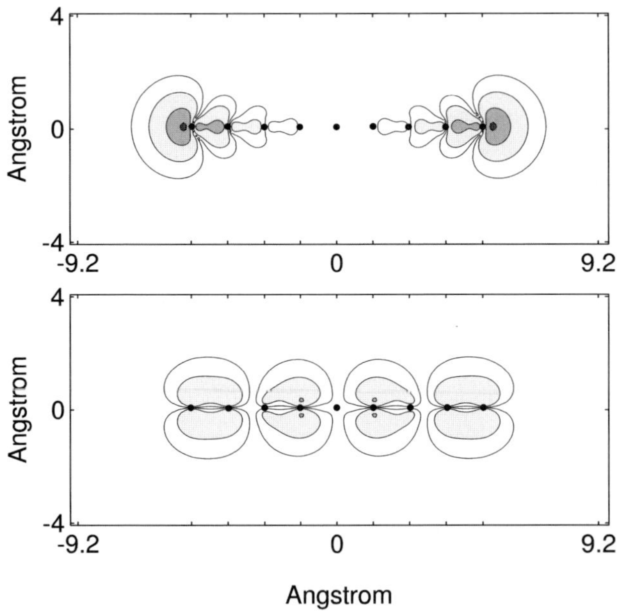

Fig. 5.9. Upper panel: Probability density $(|\psi(\rho, z, \phi = 0)|^2)$ of the highest occupied σ state (HOMO) in C_9. The vertical axis corresponds to the radial variable ρ, while the horizontal one to the z variable. The position of the carbon atoms are indicated by the *full circles*. Lower panel: the highest occupied π state (HOMO-2)

5.3.2 Optical Response and Plasmon Resonances

We discuss here the electromagnetic response of linear carbon chains, in particular the plasmon resonances. In Fig. 5.12 we show a representative plot of the electromagnetic response of the linear carbon chains C_5 and C_6.

In Fig. 5.12(a) we show the dipole longitudinal strength function $\mathcal{S}(E)$ of C_5, both for the uncorrelated (LDA) and the (correlated) TDLDA response. The strength was smeared out with a Lorentzian parameter $\eta = 0.1$ eV. In Fig. 5.12(b) the ratio between the accumulated strength, $M_1(E) = \int_0^E dE'\, E'\, \mathcal{S}(E')$, and the Energy Weighted Sum Rule is displayed as a function of the energy, for both the LDA and the full TDLDA response. The corresponding results for the transverse response are also shown. The longitudinal dipole LDA response shows a single resonance, peaked around 2.75 eV, and other very small features at higher energies, (in the range 10-15 eV and around 20 eV). From Fig. 5.12(b) one can see that the resonance of the C_5 chain exhausts about 34% of the total EWSR M_1. The inclusion of the resid-

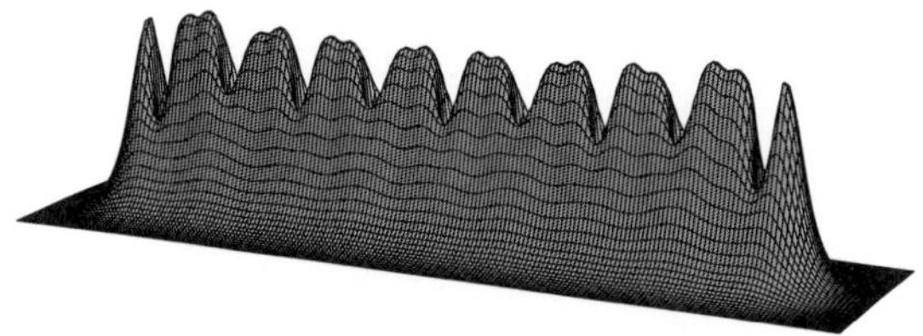

Fig. 5.10. 3D plot of the total electron number density for C_9 (Å^{-3}), in the region $[-3.5 \leq \rho \leq 3.5] \times [-7 \leq z \leq 7]\ \text{Å}^2$ (see Fig. 5.9)

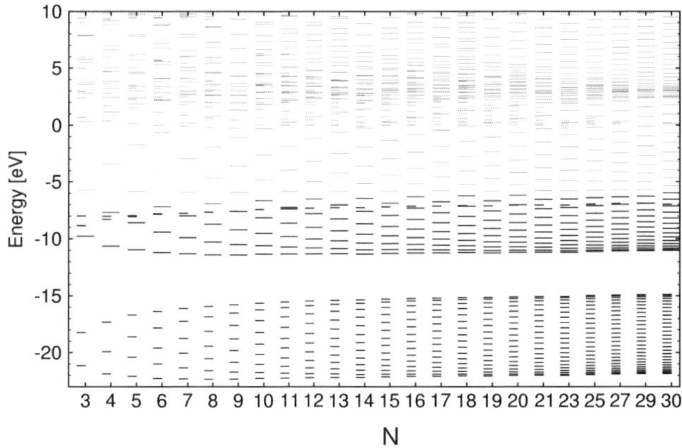

Fig. 5.11. Overview of LDA electronic spectra for C_N chains with $N = 3$ to $N = 21$ and for $N = 23, 25, 27, 29$ and 30 [18]. *Thick (thin) lines* for occupied (empty) states. *Long (short) lines* for doubly (singly) degenerate states. Only states below 10 eV are reported

ual interaction in the response shifts the LDA peak to higher energies, due to its repulsive character. The TDLDA peak is now located at 6.4 eV, taking 30% of M_1, while the remaining strength has been shifted to higher energies, up to about 45 eV. The exchange–correlation interaction is responsible for a minor red-shift of the TDLDA peak, typically of the order of 1% or less of the total shift, testifying its weak attractive character.

From Fig. 5.12(b) it is clear that the dipole transverse response, illustrated in terms of the accumulated fraction of EWSR, is completely different. It is smeared out along the spectrum and shows no distinct features. In fact, both the LDA (*dotted line*) and the TDLDA (*dot-dashed line*) accumulated EWSR do not display sharp rises, and increase considerably more slowly than the longitudinal response. In particular, the TDLDA transverse response of C_5 shows no excitations below 10 eV. Again, the residual interaction has an important overall effect in shifting the TDLDA dipole response to higher

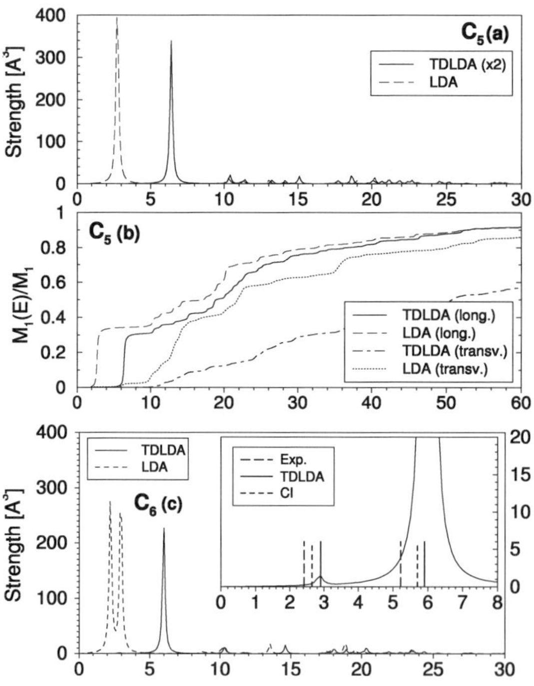

Fig. 5.12. Illustrative examples of TDLDA calculations for C_5 and C_6. **(a)** Longitudinal dipole strength $\mathcal{S}(E)$ (Å^3) for C_5 as a function of energy, shown up to 30 eV, both for single particle (LDA) and full response (TDLDA). The TDLDA strength has been multiplied by a factor 2 for clarity. **(b)** Accumulated fraction of the dipole EWSR $M_1(E)/M_1$ as a function of energy up to 60 eV, both for longitudinal and transverse responses. **(c)** Longitudinal dipole strength for C_6. In the inset the TDLDA response is enlarged and compared with configuration interaction (CI) calculations (*short dashed*) and experimental results (*long dashed*) from [102]

energies as compared to the unperturbed (LDA) response. The different features of the longitudinal and transverse dipole responses can be understood as a consequence of the different mobility (and electron spill out) along the longitudinal and transverse directions of the delocalized valence electrons of the chain in this quasi one-dimensional system. In what follows, we shall discuss only the longitudinal response. The results for C_5 agree well with the TDLDA calculations reported in [96].

Concerning the even chain C_6, Fig. 5.12(**c**), the dipole response shows the same characteristics discussed for C_5, but, because of the partial occupancy of the HOMO state, there are two low-lying uncorrelated single-particle transitions approximately of the same energy (see Fig. 5.8), and the unperturbed LDA response displays a double peak with similar strengths. The residual interaction shifts the strength almost entirely to the state at higher energy, leaving to the lower lying state a very small fraction of the EWSR. This is better appreciated in the inset of Fig. 5.12(**c**), where the low energy response is enlarged and compared with recent experimental and theoretical results [102]. The TDLDA calculations predict two peaks at 2.9 eV and 6.0 eV, with 0.04 and 7.7 units of oscillator strengths, respectively. For the higher lying state, the same theoretical result is found in [96]. The configuration interaction (CI) calculations of [102] predict two peaks at 2.7 eV and 5.7 eV, with 0.023 units of oscillator strength for the lower lying one, the oscillator strength for the higher lying state at 5.7 eV not being reported. The experimental values of the energy of the peaks are 2.42 eV and 5.22 eV. We see that the theoretical values are all appreciably blue-shifted, a feature common to all *ab initio* calculations of the linear response of atomic clusters. The main particle–hole components of the plasmon resonances for C_5 and C_6, obtained using the model (5.5), are displayed in Tables 5.5 and 5.6, respectively.

Table 5.5. Wavefunction components X_{ph} and Y_{ph} of the plasmon peak (at 6.4 eV) for C_5. The first column gives the symmetry of the transition, and the second the energy of the associated particle–hole transition

m	ε_{ph} [eV]	X_{ph}	Y_{ph}
π	2.74	-0.75	0.32
σ	8.14	-0.06	-0.01
σ	9.59	-0.08	-0.01
σ	10.25	0.18	0.04
π	10.42	-0.06	-0.02
σ	11.29	0.09	0.03
σ	13.03	0.09	0.03
σ	14.12	0.07	0.03
σ	20.18	-0.05	-0.03

Studies of the dipole response of linear carbon chains C_N with odd $N \leq$ 21 and even $N \leq 10$ have also been carried out, where a comparison with presently available experimental data is possible [101, 102]. The associated

Table 5.6. Wavefunction components X_{ph} and Y_{ph} of the plasmon peak (at 6.0 eV) for C_6. The first column gives the symmetry of the transition, and the second the energy of the associated particle–hole transition

m	ε_{ph} [eV]	X_{ph}	Y_{ph}
π	2.22	-0.47	0.23
π	2.95	-0.59	0.22
π	6.96	0.11	0.01
π	8.72	-0.07	-0.01
σ	8.94	-0.07	-0.01
σ	10.06	0.09	0.02
σ	10.27	0.13	0.03
σ	13.49	-0.07	-0.03
σ	13.52	0.06	0.02

response functions show features similar to those illustrated above in the case of C_5 and C_6. In Fig. 5.13 the TDLDA peak energies of the longitudinal dipole strength function for carbon chains C_N versus the number of carbon atoms N are displayed in comparison with the experimental results [101, 102, 103, 104]. For even N chains ($N = 6, 8, 10$) the lowest energy states are also reported, while in the inset the corresponding TDLDA peak wavelength λ are shown. The same data are reported in Table 5.7. The results reported here compare well with those reported in [96] ($N \leq 15$).

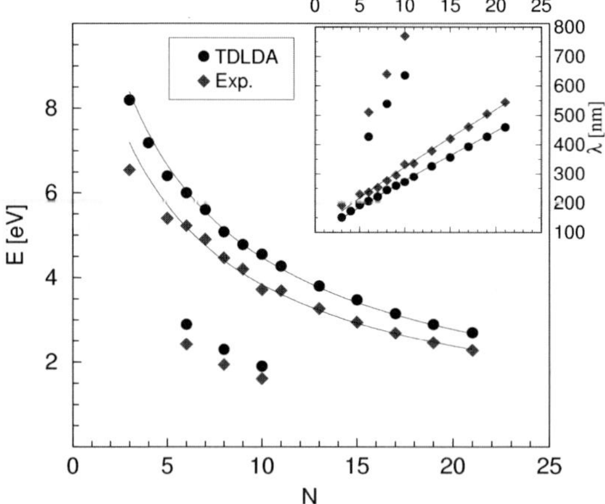

Fig. 5.13. The dipole response of carbon chains C_N. *Full circles*: TDLDA peak energies (eV) versus the number of carbon atoms N. *Full diamonds*: experimental data (from Refs. [101, 102, 103, 104]. The value of 5.4 eV for C_5 is not assigned to a longitudinal dipole transition in [103]). The values at lower energies for $N = 6, 8, 10$ corresponds to the lowest dipole transitions, having small oscillator strengths (see text and Table 5.7). Inset: the corresponding wavelengths λ (nm) (scale to the right) are plotted as a function of N (scale at the top). The lines are fits as explained in the text

Table 5.7. Dipole response of the carbon chains C_N. Col. 2: experimental values for the main absorption peaks from [104] ($N = 3$), [103] (odd N in the range 5-15, but for $N = 5$ the resonance at 5.4 eV is not assigned to a longitudinal dipole transition), [101] (odd N in the range 17-21), and [102] (even N in the range 6-10). Col. 3: the theoretical TDLDA values. Col. 4: the percentage of the EWSR M_1 associated to the TDLDA peaks. Col. 5: the corresponding oscillator strengths as defined in (4.33). Both the experimental and the theoretical spectrum of even N chains display two dipole absorption peaks with very different oscillator strengths f (the largest f in the UV, the other in the visible) [102]

N	E_{exp} [eV]	E_{TDLDA} [eV]	M_1 [%]	f
3	6.55	8.2	25	3.0
5	5.40	6.4	30	6.0
6	5.22	6.0	32	7.7
	2.42	2.9	10^{-3}	0.04
7	4.90	5.6	31	8.7
8	4.47	5.1	34	10.9
	1.94	2.3	10^{-3}	0.05
9	4.20	4.8	35	12.6
10	3.72	4.5	33	13.2
	1.61	1.9	10^{-3}	0.04
11	3.69	4.3	36	15.8
13	3.26	3.8	36	18.7
15	2.95	3.5	37	22.2
17	2.69	3.1	37	25.2
19	2.46	2.9	37	28.1
21	2.28	2.7	37	31.1

This agreement testifies to the fact that the *systematic* difference between the theoretical and the experimental values, which are always lower in energy, is a consequence of the TDLDA itself and not an artifact due to its numerical implementation. The percentual difference in wavelength tends to 17% for longer chains. Such a systematic deviation of the TDLDA has also been found in connection with the study of the optical response of other systems, like alkali metal clusters (see next Section and e.g. [5, 6, 10, 72, 105] and refs. therein). As mentioned above, taking into account the coupling to states containing one- and two-plasmons (see Chap. 8), that is going beyond the harmonic approximation, it is possible to account for most of this discrepancy.

In connection with this difference, we note that no absolute value of the cross section is reported in the experimental literature. This is an unfortunate fact, because a proper description of this quantity is as important, or even more important, than explaining the energy centroid of the photoabsorption peaks. In fact, a systematic discussion of photoabsorption cross section in comparison with the experiments has been instrumental in the understanding of the giant resonances in atomic nuclei [72]. The theoretical TDLDA oscillator strengths are shown in Table 5.7. In col. 4 we display the percentage of the EWSR M_1 exhausted by the resonance. One can see how it grows from 25% and stabilizes around 37% already at $N = 11$. In col. 5 we display the values of the f-sum rule, that is, the number of electrons involved in the resonance. They are proportional to the dipole cross section (see (4.32)).

The energy centroids of the photoabsortion peaks shown in Fig. 5.13 as a function of N, can be succesfully parametrized in terms of a (classical) expression [106] for the Mie plasmon resonance [77] of ellipsoidal metallic particles (see [18, 20, 21] for details).

5.4 Metal Clusters: the Case of Na$_8$

Small sodium clusters display a remarkable mass spectrum (Fig. 1.4). The presence of peaks in the emprical abundancies reflect the existence of "magic numbers", in analogy with atoms or nuclei: the systems characterized by closed-shell structures do have a high degree of stability in comparison with the other aggregates [5]. Initially, most of the theoretical work to study metal clusters of medium and large sizes have made use of the jellium model (see Sect. 2.2.1). More recently, however, it became clear that the actual ionic structure of the clusters must be taken into account in order to obtain reliable estimates of their electronic properties [6].

For illustration, we consider here the smallest closed shell sodium cluster, i.e., Na$_8$, and study its electronic states and plasmon spectrum in LDA and TDLDA, respectively. The phonon spectrum of this system is discussed in Sect. 7.3.1, while the consequences the interweaving of electrons, plasmons and phonons has on the photoabsorption spectrum of Na$_8$ is studied in Sect. 8.8.

The lowest energy structures of Na$_8$ are displayed in Fig. 5.14. These isomers are characterized by the symmetries associated respectively with the groups $\mathcal{D}_{2d}$ (ground state (GS)), $\mathcal{D}_{4d}$ (89 meV) and $\mathcal{T}_d$ (95 meV) – the figures in parenthesis being the total energies calculated in LDA and expressed with respect to the ground state energy (see also [107]). We refer here to results for the electromagnetic response to external fields of various multipolarities, based on the optimal $\mathcal{D}_{2d}$ ground-state structure. The Kohn–Sham energy levels for the latter are reported in Table 5.8.

The response of Na$_8$ calculated within TDLDA to external fields with multipolarity L=0, 1, 2, and 3, is shown in Fig. 5.15. Some features can be noticed in the spectra. The fragmentation of the modes increases with the angular momentum. While the dipole results from single-particle excitations which would lie at $\approx 1\hbar\omega$ without the effect of the residual interaction (in a harmonic oscillator model of the electronic average potential), and the monopole and the quadrupole would peak around $2\hbar\omega$, the octupole presents two different regions in which the strength is non negligible, clearly reminiscent of the $1\hbar\omega$ and $3\hbar\omega$ unperturbed excitations. In reality, the discrete peaks would appear as broad structures, because of the coupling of electrons with phonons and of further correlations (electron–plasmon couplings) [108]. This will be discussed in Chap. 8. Concerning the experimental situation,

Table 5.8. Kohn–Sham energy levels of Na$_8$ for its ground state configuration $\mathcal{D}_{2d}$

E[eV]	Degeneracy	Occupation
-4.75	2	2
-3.54	2	2
-3.11	4	4
-2.00	2	0
-1.84	4	0
-1.54	2	0
-1.50	2	0
-1.39	2	0
-0.65	2	0
-0.47	4	0

thermal fluctuations associated with the fact the clusters can be cooled down only to a certain extent, are superimposed to the quantal couplings we have mentioned.

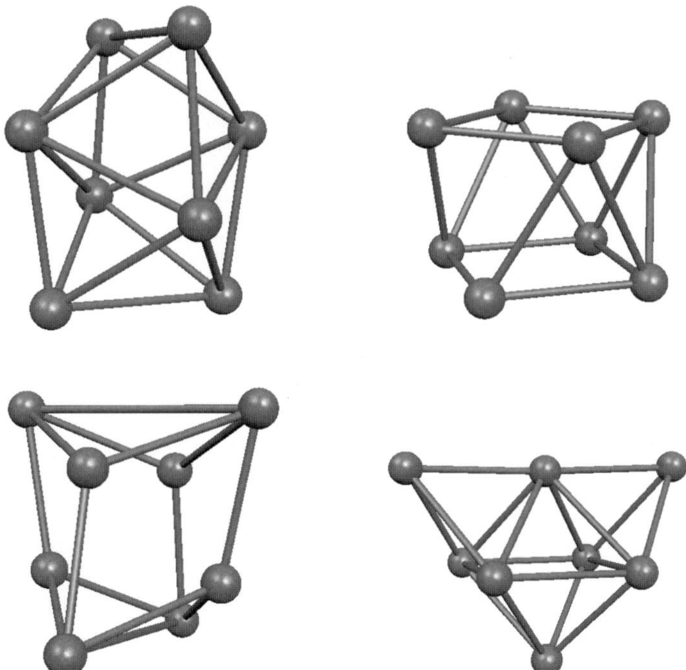

Fig. 5.14. Ionic structure of Na$_8$ in its three lowest energy isomers with symmetry: $\mathcal{D}_{2d}$ (top left), $\mathcal{D}_{4d}$ (top right), $\mathcal{T}_d$ (bottom left). Also shown, the 'close packed' structure (bottom right)

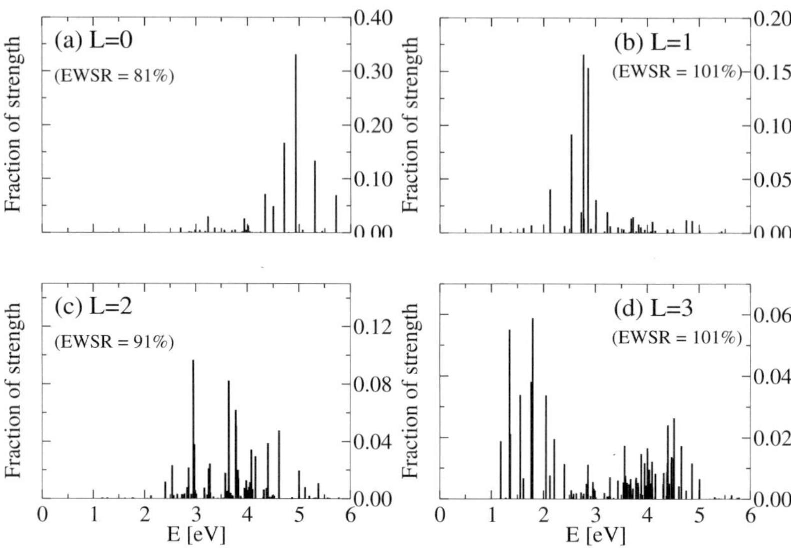

Fig. 5.15. Fraction of strength (namely $\langle n|\hat{O}|0\rangle^2$ as a function of the energy of $|n\rangle$) of Na$_8$ for the multipolarities: (a) monopole ($\hat{O} = r^2$), (b) dipole ($\hat{O} = rY_{10}$), (c) quadrupole ($\hat{O} = r^2Y_{20}$) and (d) octupole ($\hat{O} = r^3Y_{30}$). Also indicated, in parenthesis, is the fraction of the energy weighted sum rule (EWSR) exhausted by the different multipolarities (after [108])

5.5 Appendices

5.5.1 Role of Symmetries

The purpose of the present Appendix is to briefly recall the large amount of information which can be inferred concerning the properties of a quantal system like a molecule or a cluster, solely in terms of the symmetries of the Hamiltonian which describes it. We first recall the general theorems derived from group theory and we apply them to the icosahedral symmetry of C$_{60}$.

If a Hamiltonian is invariant under the action of a group G, the levels of its spectrum correspond to the irreducible representations of G, and their degeneracy is equal to the dimension of the representation. The ordinary rotational group in three dimensions has an infinite number of irreducible representations which can be characterized by the integers ℓ (non negative) and m ($-\ell \leq m \leq \ell$), which are then the quantum numbers adequate to describe finite systems which are either spherical, or whose shapes do not differ much from sphericity. In this class of systems one can include atoms, atomic nuclei, etc. When atoms come together to form a cluster, or a lattice, ℓ and m cease to be good quantum numbers, since the system has in general a different symmetry than spherical. A multiplet of levels with a definite

ℓ splits in smaller multiplets labeled by the representations of G. In what follows we shall examine how this general pattern [109] works in the case of the C_{60} cluster, which is known to display an icosahedral symmetry.

The icosahedral group $\mathcal{I}$ is formed by five classes, each one containing only rotations. We use for the classes the notation of Wilson, Decius and Cross [110]. Consequently, rotations by $2\pi/5$ and $4\pi/5$, around the five-fold axes, belong respectively to the classes $12C_5$ and $12C_5^2$ (each class contain 12 elements). Rotations around the three-fold axes are collectively denoted by $20C_3$ and the ones around the two-fold axes by $15C_2$ (there are 20 and 15 rotations, respectively). The identity operator, denoted by E constitutes a class by itself. Under a rotation of the icosahedral group $\mathcal{I}$, the spherical harmonic $Y_{\ell m}$ is transformed in a linear combination of $Y_{\ell m'}$. To find the character of such a rotation we choose a system whose z-axis coincides with the rotation axis, and call ϕ_0 the rotation angle. Then, the transformation can simply be written as

$$Y_{\ell m} \rightarrow e^{im\phi_0} Y_{\ell m}. \tag{5.7}$$

Consequently, the character of this rotation is

$$\chi^{(\ell)}(\phi_0) = \sum_{m=-\ell}^{+\ell} e^{im\phi_0} = \frac{\sin[(2\ell+1)\phi_0/2]}{\sin(\phi_0/2)}. \tag{5.8}$$

This is a well-known result first derived by Bethe [111]. From the knowledge of the rotation angle ϕ_0 (equal to $2\pi/5$, $4\pi/5$, $2\pi/3$, π, for C_5, C_5^2, C_3, C_2 respectively), the characters of operators of the classes of $\mathcal{I}$ can be calculated from (5.8), for different ℓ-representations. They are shown in Table 5.9.

Table 5.9. Characters of classes of $\mathcal{I}$ in the representations of the rotation group labeled by ℓ. Here, $\mu = (1+\sqrt{5})/2$, and $\nu = (1-\sqrt{5})/2$

	E	C_5	C_5^2	C_3	C_2
$\ell=0$	1	1	1	1	1
$\ell=1$	3	μ	ν	0	-1
$\ell=2$	5	0	0	-1	1
$\ell=3$	7	$-\mu$	$-\nu$	1	-1
$\ell=4$	9	-1	-1	0	1
$\ell=5$	11	1	1	-1	-1
$\ell=6$	13	μ	ν	1	1
$\ell=7$	15	0	0	0	-1
$\ell=8$	17	$-\mu$	$-\nu$	-1	1
$\ell=9$	19	-1	-1	1	-1
$\ell=10$	21	1	1	0	1
$\ell=11$	23	μ	ν	-1	-1
$\ell=12$	25	0	0	1	1

The ℓ-representations of the icosahedral group are reducible representations. This group has not been studied extensively, since it was thought that no physical system in nature could have icosahedral symmetry (see e.g. p. 51 of [109]). The irreducible representations are denoted usually by A, F_1, F_2, G and H, and have respectively dimensions 1, 3, 3, 4, 5. Sometimes a suffix g or u is added, to make explicit the even (gerade) or odd (ungerade) parity of the representation. The analysis of these representations can be found e.g. in [110]. We show the characters of the representations in Table 5.10 (see p. 330 of [110]).

Table 5.10. Characters of the classes of $\mathcal{I}$ in the irreducible representations named in the text

	E	C_5	C_5^2	C_3	C_2
A	1	1	1	1	1
F_1	3	μ	ν	0	-1
F_2	3	ν	μ	0	-1
G	4	-1	-1	1	0
H	5	0	0	-1	1

In order to know how many times a given irreducible representation is contained in an ℓ-representation, equation (3.150) of [109] can be used. The coefficient a_i^ℓ of the expansion of the ℓ-representation, on the basis of the irreducible icosahedral representations labeled by i, can be calculated through the relation

$$a_i^\ell = \frac{1}{g} \sum_n g_n \chi_n^{(\ell)} \chi_n^{*(i)}, \tag{5.9}$$

where $\chi_n^{(\alpha)}$ is the character of the n-th class in the representation α, as quoted in the two tables above, g_n is the number of the elements of the class, and g is the order of the group ($g = \sum_n g_n$). The complex conjugation could be omitted in the present case where all characters are real numbers. We have calculated the coefficients a_i^ℓ and displayed them in Table 5.11. The Table must be read as follows. A single-electron level with orbital angular momentum ℓ in a C_{60} cluster considered as a spherical system splits in a number of multiplets in the icosahedral field of the ions. These multiplets are labeled by the symbols of the irreducible representations of $\mathcal{I}$, and their number is indicated in Table 5.11. The degeneracy of the multiplet corresponds to the dimension of the representation. In this way one can determine the *number* of the levels, compared to the ones resulting from a spherical Hamiltonian. To get an estimate of the *position* of these levels, further considerations must be made.

Group theoretical analysis is able to give further insight concerning the single particle properties of C_{60}. In particular, concerning the question of the transitions which are allowed for an electron in the cluster, affected by an external, in general time-dependent, external field (see Chap. 4).

Table 5.11. Number of multiplets, corresponding to irreducible representations of $\mathcal{I}$, in which a level with definite ℓ split

	A	F_1	F_2	G	H
l=0	1	0	0	0	0
l=1	0	1	0	0	0
l=2	0	0	0	0	1
l=3	0	0	1	1	0
l=4	0	0	0	1	1
l=5	0	1	1	0	1
l=6	1	1	0	1	1
l=7	0	1	1	1	1
l=8	0	0	1	1	2
l=9	0	1	1	2	1
l=10	1	1	1	1	2
l=11	0	2	1	1	2
l=12	1	1	1	2	2

In particular, this can be accomplished in the case of the dipole field, which is simply the $\ell = 1$ representation of the rotation group and has a one-to-one correspondence with F_1 of $\mathcal{I}$, according to Table 5.11. The allowed transitions are the ones in which the direct product of the irreducible representations of the initial state, final state, and perturbing field, contain the identity representation. We can determine them with standard methods as follows (see e.g. p. 147 of [109]). The representation f, corresponding to the final state of the transition, can be expanded on the basis of the direct product of the representations i and e, corresponding to the initial state and external field. The coeffcents of this expansion are determined making use of (5-42) of reference [109]. One can write them in a form which resembles (5.9), that is,

$$a_{i\otimes e}^{f} = \frac{1}{g} \sum_{n} g_n \chi_n^{(i)} \chi_n^{(e)} \chi_n^{*(f)}. \tag{5.10}$$

Only in the case that these coefficients are different from 0, the transition i→f is allowed under the influence of the field e. These coefficients, in the case of dipole field, in Table 5.12. A similar calculation can also be performed in the case of other multipole fields, such as quadrupole etc.

Table 5.12. Dipole transition selection rules: 1 corresponds to an allowed transition, and 0 to a forbidden one

	A	F_1	F_2	G	H
A	0	1	0	0	0
F_1	1	1	0	0	1
F_2	0	0	0	1	1
G	0	0	1	1	1
H	0	1	1	1	1

6. Phonons: Harmonic Approximation

As illustrated in Sect. 3.1, in the Born–Oppenheimer (BO) approximation (also known as the *adiabatic approximation*), the equations which determine the electronic state are decoupled from those related to the ionic dynamics. In other words, the atomic motion of the system (molecule, cluster, or solid) is studied under the hypothesis that the electronic system always remains in the electronic ground state associated with the instantaneous geometrical configuration. We have seen in Sect. 3.1 the conditions of applicability of the BO scheme. In Sect. 6.1 we analyze the solution of the problem of atomic motion in the case of small oscillations near the equilibrium configuration (i.e. when every atom stays always near to its equilibrium position, and there is no atomic diffusion). In that case, writing down a second order expansion of the total potential felt by the ions (harmonic approximation), we show that it is possible to reduce the problem to that of a collection of independent harmonic oscillators. The general solution is then a superposition of $3N_{\mathrm{at}}$ normal modes of vibration, each of them having its own frequency and its own eigenvector, obtained by diagonalizing the dynamical matrix. The quantum description of this set of independent harmonic oscillators leads to the concept of phonons.

The harmonic approximation is usually a good one for solid state systems, molecules or atomic aggregates, at sufficiently low temperature. In the case of carbon systems (graphite, diamond, fullerenes), which are characterized by a quite strong covalent bonding, the harmonic approximation is still very good at room temperature.

In Sect. 6.2, we generalize the arguments of the previous section by discussing the general properties of the solution for the ionic motion, valid independently of the actual form of the total potential U acting on the ions. We then continue by discussing how to determine the different parameters of the model, namely the first and second derivatives of U, where the calculations range from a complete treatment of the electronic degrees of freedom (DFT and Car–Parrinello Method) to a complete neglect of them (force-constants models), going through intermediate approximations like the Bond Charge Model.

6.1 Harmonic Approximation

Let us indicate with $U(\{\boldsymbol{R}_n\})$ the total potential governing the ionic motion. According to (3.13), one can write it as

$$U(\{\boldsymbol{R}_n\}) = \mathcal{E}_0\left(\{\boldsymbol{R}_n\}\right) + V_{\text{ion,ion}}, \tag{6.1}$$

where $\mathcal{E}_0$ is the ground state electronic energy (see Sect. 3.1). Labeling the ions with indices n, m, and using vector notation, we define $\boldsymbol{R}_n = \boldsymbol{R}_n^{(0)} + \boldsymbol{u}_n$, $\boldsymbol{u}_n$ representing a small displacement of the n-th atom away from its equilibrium position $\boldsymbol{R}_n^{(0)}$. We then introduce the indices $\alpha = (x, y, z)$, $\beta = (x, y, z)$ to label the three cartesian coordinates and we write an expansion of U around its minimum value as,

$$U = U_0 + \sum_{n,\alpha} \left.\frac{\partial U}{\partial u_{n,\alpha}}\right|_0 u_{n,\alpha} + \frac{1}{2} \sum_{n,m,\alpha,\beta} \left.\frac{\partial^2 U}{\partial u_{n,\alpha}\partial u_{m,\beta}}\right|_0 u_{n,\alpha} u_{m,\beta} + ... \tag{6.2}$$

The first-order derivatives are zero since the ions are in equilibrium at a minimum of the total potential U, and the second-order derivatives correspond to a symmetric matrix (the so-called Hessian matrix [112]). If the system is in a stable minimum, all the eigenvalues of the Hessian matrix must be positive or zero. We define the *force–constant matrix* $\Phi_{n\alpha m\beta}$ as follows:

$$\Phi_{n\alpha m\beta} \equiv \left.\frac{\partial^2 U}{\partial u_{n,\alpha}\partial u_{m,\beta}}\right|_0 = \left.\frac{\partial}{\partial u_{n,\alpha}}\right|_0 \left(\left.\frac{\partial U}{\partial u_{m,\beta}}\right|_0\right). \tag{6.3}$$

Thus, $\Phi_{n\alpha m\beta}$ is proportional to the change of the component β of the *force* generated on the atom m, when the atom n is displaced by a small amount in the direction α.

If one is dealing with an isolated system, that is, a system not subject to external forces, the translational and rotational invariance conditions can be used to deduce some relations which must be satisfied by $\Phi_{n\alpha m\beta}$, hence reducing the number of independent elements in the matrix. Translational invariance yields

$$\sum_m \Phi_{n\alpha m\beta} = 0, \tag{6.4}$$

which holds for all n, α, β. This condition is intuitively understood as the fact that the force generated on a given atom as a consequence of its own displacement (diagonal term of Φ) must be equal to the opposite of the sum of all forces generated on the other atoms (third principle of dynamics). The rotational invariance yields instead:

$$\sum_m (\Phi_{n\alpha m\beta}(R_m^{(0)})_\gamma - \Phi_{n\alpha m\gamma}(R_m^{(0)})_\beta) = 0 \tag{6.5}$$

holding for all n, α, β, γ.

Making use of (6.3) one can then write

$$U = U_0 + \frac{1}{2} \sum_{n,m,\alpha,\beta} \Phi_{n\alpha m\beta} u_{n\alpha} u_{m\beta}. \tag{6.6}$$

Thus, the (classical) Lagrangian of the ionic system is

$$\mathcal{L}_{\text{ion}} = \frac{1}{2} \sum_{n,\alpha} M_n \dot{u}_{n,\alpha}^2 - \frac{1}{2} \sum_{n,m,\alpha,\beta} \Phi_{n\alpha m\beta} u_{n\alpha} u_{m\beta}, \tag{6.7}$$

the resulting Lagrange equations becoming

$$M_n \ddot{u}_{n,\alpha} + \sum_{m,\beta} \Phi_{n\alpha m\beta} u_{m,\beta} = 0 \quad \forall n, \alpha. \tag{6.8}$$

In a more compact matrix form, (6.8) reads:

$$\mathbf{M}\ddot{\mathbf{u}} + \Phi\mathbf{u} = 0, \tag{6.9}$$

where we have introduced the diagonal matrix $\mathbf{M}_{n\alpha m\beta} = \delta_{\alpha\beta}\delta_{nm}M_n$.

Equation (6.8) is a system of $3N_{\text{at}}$ coupled linear differential equations, which can be solved by the standard technique using the Fourier transform with respect to the time variable. This means, in practice, that the solution is expanded into the (complete) basis set of functions $e^{i\omega t}$. In this specific case, and to be able to formally include on equal footing in the above description both systems in which all ions have the same mass as well as those containing ions of different masses, it is convenient to introduce mass-scaled atomic displacements $w_{n,\alpha}$, defined as,

$$\mathbf{w} = \mathbf{M}^{\frac{1}{2}}\mathbf{u}. \tag{6.10}$$

The Lagrangian takes the form,

$$\mathcal{L}_{\text{ion}} = \frac{1}{2}\dot{\mathbf{u}}^\dagger \mathbf{M}^{\frac{1}{2}}\mathbf{M}^{\frac{1}{2}}\dot{\mathbf{u}} - \frac{1}{2}\mathbf{u}^\dagger\Phi\mathbf{u} = \frac{1}{2}\dot{\mathbf{w}}^\dagger\dot{\mathbf{w}} - \frac{1}{2}\mathbf{w}^\dagger\mathbf{D}\mathbf{w}, \tag{6.11}$$

where

$$\mathbf{D} = \mathbf{M}^{-\frac{1}{2}}\Phi\mathbf{M}^{-\frac{1}{2}}. \tag{6.12}$$

The Lagrange equations for $\mathbf{w}$ now reads,

$$\ddot{\mathbf{w}} + \mathbf{D}\mathbf{w} = 0. \tag{6.13}$$

By Fourier transforming with respect to time, one gets

$$\omega^2 \mathbf{w} = \mathbf{D}\mathbf{w} \tag{6.14}$$

or, writing all indices explicitly,

$$\sum_{m,\beta} \left[\delta_{\alpha\beta}\delta_{nm}\omega^2 - \frac{\Phi_{n\alpha m\beta}}{\sqrt{M_n M_m}} \right] w_{m,\beta} = 0. \tag{6.15}$$

One is then led to a canonical diagonalization problem for $\mathbf{D}$, which is known as the *dynamical matrix* of the system. It is a $3N_{\text{at}} \times 3N_{\text{at}}$ real symmetric matrix, hence it can be always diagonalized by an orthogonal transformation (i.e., by a change of the basis set[1]). The $3N_{at}$ eigenvalues of $\mathbf{D}$, ω_λ^2, are the squared frequencies of the *normal modes of vibration* for the system considered. The corresponding eigenvectors, $S_{n\alpha}^\lambda$, tell us how to build linear combinations of the mass-scaled Cartesian displacements of the single atoms setting up a *collective coordinate* associated with the vibration at frequency ω_λ:

$$q_\lambda(t) = q_\lambda(0)e^{i\omega_\lambda t} = \sum_{n\alpha} S_{n\alpha}^{\lambda\dagger} w_{n\alpha}(t) = \sum_{n\alpha} S_{n\alpha}^{\lambda\dagger} M_{n\alpha n\alpha}^{\frac{1}{2}} u_{n\alpha}(t). \qquad (6.16)$$

In other words, the matrix $\mathbf{S}$, whose columns are the $3N_{at}$ *normalized* eigenvectors of $\mathbf{D}$, is an orthogonal matrix (i.e. $\mathbf{S}^\dagger \mathbf{S} = \mathbf{1}$), and represents a change of basis such that $\mathbf{SDS}^\dagger$ is diagonal. The columns of $\mathbf{Z} = \mathbf{M}^{-\frac{1}{2}}\mathbf{S}$ are "the normal modes in Cartesian coordinates", i.e. the components of the new basis vectors with respect to the old (Cartesian) basis. Hence,

$$\mathbf{u} = \mathbf{M}^{-\frac{1}{2}}\mathbf{w} = \mathbf{Zq}. \qquad (6.17)$$

It is now possible to rewrite the Lagrangian (6.11) as a function of $\mathbf{q}$:

$$\begin{aligned}
\mathcal{L}_{ion} &= \frac{1}{2}\sum_{n\alpha} \dot{w}_{n\alpha}^\dagger \dot{w}_{n\alpha} - \frac{1}{2}\sum_{n\alpha m\beta} w_{n\alpha}^\dagger M_{n\alpha n\alpha}^{-\frac{1}{2}} \Phi_{n\alpha m\beta} M_{m\beta m\beta}^{-\frac{1}{2}} w_{m\beta} \\
&= \frac{1}{2}\sum_{n\alpha} \left(\sum_\lambda S_{n\alpha}^\lambda \dot{q}_\lambda \right)^\dagger \left(\sum_{\lambda'} S_{n\alpha}^{\lambda'} \dot{q}_{\lambda'} \right) \\
&\quad - \frac{1}{2}\sum_{n\alpha m\beta} \left(\sum_\lambda S_{n\alpha}^\lambda q_\lambda \right)^\dagger D_{n\alpha m\beta} \left(\sum_{\lambda'} S_{m\beta}^{\lambda'} q_{\lambda'} \right) \\
&= \frac{1}{2}\sum_{\lambda,\lambda'}\sum_{n\alpha} S_{n\alpha}^{\lambda\dagger} S_{n\alpha}^{\lambda'} \dot{q}_\lambda \dot{q}_{\lambda'} - \frac{1}{2}\sum_{\lambda,\lambda'}\sum_{n\alpha m\beta} q_\lambda S_{n\alpha}^{\lambda\dagger} D_{n\alpha m\beta} S_{m\beta}^{\lambda'} q_{\lambda'} \\
&= \frac{1}{2}\sum_\lambda (\dot{q}_\lambda)^2 - \frac{1}{2}\sum_\lambda \omega_\lambda^2 (q_\lambda)^2. \qquad (6.18)
\end{aligned}$$

Due to the translational and rotational invariance, it is possible to show that, for a system with N_{at} non-collinear atoms, there are always 6 vanishing eigenvalues. The six zero-frequency modes correspond in fact to the rigid translations and rotations of the system[2]. When the expansion around $\mathbf{R}^{(0)}$ reveals that this point is true quadratic minimum of the potential, all other eigenvalues ω_λ^2 of $\mathbf{D}$ are strictly positive. Extra vanishing or negative eigenvalues signal instabilities.

---

[1] In periodic systems, as bulk crystals, $\mathbf{D}$ becomes a $\mathbf{q}$-dependent complex hermitian matrix of size $3N_{cell} \times 3N_{cell}$ (the cell contains N_{cell} atoms, and atomic displacements are expanded in plane waves) which can be diagonalized by a unitary transformation.

[2] Linear molecules, in particular dimers, have only two purely rotational degrees of freedom, hence only 5 zero eigenvalues.

The system is therefore equivalent to a collection of $3N_{\text{at}}$-6 *independent* harmonic oscillators. The quantum solution is obtained from the classical one by quantizing each harmonic oscillator. For this purpose, it is convenient to scale again the coordinates, i.e. to introduce the new coordinate $Q_\lambda = \sqrt{\omega_\lambda}q_\lambda$. In this way the Lagrangian becomes,

$$\mathcal{L}_{\text{ion}} = \frac{1}{2}\sum_\lambda \left(\frac{\dot{Q}_\lambda{}^2}{\omega_\lambda} - \omega_\lambda Q_\lambda{}^2\right). \tag{6.19}$$

The conjugate momentum is then

$$P_\lambda = \frac{\dot{Q}_\lambda}{\omega_\lambda}, \tag{6.20}$$

and the Hamiltonian is

$$\mathcal{H} = \frac{1}{2}\sum_\lambda \omega_\lambda \left(P_\lambda{}^2 + Q_\lambda{}^2\right). \tag{6.21}$$

Quantization of this Hamiltonian proceeds as usual, making use of the commutation relations, $\left[\hat{Q}_\lambda, \hat{P}_{\lambda'}\right] = i\hbar\delta_{\lambda,\lambda'}$. As commonly done in the theory of the quantum harmonic oscillator, it is useful to introduce the creation and annihilation operators, $\Gamma^\dagger$ and Γ, which in the present case are defined as follows,

$$\Gamma_\lambda = \frac{1}{\sqrt{2\hbar}}(\hat{Q}_\lambda + i\hat{P}_\lambda) \qquad ; \qquad \Gamma_\lambda^\dagger = \frac{1}{\sqrt{2\hbar}}(\hat{Q}_\lambda - i\hat{P}_\lambda), \tag{6.22}$$

such that $\left[\Gamma_\lambda, \Gamma_{\lambda'}^\dagger\right] = \delta_{\lambda,\lambda'}$, and

$$\hat{Q}_\lambda = \sqrt{\frac{\hbar}{2}}(\Gamma_\lambda + \Gamma_\lambda^\dagger) \qquad ; \qquad \hat{P}_\lambda = \sqrt{\frac{\hbar}{2}}(\Gamma_\lambda - \Gamma_\lambda^\dagger). \tag{6.23}$$

One obtains hence

$$\hat{H} = \sum_\lambda \frac{\hbar\omega_\lambda}{2}\left(\Gamma_\lambda\Gamma_\lambda^\dagger + \Gamma_\lambda^\dagger\Gamma_\lambda\right) = \sum_\lambda \hbar\omega_\lambda\left(\Gamma_\lambda^\dagger\Gamma_\lambda + \frac{1}{2}\right), \tag{6.24}$$

where the operator $\Gamma_\lambda^\dagger\Gamma_\lambda$ is known as the number operator, $\hat{n}_\lambda$, having all non-negative integers as its eigenvalues (see Appendix 2.11.1).

The mean value of each $\hat{Q}_\lambda$ is of course 0, but not the mean square deviation, $\langle\hat{Q}_\lambda^2\rangle^{1/2}$. This quantity can be calculated directly (for a given quantum state) by using (6.23). It follows that

$$\langle\hat{Q}_\lambda^2\rangle^{1/2} = \sqrt{\frac{\hbar}{2}}\left\langle\left(\Gamma_\lambda + \Gamma_\lambda^\dagger\right)^2\right\rangle^{1/2}, \tag{6.25}$$

which in the ground state yields $\sqrt{\hbar/2}$.

We are now able to compute the zero-point motion or "zero point fluctuation" (i.e., the mean square deviation in the ground state) for our dynamical variables q_λ and $\mathbf{u}$ (in other words, the total mean square displacement $\langle \mathbf{u}^\dagger \cdot \mathbf{u} \rangle$ in terms of the original Cartesian coordinates):

$$\langle q_\lambda^2 \rangle^{1/2} = \sqrt{\frac{\hbar}{2\omega_\lambda}} \tag{6.26}$$

$$\langle (\mathbf{u}^\dagger \mathbf{u}) \rangle^{1/2} = \langle \sum_{n\alpha} \sum_{\lambda,\lambda'} q_\lambda Z_{n\alpha}^{\lambda\dagger} Z_{n\alpha}^{\lambda'} q_{\lambda'} \rangle^{1/2}$$

$$= \left(\sum_{\lambda,\lambda'} \langle q_\lambda q_{\lambda'} \rangle \sum_{n\alpha} S_{n\alpha}^{\lambda\dagger} M_{n\alpha n\alpha}^{-1} S_{n\alpha}^{\lambda'} \right)^{1/2}$$

$$= \left(\sum_{\lambda,\lambda'} \delta_{\lambda,\lambda'} \langle q_\lambda^2 \rangle \frac{1}{\mu_\lambda} \right)^{1/2}$$

$$= \left(\sum_\lambda \frac{\hbar}{2\omega_\lambda \mu_\lambda} \right)^{1/2} \tag{6.27}$$

where $\mu_\lambda^{-1} = \sum_{n\alpha} Z_{n\alpha}^{\lambda\dagger} Z_{n\alpha}^{\lambda}$.

When only one mode λ is excited, the mean square deviation is hence given by $\sqrt{\hbar/2\omega_\lambda \mu_\lambda}$. Relation (6.27) gives also the individual zero-point motion of specific atoms, by simply removing the n-summation. The factor μ_λ can be identified with a *reduced mass* associated to the λ-th normal mode. In the simple case of a system with equal ion masses, one recovers the usual expression $\sqrt{\hbar/2\omega_\lambda M}$. However, it is noteworthy to remark that the usual definition of reduced mass for a two-body system emerges from a *different* choice for the generalized coordinates (the position of the center-of-mass and the interparticle distance), which do not coincide with those defined above[3].

Hence, the quantum many-body problem for the ions, if adiabaticity, i.e. (3.13) and (6.6), and harmonicity conditions are satisfied, can in principle be solved exactly. The eigenvalues are simply given by (using N instead of N_{at})

$$E_{n_1,n_2,\ldots n_{3N-6}} = \sum_{\lambda=1}^{3N-6} \hbar\omega_\lambda \left(n_\lambda + \frac{1}{2}\right) \tag{6.28}$$

and the vibrational eigenstates will take the form

---

[3] In fact, in the case of a diatomic molecule, we have $q_1 = S^\dagger M^{\frac{1}{2}} = (m_1 u_1 + m_2 u_2)/\sqrt{m_1 + m_2} = \sqrt{m_1 + m_2} \cdot X_{\text{CM}}$, and $q_2 = \sqrt{m_1 m_2}(u_1 - u_2)/\sqrt{m_1 + m_2} = \sqrt{m_1 m_2/(m_1 + m_2)} \cdot d$, consistently with the Lagrangian (6.18). Our reduced masses are hence $\mu_1 = (m_1 + m_2)/2$ and $\mu_2 = m_1 m_2 (m_1 + m_2)/(m_1^2 + m_2^2)$, and not $(m_1 + m_2)$ and $(m_1 m_2)/(m_1 + m_2)$. When m_1 and m_2 are equal, the usual definition of reduced mass gives $\mu = m/2$ while in our case we have $\mu = m$. This is coherent with a choice for the collective coordinate which is $q = (u_1 - u_2)\sqrt{m/2}$.

$$|n_1, n_2, ...n_{3N-6}\rangle = (\Gamma_1^\dagger)^{n_1}(\Gamma_2^\dagger)^{n_2}...(\Gamma_{3N-6}^\dagger)^{n_{3N-6}}|0\rangle$$
$$= |n_1\rangle \times |n_2\rangle \times ... \times |n_{3N-6}\rangle \quad (6.29)$$

i.e., a state with the λ-th oscillator excited n_λ times corresponds to the state in which n_λ *phonons* of the λ-th mode have been created. The energy of the vibrational ground state, or *zero point energy* , is always different from zero and is given by $\frac{1}{2}\sum_\lambda \hbar\omega_\lambda$. In most cases, several modes will have the same frequency (degenerate modes), according to the symmetry point group under which the system is invariant (see Appendix 5.5.1). It is often convenient to label the normal modes according to the irreducible representations of the point-symmetry group; the single index λ will hence be replaced by a set of indices (e.g., n, λ, μ , where λ indicates the irreducible representation, and μ allows to distinguish between modes in degenerate subspaces).

6.1.1 A Practical Example: a Linear Triatomic Molecule

Let us consider a system made by three atoms of equal mass, and, for simplicity, take a linear and centrosymmetric configuration. One has $3\times3{=}9$ possible displacements $u_{n\alpha}$, hence the dynamical matrix is of order 9×9. If the masses are equal, one simply has

$$D_{n\alpha m\beta} = \frac{\Phi_{n\alpha m\beta}}{M}. \quad (6.30)$$

However, the number of "free" force constants is much smaller than 81. First of all, since by symmetry the atoms 1 and 3 must be interchangeable, one has 9+9+9 relations of the kind $\Phi_{1\alpha2\beta} = \Phi_{3\alpha2\beta}$, $\Phi_{1\alpha1\beta} = \Phi_{3\alpha3\beta}$, and $\Phi_{1\alpha3\beta} = \Phi_{3\alpha1\beta}$. The translational invariance conditions of the type (6.4) yield again 9+9+9 new relations: $\Phi_{1\alpha1\beta} = -\Phi_{1\alpha2\beta} - \Phi_{1\alpha3\beta}$; $\Phi_{2\alpha2\beta} = -\Phi_{2\alpha1\beta} - \Phi_{2\alpha3\beta}$; $\Phi_{3\alpha3\beta} = -\Phi_{3\alpha1\beta} - \Phi_{3\alpha2\beta}$. Finally, in keeping with the fact that D must be symmetric, the number of "free" parameters reduces further. In order to be able to solve the eigenvalue problem analytically, we will consider here only the *longitudinal* vibrations, i.e., we will neglect the degrees of freedom along the plane perpendicular to the molecular axis. In this way, the index α and β can be dropped, since they must be both equal to z, and one is left with a 3×3 dynamical matrix. Using symmetry and translational invariance, one can show that the most general form for D is

$$\begin{pmatrix} (A+B) & -A & -B \\ -A & 2A & -A \\ -B & -A & (A+B) \end{pmatrix}, \quad (6.31)$$

which yields the following equation for the eigenvalues:

$$(A+B-\Lambda)^2(2A-\Lambda) - 2A^2B - (2A-\Lambda)B^2 - 2A^2(A+B-\Lambda) = 0. \quad (6.32)$$

Solving (6.32) one gets

$$\Lambda_1 = 0; \quad \Lambda_{2,3} = (2A+B) \pm |A-B|$$

Choosing $A > B$ one has: $\Lambda_1 = 0, \Lambda_2 = 3A, \Lambda_3 = A + 2B$.

The eigenvector corresponding to Λ_1 is easily found to be any vector of the form (η, η, η): these are the rigid translations of the whole system along the molecular axis. As for Λ_2 and Λ_3, the eigenvectors can be intuitively analyzed in the particular case $B = 0$ (neglecting B means to neglect the forces generated on the atom 3 when the atom 1 is displaced; these forces are generally smaller than those generated when the nearest atom 2 is displaced. The approximation $B = 0$ is known also as the *nearest neighbors force constants* scheme). In this case, one can show that the eigenvectors corresponding to the frequencies $\sqrt{A}$ and $\sqrt{3A}$ have the form $(\eta, 0, -\eta)$ and $(\eta, -2\eta, \eta)$, respectively (Fig. 6.1). In both cases the center of mass remains fixed, and the molecule deforms according to a symmetric or antisymmetric stretching, respectively.

6.2 Phonon Calculations

6.2.1 Introduction

The essential ingredient for the calculation of phonons in an N-atoms system (we use N instead of N_{at} in the rest of this chapter) is the $3N \times 3N$ dynamical matrix $\mathbf{D}$, defined in (6.12). Its calculation for a real system with several tens of atoms is not, however, a very simple matter. In general, the total potential (6.1) includes the contribution from the electronic dynamics, in particular from valence electrons, which gives rise to many-body effects in the interatomic interaction (see the term $\mathcal{E}_0(\{\boldsymbol{R}\})$ in (6.1)). In other words, the total ionic potential U defined in (6.1), can not be written as a bare sum of two-body contributions over all the atomic pairs. The many-body effects can be exactly evaluated only via a numerical calculation. This is achieved by solving the equations of the electronic system for all the configurations in which the coordinates of one atom are slightly displaced away from the equilibrium, along one of the cartesian directions, and computing the forces generated on all the other atoms. The resulting forces, divided by the modulus of the displacement considered, are by definition the force constants (6.3) necessary to build up a given line (or row) of the dynamical matrix. A first-principles determination of the phonons of an N-atom system hence requires

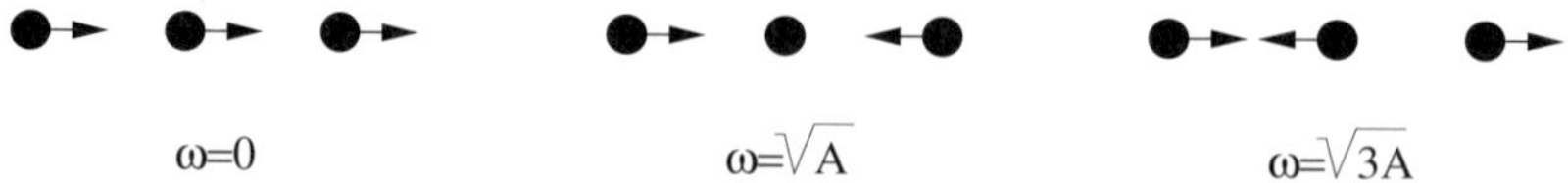

Fig. 6.1. Vibrational eigenmodes of a linear molecule of three atoms. The eigenfrequencies are $\omega = 0$, $\sqrt{A}$ and $\sqrt{3A}$, while directions of motion of the atoms are represented by the arrows

in principle $3N(3N + 1)/2$ calculations[4] of the electronic ground state (e.g., minimizations of the density functional[5]).

We will give an example of such an ab initio calculation later in Sect. 6.2.3. The large numerical effort required in such a scheme is the main reason why phonon calculations were traditionally performed within empirical or semi-empirical models, and exploiting at best the symmetry properties of the system in order to reduce the number of free elements of the dynamical matrix (DM). It is then possible to determine the DM by fitting a limited number of adjustable parameters, trying to reproduce the experimental data (e.g. the frequencies of the Raman or infrared (IR) active vibrations). This kind of approach is still useful when the size or the complexity of the system is too large to allow the use of parameter-free ab initio methods. The simplest semi-empirical model for phonon calculations is the so called "force–constants" (FC) method, in which only the FC $\Phi_{n\alpha m\beta}$ connecting pairs of nearest-neighbors, or second-neighbors atoms are considered, and fitted as adjustable parameters. It is also possible to start from an expression of the total potential approximated with a sum of two-body (pair) contributions,

$$U \simeq \sum_{n} \sum_{m \neq n} U_{\mathrm{p}}(|R_m - R_n|), \tag{6.33}$$

and to write the force constants Φ as functions of the values of the first and second order derivatives of those pair potentials. Approximation (6.33) works well only in those systems, such as Van der Waals crystals, where the chemical bonding is very much isotropic. It is easy to verify that from (6.33) one obtains:

$$\Phi_{n\alpha m\beta} = \left. \frac{\partial^2 U_{\mathrm{p}}(|\boldsymbol{R}_m^{(0)} + \boldsymbol{u}_n - \boldsymbol{R}_n^{(0)} - \boldsymbol{u}_n|)}{\partial u_{n\alpha} \partial u_{m\beta}} \right|_0$$

$$= \left\{ \delta_{\alpha\beta} \frac{U_{\mathrm{p}}'(d_0)}{|d_0|} + \frac{(d_0)_\alpha (d_0)_\beta}{|d_0|^2} \left[U_{\mathrm{p}}''(d_0) - \frac{U_{\mathrm{p}}'(d_0)}{|d_0|} \right] \right\}, \tag{6.34}$$

where $\boldsymbol{d}_0 = \boldsymbol{R}_m^{(0)} - \boldsymbol{R}_n^{(0)}$ is the vector directed from the atom m to the atom n.

The values of U_{p}' and U_{p}'' in correspondence to each equilibrium distance, d_0, become then the parameters of the model, and are fitted on the basis of the experimental data for the phonon frequencies. The first order derivative of the pair potential, U_{p}', does not need to be zero even if the global equilibrium condition $(\partial U/\partial u_{n\alpha}|_0 = 0)$ must be satisfied, because repulsive interactions between first neighbors can be compensated by attractive interactions with more distant atoms (see the example on the three-atom molecule

---

[4] Remember that the force constant matrix is symmetric. To be precise, the actual number of independent elements is slightly smaller than that, due to the translational and rotational invariance conditions discussed in Sect. 6.1.

[5] Or, as an alternative, solutions of the set of the corresponding equations within the linear response theory [113].

in Sect. 6.1.1). However, the force constants models, like the one described here, have several limitations. First of all, the number of free parameters can become very large if the system has only but a few symmetries. Hence, their determination can become difficult and physically neither very significant nor transparent. Such large sets of parameters usually have poor transferability between different systems. Moreover, in the case of covalent bonding (e.g., in semiconductors as Si and Ge, or in carbon structures) the interaction associated with a given pair of atoms normally depends also on the *orientation* of the bond with respect to other atoms or bonds, and not only on the distance between the atoms n and m. One has then *non-central* interactions, arising from strongly directional bonds (let us think of the sp^3 bonding of silicon or diamond), which would require, to be correctly modeled, the introduction of three-body or many-body interatomic potentials. To describe correctly this kind of interaction between ions, and the related vibrational properties, it is then necessary to take into account the contributions of the *electrons*, i.e., the response of the electronic system to the ionic motion.

This response is naturally included (and exactly calculated) in the case of ab initio methods (e.g., in the Car–Parrinello scheme, see Chap. 7), where the whole electronic structure of the system is obtained explicitly.

6.2.2 Bond Charge Models

In what follows, we discuss an intermediate scheme for the calculation of the phonon spectrum, in which the effective ion–ion interactions mediated by the electrons are neither completely neglected (as, e.g., in the force–constants models), nor fully taken into account (as in ab initio methods), but are described in a phenomenological way. This is done by introducing some kind of electronic degrees of freedom which, without requiring the explicit knowledge of the electronic structure, allow one to parameterize in an efficient way the response of the "electronic cloud" to the ionic motion. The so called *shell models* for the study of the vibrational properties of ionic solids belong to this class [114]. Another model of this kind is the *bond charge model* (BCM), developed for the case of covalent systems.

In practice, these schemes are all based on multipolar expansions of the valence electron charge density around high-symmetry points (the atomic positions in the case of ionic solids, or the center of the bonds in the case of covalent systems). The expansion is usually carried out only to the first term (monopole), but additional terms have sometimes been included. In the case of covalent systems, this approach leads one to consider a set of *point charges*, sitting near the center of the bonds. These charges, called the *bonding charges* or simply *bond charges*, carry their own spatial degrees of freedom, i.e., they can move in the neighborhood of their equilibrium positions. Clearly this scheme is only valid when small oscillations of the ions are considered (that is, for the study of the vibrational properties in the harmonic approximation). The bond charges are coupled to each other and to the (positively charged) ions by the Coulomb interaction, and, of course, by one or more phenomenological repulsive potentials, in analogy to what is done in force–constant models. The total energy of the system is hence considered as a function of the $3N$ ionic coordinates, and of a set of $3S$ "electronic degrees of freedom", which correspond to the positions of the bond charges. This kind of description finds a rough justification, in principle, within the Density Functional Theory, since we know that the total energy of a system can be expressed as a functional depending, besides the ionic positions, also on the ground state charge distribution, $\varrho_0(r)$. A BCM calculation is not, of course, an ab initio one, since this dependence is simply parameterized by a phenomenological model, in which the values of the parameters are found on the basis of a fit to the experimental data. On the other hand, this class of phenomenological models diplays a number of attractive features:

1. they are computationally much simpler than an ab initio calculation. In fact, at least one order of magnitude is gained in speed and spared in memory requirements,
2. if the system under study is sufficiently symmetric, the calculation requires only a small number of adjustable parameters, bearing a clear physical interpretation, and being well transferable from system to system.

Vibrational spectrum of C_{60}. We summarize the method in the following for the practical example of the C_{60} (buckminsterfullerene) cluster (see Fig. 1.3). In the calculations, besides the 60 carbon atoms which build the molecule, one introduces as many bond charges as the number of covalent bonds (in this case, 90 but we use in the following the more general notation N_{bc}). One obtains hence a system with 180 ionic and 270 electronic degrees of freedom. Considering the Coulomb and the phenomenological potentials, one then computes the force–constant matrix $\Phi_{n\alpha m\beta}$. This is a $450{\times}450$ symmetric matrix, which can be written as

$$
\begin{pmatrix}
\begin{matrix} R \\ (180 \times 180) \end{matrix} & \begin{matrix} T \\ (180 \times 270) \end{matrix} \\
\hline
\begin{matrix} T^\dagger \\ (270 \times 180) \end{matrix} & \begin{matrix} S \\ (270 \times 270) \end{matrix}
\end{pmatrix},
\tag{6.35}
$$

where the blocks related to ion–ion (R), ion–bond charge (T), and bond charge–bond charge (S) interactions are set into evidence. To be noted that in a bare force–constants calculation only the $(180{\times}180)$ block labeled by R is considered.

Indicating with $\mathbf{u}$ and $\mathbf{v}$ the displacement of ions and bond charges, respectively, one can write

$$
U = U_0 + \frac{1}{2} \sum_{nm\alpha\beta} R_{n\alpha m\beta} u_{n\alpha} u_{m\beta} + \sum_{nm\alpha\beta} T_{n\alpha m\beta} u_{n\alpha} v_{m\beta}
$$

$$+\frac{1}{2}\sum_{nm\alpha\beta} S_{nm\alpha\beta} v_{n\alpha} v_{m\beta} \tag{6.36}$$

where

$$R_{nam\beta} = \frac{\partial^2 U}{\partial u_{n\alpha}\partial u_{m\beta}}, \qquad T_{nam\beta} = \frac{\partial^2 U}{\partial u_{n\alpha}\partial v_{m\beta}},$$

$$S_{nam\beta} = \frac{\partial^2 U}{\partial v_{n\alpha}\partial v_{m\beta}}, \tag{6.37}$$

where in each term both the long-range Coulomb contribution and a short-range (first-neighbors) phenomenological interaction have been included[6].

Indicating with M_n the ionic masses, and with m the mass of the bond charges, the equations of motion in the harmonic approximation become, after Fourier transforming to the frequency domain,

$$M_n\omega^2 u_{n\alpha} = \sum_{m=1}^{N_{\mathrm{ion}}}\sum_{\beta=1}^{3} R_{nm\alpha\beta}\cdot u_{m\beta} + \sum_{m=1}^{N_{\mathrm{bc}}}\sum_{\beta=1}^{3} T_{n\alpha m\beta}\cdot v_{m\beta}, \tag{6.38}$$

$$m\omega^2 v_{n\alpha} = \sum_{m=1}^{N_{\mathrm{ion}}}\sum_{\beta=1}^{3} (T_{n\alpha m\beta})^{\dagger}\cdot u_{m\beta} + \sum_{m=1}^{N_{\mathrm{bc}}}\sum_{\beta=1}^{3} S_{n\alpha m\beta}\cdot v_{m\beta}. \tag{6.39}$$

Let us now take the limit of these expressions for $m \to 0$, and use (6.39) to express **v** as a function of **u**, that is,

$$\mathbf{v} = -S^{-1}T^{\dagger}\mathbf{u}. \tag{6.40}$$

Substituting (6.40) in (6.38), one gets

$$M_n\omega^2 u_{n\alpha} = \sum_{m=1}^{N_{\mathrm{ion}}}\sum_{\beta=1}^{3}\left(\Phi^{(\mathrm{eff})}_{n\alpha m\beta} u_{m\beta}\right), \tag{6.41}$$

where we have introduced an *effective* force–constant matrix,

$$\Phi^{(\mathrm{eff})}_{n\alpha m\beta} = R_{n\alpha m\beta} - \sum_{rs\delta\gamma} T_{nr\alpha\delta} S^{-1}_{rs\delta\gamma} T^{\dagger}_{sm\gamma\beta}. \tag{6.42}$$

The corresponding effective dynamical matrix is hence

$$D^{(\mathrm{eff})} = \frac{\Phi^{(\mathrm{eff})}}{\sqrt{M_n M_m}} = \frac{R - TS^{-1}T^{\dagger}}{\sqrt{M_n M_m}}. \tag{6.43}$$

---

[6] In practical calculations, the Coulomb part appears through the single parameter Z^2/ε (screened interaction acting among all ions and bond charges). The short-range potential is parametrized by the second derivative of the repulsive term (Φ'') at the nearest-neighbors distance, while the first derivative (Φ') is fixed by the requirement of the global equilibrium of ions and bond charges. A possible angular (three-body) contribution of the Keating form (controlled by the single parameter β) is often introduced to mimic the strongly directionality of the sp^3 bonding [114].

In (6.43), the electronic degrees of freedom do not appear explicitly any more. They have been in fact eliminated through the "adiabatic condition" ($m = 0$), which means that the bond charges follow instantaneously the vibrational motion of the ions. When applied to the C_{60} molecule mentioned before, the BCM allows one to describe the vibrational spectrum within a few % of mean square relative deviation (MSRD) from the experimental one.

The C_{60} molecule has the full symmetry of the icosahedral group with inversion, $\mathcal{I}_h$ [115] (see Appendix 5.5.1). Of the 174 normal modes with nonzero frequency, only four are IR active (three-fold degenerate, belonging to the irreducible representation F_{1u}). The Raman active modes are ten: two non-degenerate, totally symmetric (the "breathing" mode and the "pentagonal pinch" mode, both of A_{1g} symmetry), and eight five-fold degenerate modes of H_g symmetry, the lowest of them corresponding to the "quadrupolar" or "squashing" deformation of the cluster. Despite the small number of adjustable parameters (i.e. four, see Table 6.1), the fourteen experimental Raman and infrared-active frequencies are reproduced very accurately. Good agreement is found also for other optically inactive modes, detected by neutron scattering and reflection electron energy loss spectroscopy.

Moreover, the values of the model parameters are similar to those determined for similar carbon structures, e.g., for bulk graphite [115]. The BCM offers hence a simple way to compute the vibrational properties of many-body systems, taking a step beyond the bare force constant models, since it allows one to include some of the physics connected with the electronic response to the atomic motion.

The BCM has the advantage of using a very reduced set of free parameters, and demanding a low computational cost. However, it is still a simplified model, and simplicity is payed by the fact that it is not possible to describe low-symmetry systems with the same degree of accuracy and with the same small set of parameters. To give an example, remaining in the field of carbon clusters, BCM allows one to describe in an excellent way the phonons of C_{60} and C_{70}, which are high-symmetry molecules, or those of planar carbon rings and carbon linear chains in their most symmetric configuration, but can be hardly applied to lower symmetry structures, as in the case of the "bowl" isomer of C_{20}, or chains and rings in distorted configurations. In the latter cases, the number of free parameters tends to grow in an uncontrollable way.

In the rest of this section we describe in more detail the BCM calculations for C_{60}. To start with, it is to be noted that the rest position of atoms and BC's are not uniquely determined by symmetry requirements. There are indeed two degrees of freedom in positioning atoms, and three in choosing the BC positions. The atomic positions are determined by the known bond lengths r_p and r_h, provided by a calculation making use of the Car–Parrinello method ($r_h = 1.39$ Å and $r_p = 1.45$ Å [107]. These values agree within the experimental error (± 0.015 Å), with those obtained from a best fit of NMR data (i.e. $r_h = 1.40$ Å and $r_p = 1.45$ Å) [116].

We then start from a configuration with all the BCs located exactly at the midpoints of the bonds. Since the thirty BCs located along shorter bonds lie on a two-fold symmetry axis, their rest position can only vary in radial direction, while the remaining sixty BCs can also be displaced along the tangential direction perpendicularly to the bond. With the atomic positions fixed, we let the BCs relax to their equilibrium position, determining simultaneously the equilibrium condition on the short-range ion–ion potential.

As a result of the relaxation, the BCs are found to move radially toward the spherical surface defined by the C atoms, and the sixty BCs lying on longer bonds are found to move slightly toward the hexagon center. The distance of BCs from the cluster center changes, after relaxation, from 3.462 (3.467) Å to 3.507 (3.508) Å for long (short) bonds, and the BC–atom–BC angles, which in the ideal case are 108 and 120 degrees for pentagonal and hexagonal rings respectively, become 112.6 and 120.7 degrees. The BC positions are independent of the value of the model parameters.

Table 6.1. BCM parameters. Units for β, $\Phi_{\mathrm{p}}^{''}$ and $\Phi_{\mathrm{h}}^{''}$ are 10^4 dyne·cm^{-1}

	Graphite	C_{60}, set 1	C_{60}, set 2
Z^2/ϵ	1.31	1.31	1.628
β	32.8	32.8	50.5
$\Phi_{\mathrm{p}}^{''}$	136.1	128.5	144.1
$\Phi_{\mathrm{h}}^{''}$	136.1	145.3	171.6

In [115], a calculation was made starting from the three graphite parameters (see Table 6.1), which were obtained by fitting the experimental frequencies of intralayer graphite phonons at the Γ point and along the Γ–M and Γ–K directions. The parameters Z^2/ϵ and β were kept unchanged, and $\Phi_{\mathrm{h}}^{''}$ and $\Phi_{\mathrm{p}}^{''}$ were obtained by scaling $\Phi_0^{''}$ scaled through the ratios $(r_0/r_h)^3$ and $(r_0/r_p)^3$, respectively, where r_0 is the bond length in graphite. With this choice, the effective C–C stretching force constant, which is given by the sum of $\Phi_{\mathrm{p}}^{''}$ ($\Phi_{\mathrm{h}}^{''}$) and the (negative) contribution coming from the Coulomb interactions, is roughly the same for shorter and longer bonds (in the case $\Phi_{\mathrm{h}}^{''} = \Phi_{\mathrm{p}}^{''}$ the short bond would be unphysically softer than the long one). The calculated frequencies for this set of parameters (set 1) are shown in Table 6.2 and compared with the experimental IR and Raman frequencies, labeled by the irreducible representations of the Y_{h} point group. The global agreement with measured spectra is good, with a maximum deviation of 17% and a mean-square relative deviation (MSRD) of 7.2% with respect to the whole set of 14 optically active frequencies. This result almost parallels that obtained with the Car–Parrinello molecular dynamics (MSRD = 6.0%, column c [107]) and represents an improvement over the other methods listed in Table 6.2 (MSRD $\geq$ 11%, columns d–f [117, 118, 119]), except for the QCFF/PI method [120] which gives a MSRD of 4.0% (column g), and the ab initio calculation made using the frozen phonon method [121] (see also Sect. 7.3.1) which gives a MSRD as good as 2.5%.

The 7.2% agreement obtained with parameters scaled directly from graphite should give confidence in the transferability of BCM to more complex fullerene and fullerite structures. We can however learn more about the peculiarities of C_{60} with respect to graphite by letting the four parameters vary up to the best fit of the 14 experimental frequencies [122]. This gives a new set of parameters (Table 6.1: set 2) and the calculated frequencies (Table 6.2, set 2) now agree with experiment within a 3.0% MSRD (maximum deviation is 6.6%). One can obtain interesting physical insight from these new parameters. In fact:

(i) The average charge along the bonds is increased with respect to graphite by 11.4%, which suggests that p_z electrons participate in the bond charge, differently from the case of planar graphite.

(ii) The stiffening of β is also likely to be an effect of the distortion of the sp^2 orbitals into a non-planar configuration, and of the increased bond charge. One can conjecture that the angular force constants, depending on both interactions between the vertex atom and the two BCs, scale as Z^4, as suggested by the 54% stiffening of β with respect to graphite.

(iii) In this BCM version, where direct ion–BC repulsive force constants are not considered, the ion–ion force constants $\Phi''_{h,p}$ depend on the kinetic energy of the interposed BC electrons. Thus, for a scaling $\Phi''_{h,p} \sim Z^{5/3}$, they should increase by the 19% with respect to the values of set 1. Actually the fitted increments are of 18% for Φ''_h and 12% for Φ''_p, and indicate that the long bonds have less charge than the short ones, as expected. The inequality of the two bonds suggests a six-parameter fit, with unequal values of Z and β, which is likely to reduce further the MRSD. For the present purpose, however, we considered such a complication unnecessary.

The frequencies and symmetries of the 46 vibrational levels for set–2 parameters are listed in Table 6.3. Superscripts r and t label the modes whose eigenvectors have a radial component larger than 90% or smaller than 10%, respectively. The lines at 1568 and 1062 cm^{-1} have been assigned to contamination by different fullerenes (in particular, C_{70}), and the present calculation strongly supports this assignment [126].

The calculated frequencies also compare well with the experimental data for some optically inactive modes as provided by reflection electron energy loss spectroscopy (REELS) and neutron scattering experiments, particularly in the lower part of the spectrum. The REELS spectrum of a C_{60} thin film deposited on Si(100) has been measured [127]. The four principal peaks they found at 532, 758, 1258 and 1565 cm^{-1} correlate well with the modes we calculated at 527 (F_{1u} symmetry), 746 (H_u), 1216 (H_u), 1574 (G_g) cm^{-1}. More recently, 28 normal frequencies of C_{60}–fullerite have been measured by neutron scattering [128]. All their results agree well with the calculations

Table 6.2. IR and Raman active modes (cm^{-1}) for C_{60}. Experiment: (a) Gas phase [123], (b) Solid [122]. Theory: (c) ab initio Car–Parrinello MD [107], (d) ab initio real-space MD [117], (e) Classical force field [118], (f) MNDO [119], (g) Quantum Chemical Force Field /PI [120], (h) Frozen-phonon LDA calculation [121]. Other calculations: Force–constant model (MSRD = 21% [124]), Quantum Chemical Force Field/AM1 (MSRD = 19% [125]). The last line contains the MSRD (%)

	Experiment		Theory							
	a	b	set 1	c	d	e	f	g	h	set 2
F_{1u}	527	527	463	530	494	478	577	544	525	527
F_{1u}	570	577	569	555	643	618	719	637	573	586
F_{1u}	1170	1183	1178	1105	1358	1462	1353	1212	1170	1177
F_{1u}	1407	1428	1424	1345	1641	1868	1628	1437	1401	1423
A_g		496	479	454	537	510	610	513	475	496
A_g		1470	1433	1368	1680	1830	1667	1442	1442	1435
H_g		273	239	246	249	274	263	258	257	265
H_g		437	361	$405^{\dagger}$	413	413	447	440	426	408
H_g		710	641	$688^{\dagger}$	681	526	771	691	710	726
H_g		774	757		845	828	924	801	766	786
H_g		$1099^{\dagger}$	1108	$1110^{\dagger}$	1209	1292	1261	1154	1088	1149
H_g		$1250^{\dagger}$	1230	$1314^{\dagger}$	1453	1575	1407	1265	1224	1207
H_g		1428	1427	$1500^{\dagger}$	1624	1910	1596	1465	1389	1433
H_g		1575	1554		1726	2085	1722	1644	1530	1651
(%)			7.2	6.0	11	21	14	4.0	2.5	3.0

(†)Tentative assignment

(MSRD is 3.0% and the maximum deviation is 6.3%). Information about the symmetry of their optically inactive modes can be extracted from Table 6.3.

Table 6.3. Calculated vibrational frequencies (cm^{-1}) for C_{60} within the BCM, rounded to integer

A_g	496^r	1435^t						
A_u	1343^t							
F_{1g}	568^r	992	1461^t					
F_{1u}	527^r	586	1177	1423^t				
F_{2g}	640^t	967^r	986	1480^t				
F_{2u}	355^r	746^r	1011	1201	1606^t			
G_g	430^r	660	861^r	1121	1533^t	1574^t		
G_u	330^r	903	954	991	1428^t	1513^t		
H_g	265	408^r	726^r	786	1149	1207	1433^t	1651^t
H_u	417	488	746	848	1216	1524^t	1643^t	

^{r}Radial mode, ^{t}Tangential mode (see text)

The incoherent neutron scattering spectrum measured for the solid fcc fullerite [129], is displayed in Fig. 6.2 together with the calculated total density of vibrational states for C_{60} (set 2). The low-energy neutron peaks at

274, 355, 403, 436, 492, ~ 525, and ~ 560 cm^{-1} agree with the BCM calculation to within 6.4%. By contrast, the large gap centered at about 1300 cm^{-1},
which is shown by the BCM calculation, correspond to only a small kink
in the neutron spectrum of [129]. A similar gap is also found in QCFF/PI
calculations [120]. On the other hand, the IR active modes measured in the
free molecule show the largest discrepancy (21 cm^{-1}) with those in the solid
just for the F_{1u} mode at 1407 cm^{-1} (Table 6.2, first column). This provides
an estimate for the dispersion effects, which look not-at-all negligible in this
spectral region of the solid.

Vibrational spectrum of C$_{70}$. In the case of C$_{60}$, the number of independent optical active frequencies is limited by the severe selection rules of
the icosahedral symmetry, and no difficulty is met in assignment. The situation becomes more complex in clusters such as C$_{70}$, where selection rules are
relaxed by the lower symmetry ($\mathcal{D}_{5h}$) and quite a large number of IR and Raman active modes is allowed. The vibrational spectrum of the C$_{70}$ molecule is
much much richer than that of C$_{60}$, having 122 distinct frequencies, of which
53 Raman and 31 IR-active. The 210-dimensional representation of $\mathcal{D}_{5h}$ given
by the cartesian displacements of the 70 atoms of the cluster decomposes as:

$$12A'_1 + 10A'_2 + 22E'_1 + 22E'_2 + 9A''_1 + 11A''_2 + 20E''_1 + 20E''_2 \qquad (6.44)$$

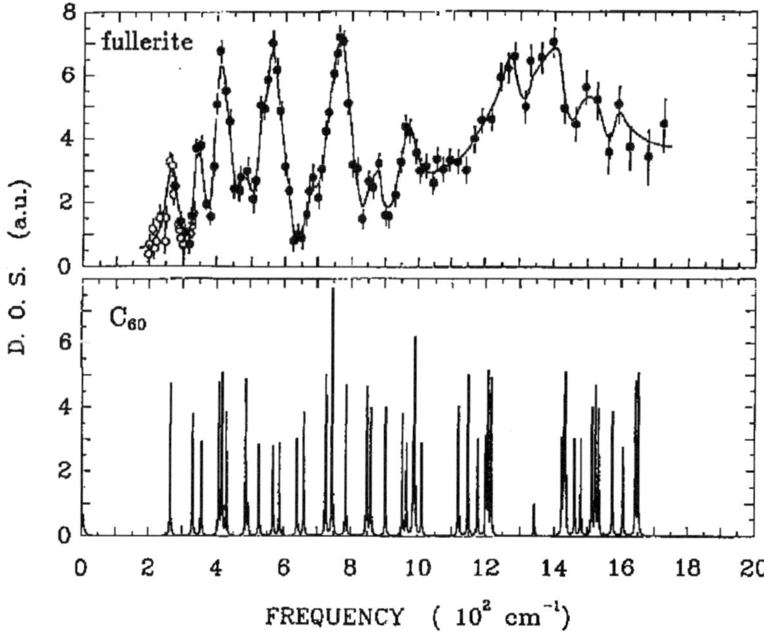

Fig. 6.2. Calculated density of vibrational states for C$_{60}$, (bottom panel) compared
with the measured incoherent neutron scattering spectrum (upper panel) [129]

where A_1', A_2', $\cdots$, are the irreducible representations of the $\mathcal{D}_{5h}$ point symmetry group. E and A irreducible representations are two and one-dimensional, respectively. The zero frequency modes associated to rigid translations belong to A_2'' and E_1'; the rigid rotations to A_2' and E_1''. The IR active modes are then 21 E_1' two-fold degenerate modes and 10 A_2'' nondegenerate ones, giving a total of 31 different peaks in the spectrum. The Raman active vibrations are 12 nondegenerate modes belonging to A_1', and 41 degenerate pairs belonging to either E_2' (22 pairs) or E_1'' (19 pairs). Theory predicts then the existence of 53 distinct peaks in the Raman spectrum. As a consequence, the assignment of the experimental peaks to the calculated frequencies is not straightforward.

Since the largest common subgroup of $\mathcal{I}_h$ (C_{60}) and $\mathcal{D}_{5h}$ (C_{70}) is the cyclic group $\mathcal{C}_{5v}$, normal modes of both systems can be classified according to the $\mathcal{C}_{5v}$ irreducible representations, and are compatible if belonging to the same one. For instance, among the five (degenerate) H_g C_{60} modes, one is compatible with an A_1' C_{70} mode, two with an E_2' doublet, and two with an E_1'' doublet.

Several semiempirical calculations have been performed based on different geometrical prescriptions and on different models for the interatomic interactions, ranging from MNDO [130] to the BCM [131] applied to the LDA optimized structure [132], to the more sophisticated QCFF/PI scheme [133] and to empirical force field models [125, 134].

For the IR active modes of C_{70} the comparison between the BCM theoretical results [132] and experimental intensities is moderately good. Theory predicts a multiplet of high-frequency modes in a region around 1600 cm^{-1} where experiment gives only a set a weak features. Altogether the experimental IR-active modes in the intermediate frequency region appear to be softer than the theoretical ones.

C$_{20}$. Many theoretical studies of the vibrational properties of C_{20} in its dodecahedral cage structure, which makes it the smallest fullerene, have been published [135, 136, 137, 138]. The highly symmetric Y_h configuration has been chosen in a BCM calculation performed in [98]. This choice simplifies both the comparison with other calculations, and allows one to establish a meaningful link to the normal modes of C_{60}, the paradigmatic fullerene. For example, it is easy to identify the (single) A_g totally symmetric mode which corresponds to the "breathing mode" of the molecule, at 841 cm^{-1}. In Table 6.4 we report the BCM results for the 14 different nonzero frequencies of the 54 normal modes of the molecule, together with their symmetry classification. These values were obtained with the same model parameters as in [115] and with $\Phi_{\mathrm{ion-ion}}'' = 144 \cdot 10^4$ dyne/cm, corresponding to a single bond length of 1.45 Å. The geometrical relaxation of BC's positions, performed in the same way as for other fullerenes [131], yields equilibrium positions which are radially displaced by 0.08 Å with respect to the bond centers.

Table 6.4. Vibrational eigenfrequencies (cm^{-1}) of C_{20} in the Y_h cage structure, obtained with BCM. Degeneracies are 1, 3, 4 and 5, for A, T, G and H modes respectively. The bond length is set to 1.45 Å; parameter values are: $\Phi''_{ion-ion} = 144 \cdot 10^4$ dyne/cm, $\beta = 50.5 \cdot 10^4$ dyne/cm; $z^2/\varepsilon = 1.63$

A_g	T_{1u}	T_{2g}	T_{2u}	G_g	G_u	H_g	H_u
841	925	1065	990	970	682	525	736
			1195	1162	1313	1089	1464
						1510	

The Raman active modes are predicted at 525, 841, 1089 and 1510 cm^{-1}, while the (single) IR line is found at 925 cm^{-1}. These values coincide within $\simeq 15\%$ with those obtained in [135] for the same structure, the agreement being better for the higher-frequency modes. A comparison with the IR spectra predicted by recent, ab initio calculations for different Jahn–Teller distorted structures is also possible, keeping into account that the modes generated from the splitting of our G_u, H_u, and T_{2u} modes will also become IR allowed. In the IR spectrum we expect, besides a triplet at about 925 cm^{-1}, additional multiplets at about 680, 740, 1000, 1200, 1300 and 1460 cm^{-1}. Such a distribution of the IR peaks agrees with the predictions of [137], with larger splittings ($\simeq 100$ cm^{-1}) in the lower frequency range, and smaller ones ($\simeq 10$ cm^{-1}) for the higher frequencies.

Linear C_n chains. Some of the optically active vibrational frequencies of odd-numbered linear C_n are experimentally known, both for clusters in the gas-phase, and deposited on inert matrices [139, 140, 141, 142, 143, 144]. In some cases, also the isotopic effects on the highest stretching mode have been experimentally investigated [140, 145]. It has been shown that it is possible to reproduce correctly the vibrational spectra of these molecules using the BCM model, which, in contrast to ab initio calculations, remains computationally cheap even for systems of many atoms. BCM has been shown to be an efficient tool for the study of C_n linear chains, being substantially more precise that a trivial force–constants model.

The interactions included in our BCM parametrization for sp bonded Carbon atoms are reduced to the screened Coulomb interaction, plus a first-neighbor repulsion. Due to the lower coordination number and to the absence of sp^2 or sp^3 bonds, the three-body Keating potential can be dropped off. In this simple scheme, the model contains only two free parameters (the value of Z^2/ε and that of $\Phi''_{ion-ion}$), $\Phi'_{ion-ion}$ being determined by the equilibrium condition. For simplicity, we have considered an unique bond length of 1.3 Å, neglecting the small differences (which are always less than 3%) which high level calculations report for different bonds within the chains, and between different chains. However the bonds are not completely equivalent in our calculation, since the relaxed bond–charge positions are generally different for the different bonds. The small differences in the resulting frequencies obtained by also taking bond length differences into account cannot be considered very significant, due to the semi-empirical character of the calculation. Having eliminated the angular potential, the simplest approach of rescaling the parameter values used for fullerenes (where the Keating potential plays an important role) becomes less effective. On the other hand, since reliable and precise experimental data are available for the highest vibrational mode of some well-characterized chain, namely C_5 and C_7, the model parameters can be easily refitted. We have then simply fixed Z^2/ε at the graphite value [146], and fitted the (unique) value of $\Phi''_{\text{ion}-\text{ion}}$ in order to reproduce exactly the highest C_5 frequency. These parameter values have then been used to compute the full vibrational spectra of C_5 *and* of the other C_n chains, yielding very satisfactory results. The results for C_5 and longer chains are shown in Tables 6.5 and 6.6.

Table 6.5. Longitudinal modes for odd-numbered C_n linear chains up to 19 atoms (cm^{-1}). Experimental data are reported in parenthesis, where available

n=5	n=7	n=9	n=11	n=13	n=15	n=17	n=19
2164	2160	2150	2142	2135	2129	2125	2121
(2164[a])	(2128[a])						
(2169[b])	(2138[c])						
1860	2014	2065	2086	2096	2100	2102	2103
	(1893[d])	(1997[d])					
1372	1764	1918	1990	2028	2050	2064	2073
739	1424	1711	1853	1931	1978	2008	2029
798							
(±45[e])							
	1006	1450	1678	1806	1885	1937	1972
				(1809[f])			
	523	1141	1466	1654	1771	1848	1901
	548	1258					
	(±90[e])	(±50[e])					
		788	1221	1476	1637	1744	1818
		404	946	1274	1483	1624	1722
		484					
		(±48[e])					
			646	1050	1311	1489	1613
			329	806	1122	1339	1493
				547	918	1176	1361
				277	701	1001	1218
					474	815	1065
					239	620	903
						418	732
						210	556
							373
							188

(a) IR in matrix [140, 147], (b) gas phase [139, 141], (c) gas phase [142], (d) [147] as reassigned in [145], (e) photoelectron spectroscopy [144], (f) [148]

In parallel, we have performed a force–constant (FC) model calculation with the same number of parameters (two), trying to fit them to the experimentally known data, in a way similar to that of [140].

Table 6.6. Transversal modes for odd-numbered C_n linear chains up to 19 atoms (cm^{-1}). Experimental data are reported in parenthesis, where available

n=5	n=7	n=9	n=11	n=13	n=15	n=17	n=19
509	553	566	569	569	568	567	566
(512 ± 45[a])	(496 ± 110[a])						
302	433	491	519	534	542	547	550
(222 ± 45[a])							
127	292	394	451	484	504	517	526
(101 ± 45[a])							
(118[b])							
	161	288	370	423	457	480	496
	65	186	285	355	402	435	459
	($\simeq 70$[c])						
		100	202	283	343	386	418
		40	128	214	282	334	374
			68	149	222	281	327
			27	93	165	228	280
				49	114	178	233
				19	71	131	188
					37	90	145
					14	55	106
						29	73
						11	45
							23
							9

(a) photoelectron spectroscopy [144], (b) gas phase [141], (c) [149]

Even if, at a first sight, the main features of the spectra are almost equally well fitted in the FC and BCM approach, the advantages of the latter are evident. First of all, in the FCM the two model parameters (a stretching force–constant K_{str} analogue to our $\Phi''_{ion-ion}$, and a bending force–constant K_{bend}, corresponding to our factor $\Phi'_{ion-ion}/r_0$, where r_0 is here the equilibrium bond length) determine, respectively and *independently*, the frequencies of the longitudinal and transverse modes. Since the experimental data on transverse frequencies are affected by large error bars, the bending force constant remains essentially undetermined. Moreover, in order to satisfy the equilibrium condition, it is necessary to consider short-range interactions extended at least to second neighbors, while in the BCM the only interaction acting beyond first neighbors is the screened Coulomb potential. In the BCM, stretching and bending frequencies cannot be varied independently; a natural interplay between them exists due to the consideration of the Coulomb interaction.

Another important difference between FC and BCM approaches is found in the different behavior of the highest vibrational frequency as a function of the number of atoms. In the BCM, the limiting value for $n \to \infty$ is approached from above. In the FC model, instead, the highest frequency in the $n \to \infty$ limit, although similar to the BCM one, is approached from below. The experimental evidence supports the BCM predictions, at least for the shorter chains, where the highest frequency in C_5 is 20–30 cm^{-1} higher than in C_7 [140, 141, 142]. Similar differences between the FCM and the BCM results are found in the behavior of the highest transversal frequency. Lowest and highest longitudinal and transversal frequencies, computed in both the FC and BCM approaches, are reported in Table 6.7, for all the chains considered.

Table 6.7. Highest longitudinal (L) and transversal (T) vibrational frequencies for C_n linear chains (cm^{-1}), computed with the present BCM and FCM (see text)

L.	n=5	n=7	n=9	n=11	n=13	n=15	n=17	n=19
BCM	2164	2160	2150	2142	2135	2129	2125	2121
FCM	2164	2218	2240	2252	2258	2262	2265	2267

T.	n=5	n=7	n=9	n=11	n=13	n=15	n=17	n=19
BCM	509	553	566	569	569	568	567	566
FCM	508	547	562	569	573	576	577	579

Cyclic C_n. Linear isomers of C_n with 11 or more atoms are predicted, by ab initio calculation, to lie higher in energy than the corresponding cyclic structures. However, high temperature preparation conditions, followed by a fast temperature decrease, allow the experimental detection of linear chains up to 13 atoms [148]. Despite the fact that no experimental data on cyclic C_n vibrational frequencies has been reported to date, it is clearly of interest to extend the BCM study to the ring isomers of C_{11}–C_{19}. Since in this case no experimental input can be used to fit the model parameters, the latter must be obtained by transferring or scaling those used for other structures. Similarly to the case of the linear chains, a single bond length of 1.3 Å has been considered for all the rings (C_{11}–C_{19}) within the full D_{nh} symmetry. This approximation (which, according to ab initio calculations of [95], is excellent for C_{19}–C_{15} and becomes gradually less valid for smaller rings) simplifies the equilibrium condition (relaxation of the BC and determination of $\Phi'_{\mathrm{ion-ion}}$), and allows one to keep the minimum number of model parameters (two). For our calculations, we have then simply used the values of Z^2/ε and $\Phi''_{\mathrm{ion-ion}}$ used for the linear chains. As for the other structures the bond–charge positions have been relaxed, and the $\Phi'_{\mathrm{ion-ion}}$ determined by the equilibrium condition. The resulting frequencies, for the 11 to 19 atoms rings, are reported in Table 6.8. A comparison with Tables 6.5 and 6.6 shows that going from the open (linear) to the closed (ring) topology yields, as expected, a reduction of the spectrum width, with larger effects on the smaller clusters.

The width of the closed rings vibrational spectrum is hence intermediate between those of the fullerenic cage and of the open linear chains. In particular the highest ring mode is lowered by 160–70 cm^{-1} with respect to the linear chain one, and the lowest ring mode is found at about 4.5 times the frequency of the lowest mode of the corresponding linear chain. In the case of linear C_5–C_{15} and ring C_{11}–C_{15} the BCM results can be compared with those of B3LYP calculations: BCM frequencies are generally slightly higher than the B3LYP ones, both for linear chains and planar rings.

Table 6.8. Vibrational frequencies for odd-numbered C_n rings up to 19 atoms (cm^{-1})

n=11	n=13	n=15	n=17	n=19
1977	2007	2026	2039	2048
1854	1914	1954	1982	2002
1605	1727	1809	1867	1908
1246	1447	1590	1693	1767
846	1094	1305	1462	1579
628^a	728	971	1182	1347
546	547	638	870	1078
545	546	547	567	787
455	533^a	546	547	547
448	481	497	546	545
305	478	496	508	515
287	367	463^a	508	515
144	358	409	437	510
117	232	404	435	457
	215	297	409^a	456
	105	287	344	380
	84	181	338	376
		167	242	366^a
		80	233	291
		64	144	285
			132	200
			63	192
			50	117
				108
				50
				40

(a) breathing mode

The case of ring–C_{20}. The successful application of BCM to planar odd-numbered rings suggests a further extension. The sp-bonded isomer of C_{20}, which is also a planar ring but, at difference with the odd-numbered C_n clusters, has a polyacetylenic (i.e., partially dimerized) structure. This structure is predicted to be the most stable C_{20} isomer at both the HF and DFT Generalized Gradient Approximation (GGA) levels [97, 136, 150, 151], and, together with the "bowl" isomer, is likely to be present at the experimental conditions. In these calculations, the bond distances 1.24 Å and 1.35 Å has been taken from [137]. In order to facilitate the comparison of the results with those given in Table 6.4, BCM calculations are reported for two different parameter sets: the "fullerene" set (rescaling the icosahedral C_{20} value of $\Phi''_{ion-ion}$ with the actual ring bond lengths), and the "linear chain" set used above for odd-numbered linear chains and cumulenic rings. The results of such calculation are shown in Table 6.9.

The vibrational spectrum contains 32 frequencies (22 doubly degenerate and 10 non-degenerate modes), distributed from $\simeq 40$ to $\simeq 2200$ cm^{-1}. A comparison with the spectra computed by ab initio methods [136, 137, 152], also reported in Table 6.9, indicates overall agreement, with the "linear chain" parameters set yielding frequencies which are systematically lower by about 10% with respect to those obtained using the "fullerene" parameters. The absence of modes in the region $\simeq 800 - 1100$ cm^{-1} (a "pseudogap" in the spectrum), predicted by BCM, is similarly present in HF, LDA and B3LYP calculations. On the other hand, the width of a similar pseudogap around 1700 cm^{-1} shows a stronger dependence on the method of calculation employed. As was pointed out in [137], a basic difference distinguishing the vibrational properties of the cage and the ring-C_{20} is the presence, in the phonon spectrum, of low-frequency bending modes which are absent in the case of the cage (see Table 6.4). In [137], it was found that the entropic contributions of these low-lying modes are responsible for a large relative abundance of the ring structure at the usual experimental conditions ($T \geq 1000$ K).

Summary on the BCM calculations. In summary, we have shown in this Section that it is possible to compute the vibrational properties of a large class of pure carbon clusters -ranging from linear C_5 to fullerenes- using a reliable and computationally inexpensive method based on the Bond-Charge Model approach, obtaining an overall agreement with both experimental data and with ab initio results. This approach, due to the inclusion of the (classical) screened Coulomb interaction effects, is more efficient than traditional force–constant models with the same number of adjustable parameters. In particular, in the case of the linear chains, the BCM ensures that: a) a natural interplay between the frequencies of stretching and bending modes is present; b) the experimental trend of the highest frequency with respect to the chain length is reproduced. Both these features are missing in the case of the second-neighbors force–constant model.

6.2.3 Phonons from ab initio Calculations

Empirical and semi-empirical methods often fail when applied to low-symmetry systems, since the number of adjustable parameters becomes too large. In those cases, the use of parameter-free ab initio methods becomes essential. The same is true when quantitative, high-precision theoretical results are required. Possible approaches for computing vibrational properties in an ab

Table 6.9. Vibrational modes for the polyacetylenic ring isomer of C_{20} as obtained in BCM. Bond lengths are 1.24 and 1.35 Å [137]. Results from other calculations are reported for comparison

Degen.	LDA [152]	HF [136]	B3LYP [137]	BCM^a	BCM^b
2	2189	2522	2240	(2260) 2255	2075
2	2177	2506	2231	(2204) 2186	2011
1	2111	2495	2143	(2278) 2277	2096
2	2076	2477	2119	(2106) 2073	1906
2	2025	2393	2101	(1965) 1918	1763
1	1994	2346	2078	(1843) 1787	1643
1	1589	1519	1563	(1636) 1584	1455
2	1450	1417	1440	(1464) 1411	1295
2	1175	1173	1174	(1180) 1131	1038
2	860	869	860	(864) 825	757
1	561	761	751	(662) 616	553
2	539-558^c	757	731	(644) 600	539
2	531	702	640	(599) 599	538
2	503	618	566	(592) 554	497
2	493	615	547	(560) 553	496
2	484	577	493	(554) 533	489
2	480	544	492	(513) 484	434
1	474	536	468	(615) 615	552
2	440-471^c	527	460	(484) 482	432
2	422	514	435	(416) 397	356
2	415	463	421	(397) 393	352
1	373	425	387	(404) 383	351
1	350	396	370	(320) 311	279
1	343	395	368	(311) 303	272
1	329	369	335	(300) 290	261
1	302	340	307	(290) 284	255
2	262	273	251	(206) 205	184
2	239	252	233	(205) 196	176
2	166	161	149	(119) 119	107
2	152	143	133	(114) 109	97
2	83-96^c	69	64	(51) 51	46
2	56-73^c	53	49	(42) 40	36

(a) BCM parameters, scaled from fullerene calculations:
$\Phi''_{\mathrm{ion-ion}} = 230.2$ (short bond) and 178.5 (long bond) 10^4 dyne/cm, $\beta = 0$ (in parenthesis, results for $\beta = 50.5 \cdot 10^4$ dyne/cm), $Z^2/\varepsilon = 1.63$

(b) BCM parameters, scaled from chains calculations:
$\Phi''_{\mathrm{ion-ion}} = 183.8$ (short bond) and 147.1 (long bond) 10^4 dyne/cm, $\beta = 0$, $Z^2/\varepsilon = 1.31$

(c) The LDA calculations [152] are performed for a lower symmetry than the C_{10h} one

initio approach are the so-called "frozen-phonon" method (see Sect. 7.3.1), and methods based on the Density Functional Perturbation Theory (DFPT) (see e.g., [153]). Several recent applications to infinite and finite systems can be found in the literature [153]. We mention, in the next chapter, some applications based on the "frozen-phonon" approach within the Car–Parrinello method.

7. The Car–Parrinello Method

The Car–Parrinello (CP) method has been briefly mentioned in previous chapters. We will describe it here in an introductory way. Both the CP and Norm-Conserving Pseudopotentials (NCPP, see Sect. 3.3.2) methods are currently implemented in Fortran codes, and public domain versions exist on the Internet[1].

7.1 An Efficient Tool for Solving the Electronic Problem

As the title of the original Car and Parrinello's work states [54], the method consists in the unification of the Density Functional Theory (DFT) with the Molecular Dynamics (MD) technique. By (classical) Molecular Dynamics one understands the techniques for the numerical solution of the equations of motion of a many-body system interacting through a classical potential. One of the most stable and simple algorithms used to integrate classical MD equations is the Verlet algorithm: given the initial positions of the particles of the system, $\boldsymbol{X}(t)$, and the forces acting on them, $\boldsymbol{f}(\boldsymbol{X}) = -\boldsymbol{\nabla}V(\boldsymbol{X})$, the Newton equations are numerically integrated by using small *time steps* Δt according to the equation

$$\boldsymbol{X}(t + \Delta t) = 2\boldsymbol{X}(t) - \boldsymbol{X}(t - \Delta t) + \frac{(\Delta t)^2}{M}\boldsymbol{f}(\boldsymbol{X}(t)). \tag{7.1}$$

This equation can be derived by summing the Taylor expansions of $\boldsymbol{X}(t - \Delta t)$ and of $\boldsymbol{X}(t + \Delta t)$ up to the fourth order, to obtain:

$$\boldsymbol{X}(t - \Delta t) + \boldsymbol{X}(t + \Delta t) = 2\boldsymbol{X}(t) + (\Delta t)^2\ddot{\boldsymbol{X}}(t) + \mathcal{O}(\Delta t)^4. \tag{7.2}$$

Equation (7.1) follows directly from (7.2) and Newton equations. In the present case, $\boldsymbol{X}$ means the set of ionic positions, $\{\boldsymbol{R}\}$. On the other hand, DFT is the theory described in previous chapters, for the solution of the ground state problem of a many-electrons system with fixed ions. In DFT the problem of the interacting electrons is reduced to a system of coupled integro-differential equations (the Kohn–Sham equations) for a system of non-interacting particles which feel an effective, self-consistent potential.

---

[1] See e.g., the Mattias Scheffler's group codes available at the Fritz-Haber Institut of Berlin.

The first goal of the Car–Parrinello (CP) method is to substitute the traditional method for the solution of the Kohn-Sham (KS) equations with a method which we can call "dynamical minimization" of the energy functional. In the traditional approach, the KS equations are solved by repeated diagonalizations of the KS Hamiltonian, until reaching full self-consistency between wavefunctions, electronic charge and effective potential. In the CP approach, there is no need to diagonalize the KS Hamiltonian, instead the latter needs only to be applied to the electronic states many times. This is a very important advantage when large basis sets are used (e.g., with a plane-wave basis, where the KS Hamiltonian is usually a matrix of the order of $10^4 \times 10^4$). Moreover, as we will see, the self-consistency between charge, potential, and wavefunctions comes out automatically.

This is only the first of the two important novelties which have been introduced by the CP method. The second is the fact that, by coupling together the electronic and ionic degrees of freedom, the CP method allows one to simulate the actual motion and structural changes of the system (i.e., the atomic displacements), in the spirit of the Born–Oppenheimer separation. Within this context it is important to stress that the interatomic forces, which produce the atomic motions, are computed from first-principles, exploiting the knowledge of the electronic ground state of the system corresponding to the actual ionic configuration. Moreover, in a CP calculation the electronic ground state follows adiabatically the ionic motion. This fact, as it will become clear in the following, allows one to avoid solving repeatedly the electronic problem after each atomic move. In summary, the two main goals of CP can be summarized as follow:

1. solution of the quantum problem of the electronic ground state by means of the techniques of classical molecular dynamics,
2. calculation of the atomic motion by computing the interatomic forces from first-principles, hence realizing a parameter-free molecular dynamics simulation.

In the present section we consider point 1), demonstrating how the CP method can be used to perform static DFT calculations (i.e. without moving the ions). Point 2) will be discussed in Sect. 7.3.

We point out that the physical information which can be obtained using the CP method with fixed atoms are exactly the same as those supplied by a standard DFT calculation: total energy, charge density, ionicity (which is the charge transfer between different atoms in a molecule or a compound), and total energy derivatives, as the phonon frequencies and so on. The difference with a standard DFT calculation concerns only the method employed to minimize the energy functional.

In the minimization process, the possibility to work simultaneously on the electronic and ionic degrees of freedom can be exploited to implement an efficient scheme to search for the equilibrium geometry of the system. A combined minimization, with respect to all the degrees of freedom, can be eventually associated to simulated annealing techniques, in order to obtain the global minimum energy configuration if several minima are present.

The advantages of the CP method, beyond those which are common to any DFT calculation (i.e., the absence of any adjustable parameter and hence the reduction of transferability problems), are essentially computational: one can treat larger and more complex systems than using standard DFT codes. The use of a plane-wave (PW) basis , moreover, allows one to avoid symmetry constraints, which are often a limit in quantum-chemistry codes. Using PW, each KS state is typically expanded on as many as $10^5 - 10^6$ basis functions. A system containing 10^2 atoms, i.e., about 10^8 electronic degrees of freedom, can be treated on a table workstation equipped with 512 Mb of RAM.

The computational cost is anyway large with respect to any semi-empirical calculation for a system of the same size. In particular, the number of atoms included in the system cannot be much larger than 10^2, and, in the case of actual molecular dynamics simulations, the simulation time is limited to some picoseconds. On the other hand phenomenological, as opposed to ab initio methods, can be used with ease to treat up to 10^5 atoms, and intervals of time of the order of a nanosecond.

7.2 Car–Parrinello Molecular Dynamics

We shall now show how Car and Parrinello proposed to unify DFT with MD. They started with the definition of the (classical) Lagrangian,

$$\mathcal{L}_{\mathrm{CP}} = T_{\mathrm{ions}} + T_{\mathrm{fake}} - [U - \sigma]. \tag{7.3}$$

In it, besides the standard kinetic energy of the ions

$$T_{\mathrm{ions}} = \frac{1}{2} \sum_{n=1}^{N_{\mathrm{at}}} M_n \dot{\boldsymbol{R}}_n^2, \tag{7.4}$$

an additional term was introduced, called a "fictitious (fake) kinetic energy",

$$T_{\mathrm{fake}} = \tilde{\mu} \sum_{i \in occ} \int \dot{\varphi}_i^* \dot{\varphi}_i, \tag{7.5}$$

where the index i runs over *occupied* Kohn–Sham states φ_i. This kinetic energy has noting to do with the real kinetic energy of the electron system, nor with that of the auxiliary system of independent Kohn–Sham particles. In (7.5), $\tilde{\mu}$ is an arbitrary parameter whose role will become clear in what follows, and which is called the *fictitious mass*. In the same equation, $\dot{\varphi}_i(r)$ indicates the derivative with respect to a time parameter t which is attached to the Kohn–Sham orbitals $\varphi_i = \varphi_i(r; t)$. Hence, the temporal dependence of φ introduced in this way has nothing to do with the real time evolution of a quantum system. It must be considered simply as a parametric dependence, attached to the Kohn–Sham orbitals.

In the potential energy part, Car and Parrinello considered the energy functional of the DFT, including the orthogonality constraint. In other words, the potential part of the CP Lagrangian coincides with the quantity which one must minimize to obtain the ground state within the Density Functional Theory:

$$U = E^{\text{DFT}}_{\{V_{\text{ext}}\}} \left[\{\varphi_i\}, \{\boldsymbol{R}_n\} \right], \tag{7.6}$$

as in (3.20), and

$$\sigma = \sum_{ij} \lambda_{ij} \int \varphi_j^* \varphi_i - \delta_{ij} \tag{7.7}$$

(see (3.21)). It must be noticed that in the expression of U also the ionic positions appear, through the external potential $V_{\text{ext}} = V_{\text{ext}}(\{\boldsymbol{R}_j\})$. If more than one nucleus or ion is present, also the direct Coulomb repulsion acting between every pair of them is included in U.

It is instructive to write down explicitly the equations of motion generated by this Lagrangian. Using the relations

$$\frac{\delta \mathcal{L}}{\delta \dot{\varphi}_i^*} = \tilde{\mu} \dot{\varphi}_i, \tag{7.8}$$

and

$$\frac{\delta \mathcal{L}}{\delta \dot{\boldsymbol{R}}_j^*} = M_j \dot{\boldsymbol{R}}_j, \tag{7.9}$$

one obtains the equations

$$\tilde{\mu} \ddot{\varphi}_i(r; t) = -\frac{\delta}{\delta \varphi_i^*} E_{\{V_{\text{ext}}\}} \left[\{\varphi_i\}, \{\boldsymbol{R}_n\} \right] + \sum_j \lambda_{ij} \varphi_j(r; t) \tag{7.10}$$

which governs the time evolution of the electronic degrees of freedom, and

$$M_j \ddot{\boldsymbol{R}}_j(t) = -\frac{\partial}{\partial \boldsymbol{R}_n} E_{\{V_{\text{ext}}\}} \left[\{\varphi_i\}, \{\boldsymbol{R}_n\} \right] \equiv \boldsymbol{F}_j(t) \tag{7.11}$$

governing the time evolution of the ionic degrees of freedom. We notice that $\boldsymbol{F}_j(t)$ represents the actual force on the n-th ion, *only* if the energy functional E is evaluated *at the density of the electronic ground state*, $\varrho_0(r)$. That is if, and only if, the $\varphi_i(r)$ are the solutions of the Kohn–Sham equations for the given external potential. In general, for an arbitrary set of $\{\varphi_i\}$ which are not the solution of the electronic ground state problem, $\boldsymbol{F}_j(t)$ have no direct physical meaning. We will come back to the interpretation of (7.11) later.

Let us first assume the ions to be *fixed*, forgetting the variables $\{\boldsymbol{R}_j\}$, and concentrating our attention on (7.10). This is an entirely *classical* equation describing the time-evolution of the Kohn–Sham orbitals $\varphi_i(r; t)$, which in turn determine the density $\varrho(r) = \sum_{i=1}^{N} |\varphi_i|^2$. Let us suppose we start at time t_0 with a given initial set of orthogonal orbitals $\{\varphi_i^{\text{Random}}(r; t_0)\}$, chosen at random.

One then allows the orbitals to evolve, by integrating the equation of motion (7.10). This can be done numerically, using standard classical molecular dynamics techniques. That is, starting from $\{\varphi_i(r; t_0)\}$ one computes $\{\varphi_i(r; t_0 + \Delta t)\}$, and so on, by using, e.g., the Verlet algorithm described above (see (7.1)). We must moreover do something in order to preserve, during the time evolution, the orthonormality constraints (3.21).

This can be done in several ways, and again the numerical techniques can be borrowed from classical molecular dynamics, where they have been developed to simulate the motion of molecules by keeping fixed some substructure, or bond, whose dynamic is not relevant (constrained molecular dynamics, [154]). However, as it will become clear, the choice of a specific technique ensuring the respect of the orthonormality constraints will be of importance only when the ions are also displaced: at the present stage, when only the electronic degrees of freedom are considered, the simple Grahm–Schmidt procedure is perfectly adequate.

The orbitals $\varphi_i(r; t)$, which from now on we will simply call the *electronic degrees of freedom*, will then start to "move", dragging with them the electronic charge distribution, hence the Hartree and exchange correlation potentials of the system. Their initial motion will clearly be toward regions where the potential part of the Lagrangian is smaller. If the time evolution is continued, however, they will gain kinetic energy, and eventually they will return to regions of larger potential energy, since equation (7.10) is a second–order, Newtonian equation. If instead their dynamics is "cooled down", i.e., if their kinetic energy is artificially removed, for example by zeroing the velocities at fixed time intervals, or each time they have a negative projection on the direction of the forces, it is possible to make the system evolve toward the minimum of the potential part of the Lagrangian.

This technique, known also as "simulated annealing", can be implemented in a number of refined ways, e.g. by introducing a dissipative term in the Lagrangian, a technique which has proved useful in the search of the global minimum in multi-minima potential energy surfaces. However, 'simulated annealing' is not necessary in the present case, since the DFT potential energy surface has no additional local minima unless also the atomic positions are moved. Hence, when only the electronic system is considered, any minimization scheme will eventually lead to the global minimum of $U - \sigma$ (see (7.6) and (7.7)). Reaching this minimum is equivalent to solving the Kohn–Sham equations. This is analogous to the problem of finding the stationary point of a function $f(x)$, which may be solved either by calculating the roots of the equation $f'(x) = 0$, or by moving along the x-axis by small steps along the steepest descent path.

One must notice, however, that the case of the DFT functional is complicated by the fact that the electronic charge must be self-consistently updated at every step, since the potential depends on it. By letting the system evolve toward the minimum one finally reaches a state where:

1. the electronic degrees of freedom do not change any more with time (i.e., $\dot{\varphi} = 0$ and $\ddot{\varphi} = 0$),
2. the KS wavefunctions $\varphi_i(\boldsymbol{r})$, the electronic charge density $\varrho(\boldsymbol{r})$, and the Hartree and exchange-correlation potential are all self-consistent with each other.

Consequently, the expression given by (7.10) is equal to zero. From this result, and making use of the fact that $\delta E^{\mathrm{DFT}}/\delta\varphi_i^* = H^{\mathrm{KS}}\varphi_i$, one obtains

$$H^{\mathrm{KS}}\varphi_i(\boldsymbol{r}) = \sum_{j\in occ} \lambda_{ij}\,\varphi_j(\boldsymbol{r}). \tag{7.12}$$

The sum over j runs over the occupied Kohn-Sham states only. We recognize in (7.12) the Kohn–Sham equations (3.25), except for the fact that λ can be (and generally is) a non-diagonal (hermitian) matrix with matrix elements λ_{ij}. Hence, it can be diagonalized by a unitary transformation. Anyway, since the total charge (see (3.19)) and the sum of the Kohn–Sham eigenvalues (see (3.28)) are invariant under unitary transformations, (7.12) yields the DFT solution for the ground state of the electronic system.

The advantage of using this "dynamic" procedure to minimize the DFT functional, is that while we solve the KS equations, we *never* diagonalize H^{KS}. Instead, we simply perform the product $H^{\mathrm{KS}}\varphi_j$ many times, in order to obtain the "forces" driving the electronic degrees of freedom toward the minimum. This point is of crucial importance, since the size of the Hamiltonian matrix is $M \times M$ where M is the number of basis functions used (e.g., plane waves, or spherical wavefunctions as in (3.46), a number which can be as large as 10^5, while the number of Kohn–Sham states one is usually interested in is much smaller, comparable to the number of electrons in the system).

Instead of the huge amount of computer memory which is needed to store the full KS Hamiltonian, in the CP scheme it is sufficient to store just *one row at a time*, and perform the matrix-vector multiplications as a series i of scalar products (for details see [155]). The traditional method to solve the KS system requires one instead, to build and to diagonalize the full H^{KS} matrix. Moreover this must be done several times, until the full self-consistency between wavefunctions and potential is reached.

One is now prepared to study what happens when also the ionic positions, $\{\boldsymbol{R}_j\}$, are allowed to move.

7.3 Ionic Motion: Parameter-Free Molecular Dynamics Simulation

In this Section we will focus on the second of the two novelties introduced by the CP method: the fact that, by coupling together the electronic and ionic degrees of freedom *within* the ab initio framework, one can carry out the calculation of the actual motion of the atoms in the system fully from first principles. The only condition assumed is the validity of the Born–Oppenheimer separation (see Sect. 3.1), but without being limited to the harmonic regime (see Chap. 6). The method allows one to compute the forces acting on all atoms without introducing any phenomenological two-body interaction or mean potential. Moreover, it will be shown how the method can be applied to realize a (ionic) molecular dynamics simulation, *without* solving *ex novo* the electronic problem at each ionic "step".

We return again to the CP equations of motion (7.10) and (7.11). We recall that $\tilde{\mu}$ is an arbitrary parameter, having no effect on the results (at the conditions described in the following), and that the time dependence of the orbitals must be considered as a parametric dependence. We remember finally that $\delta E^{\mathrm{DFT}}/\delta\varphi_i^* = H^{\mathrm{KS}}\varphi_i$.

We have seen previously that (7.10) represents a purely classical motion of a set of particles in a given potential $U - \sigma$ (see also (7.6) and (7.7)), and that letting the system evolve and "cool down", until it reaches the potential energy minimum, one reaches a situation where $\ddot{\varphi} = 0$. In these conditions, (7.10) becomes equivalent to the Kohn–Sham equations (3.25), i.e., one reaches the solution for the electronic ground state avoiding any explicit diagonalization of H^{KS}.

Up to this point, we have kept fixed the ionic coordinates $\{\boldsymbol{R}_i\}$. Now, we look at equation (7.11). We know that the right–hand side represents the actual force on the n-th ion, *provided* that the $\{\varphi_i\}$ are those which minimize E^{DFT} for the ionic configuration $\{\boldsymbol{R}_n^0\}$. Our first action must then be to prepare the system in the minimum of the potential energy with respect to the electronic degrees of freedom. We are hence in a situation where $\ddot{\varphi} = 0$, and the $\{\varphi_i\}$ are "the right ones" for the given ionic configuration.

In these conditions, the evaluation of $\boldsymbol{F}_j$ in (7.11) yields the forces acting on the ions. If the procedure "electronic minimization + evaluation of F_j" is repeated for many ionic configurations slightly displaced away from the equilibrium geometry, the scheme can be used to build up the dynamical matrix, as described in Sect. 7.3.1. However, sometimes the most interesting applications require one to solve the (classical) ionic equations of motion *beyond* the harmonic approximation, that is to integrate numerically (7.11). In this case there is no need for the ionic displacements to be restricted to small oscillations, and any kind of motion can be studied, including ion diffusion. In practice, the Verlet algorithm can be used to integrate (7.11) as well.

If (7.10) and (7.11) are considered separately, after any ionic move one would expect, in principle, that a new electronic minimization should be performed, since the new $\{\boldsymbol{R}_n\}$ are no more compatible with the old $\{\varphi_i\}$. Such a scheme is possible, but very expensive from the computational point of view. If, instead, equations (7.10) and (7.11) are integrated simultaneously and no electronic re-minimization is performed, one would expect that, after some step, the condition $\ddot{\varphi} = 0$ will not be satisfied any more, hence the electronic ground state will not be described correctly, nor the forces on the ions, $\boldsymbol{F}_i$.

To solve this problem, Car and Parrinello observed that the coupled system, (7.10) plus (7.11), is an entirely classical system constituted by the real ions coupled to a collection of "particles" (the Kohn–Sham orbitals, or the coefficients of their expansion on a given basis) moving in their own configuration space, and having their own "mass" $\tilde{\mu}$. Since $\tilde{\mu}$, the "inertia" of the coefficients, is completely arbitrary, it is possible to choose it *much smaller than the ionic masses* (in practice, much smaller means at least 100 times smaller).

In these conditions, it is easy to verify that the coupled system evolves in time without appreciable energy transfer from the ionic to the electronic degrees of freedom or viceversa. More precisely, the energy transfer occurs only on a very long timescale (since after an infinite time the energy will be in any case equi-partitioned between all degrees of freedom), much longer than the actual duration of the MD simulation. Consequently, if one starts from a situation in which the kinetic energy of the coefficients is very small (that is, $\dot{\varphi} \simeq 0$ and $\ddot{\varphi} \simeq 0$), the same will remain true along the simulation time.

In practice, the system will be in a metastable, non-equilibrium situation, in which the "temperature of the electronic coefficients" is very small, and rises extremely slowly, no matter how much the ions "heat up". This is a very well known property in classical molecular dynamics, where it represents sometimes an inconvenient, because a system containing atoms of very different masses can take a very long time to reach the thermodynamic equilibrium.

The following example can help us to understand the phenomenon. One may think of a long pendulum, of large mass M, with several short pendula of mass $m \ll M$ attached to it. During the motion of the system, the short pendula always oscillate quickly around their instantaneous potential energy minimum determined by the position of the long pendulum. However, the long pendulum will transfer only a negligible amount of kinetic energy to the short ones. In fact, the short and long pendula are (essentially) energetically decoupled on a very long time scale.

In the Car–Parrinello scheme, hence, it is possible to exploit this energetic decoupling in such a way that the electronic system evolves in time, keeping the relation $T_{\text{fake}} \simeq 0$ (see (7.5)) always verified. This means that the KS equations are instantaneously always satisfied to a high degree of accuracy, the set of the electronic coordinates $\{\varphi_i\}$ are always those which minimize E^{DFT} for the instantaneous geometrical configuration $\{\boldsymbol{R}_n\}$, and the forces on the ions, $\boldsymbol{F}_i$, are correctly given by the right-hand side of (7.11). In this way it is possible to realize a MD simulation for the complete system. This MD simulation will be, in particular, a *microcanonical* one, i.e. the quantity $T_{\text{ions}} + U$ will be almost exactly conserved (by contrast, the quantity which is conserved in general, no matter how large is $\tilde{\mu}$ with respect to M, is $T_{\text{ions}} + U +$

T_{fake}). As a technical detail, we mention that the numerical procedure used to enforce the orthogonality condition (3.21) becomes relevant, as illustrated, e.g., in [156].

The CP method allows one to simulate molecules, solids, surfaces and clusters *at finite temperature*, even if the atomic displacements are very large (e.g., in liquid dynamics). The only necessary conditions are: a) always start from a situation where $T_{\text{fake}} \simeq 0$, i.e., after a "good" electronic minimization, b) choose $\tilde{\mu} << M$. It must be noticed that the size of the time-step Δt used in the numerical integration algorithm (7.1) is governed by the components having the smallest mass. This means that the smaller is $\tilde{\mu}$, the smaller must be Δt. This is one of the reasons why CP–MD simulations are very CPU time consuming calculations. The small errors due to the fact that the condition $\ddot{\varphi} = 0$ is not *exactly* satisfied, tend to average to zero: these errors lead to both positive and negative contributions. This has been discussed extensively in the literature (see, e.g., [156]).

In an attempt to make the above discussion more didactic, we show in Fig. 7.1 the typical time-evolution of different quantities during a CP molecular dynamics simulation, performed for the case of bulk silicon. The upper curve (U) represents the total potential energy associated with the ions (i.e., the "total energy" computed in DFT for a given set of ionic positions). When it is added to the ionic kinetic energy (T_{ions}), which is proportional to the system's temperature, one obtains the quantity which should be physically conserved in a microcanonical simulation (plotted as the third curve from above). However, enlarging the energy scale (see right part of the plot), the latter curve reveals small and fast oscillations, which originate from changes in the "orbital kinetic energy" due to the "motion" of the Kohn–Sham wavefunctions (T_{fake}, see (7.5)). The quantity which is rigorously conserved by the Car–Parrinello Lagrangian (independently of the ratio $\tilde{\mu}/M$) is hence the sum of $T + U$ plus T_{fake}, plotted as the bottom curve. In the right part of the figure, the energy scale is expanded by a factor of 100.

When the motion is limited to small, harmonic oscillations around a stable configuration (i.e., around a minimum of the potential U), a molecular dynamics trajectory obtained in a full CP run can be also used to extract the vibrational frequencies and the normal mode eigenvectors. C_{60} phonons have been computed ab initio for the first time in [157, 158].

Following the same method used to calculate structural and vibrational properties of C_{60}, and which successfully reproduced all available experimental data. The more complex case of C_{70} has also been investigated [159]. The basis of this analysis was a molecular dynamics low temperature trajectory calculated with the LDA-based Car–Parrinello method, previously applied to optimize the molecular structure [132]. The derivation of the vibrational spectrum follows a procedure which, in the case of a harmonic system such as C_{60} and presumably C_{70} , allows one to obtain very accurate results in spite of the short times the simulation is carried out, times in which the system is not able to reach thermal equilibrium [159].

In [159] frequencies of the *optically active* vibrational modes of C_{70} are also reported. Those modes consist into infrared active ones (modes which couple to the electromagnetic field in the dipole approximation, allowing absorption or emission of infrared light), and Raman active ones (modes which modulate the polarizability is such a way that Raman lines, i.e. inelastically scattered light, is emitted when the sample is illuminated by an intense laser source). Comparison with experimental data is not obvious since, in the measured spectra, not all the 31 infrared and 53 Raman active modes have been identified. We can safely attribute only the lowest Raman active mode which is a doubly-degenerate E_2' mode at 222 cm^{-1} in our calculations and observed at 226 cm^{-1}. Two accidentally degenerate modes follow in the Raman spectrum, one non degenerate (A_1') and one doubly degenerate (E_1'').

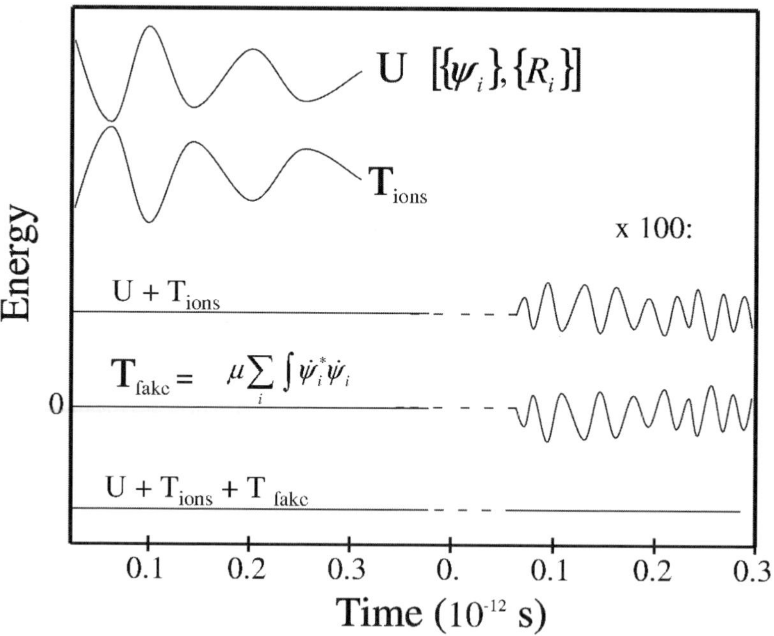

Fig. 7.1. Plot of the total potential energy (U) and the ionic kinetic energy (T_{ions}) as a function of time during an ab initio Car–Parrinello molecular dynamics simulation. The sum of those energies, $U + T_{ions}$, remains constant except for a small oscillatory behavior limited to less than 1 meV (visible only on the expanded y–scale on the right part of the plot). The latter oscillations are due to the small *orbital* kinetic energy, appearing in the CP Lagrangian and plotted as T_{fake}. The sum of T_{fake} with ($U + T_{ions}$) is the energy which is exactly conserved by the CP Lagrangian, and which should remain constant also on the expanded energy scale. It corresponds to the bottom curve in the graph and allows one to monitor the error introduced by the numerical integration of the equations of motion

The eigenvectors corresponding to the three lowest Raman active modes (having symmetries E_2', E_1'' and A_1' respectively) are easily recognized as "squashing" modes, and closely resemble the lowest, five-fold degenerate Raman active mode of C_{60}. The splitting of this mode into a 2+2+1 multiplet is clearly due to the symmetry reduction. Our softest mode is an E_2' one (see Appendix 5.5.1), and it corresponds to a deformation along one of the short axis of the molecule.

A good agreement is found between the observed major spectral features and the ab initio results; this agreement, on the other hand, is appreciably more satisfactory than with the results of model calculations.

More efficient computational strategies can be used to obtain these ab initio results if one is only interested in the harmonic regime. In such a case, it is sufficient to carry out *static* CP runs (i.e., with fixed ionic positions) to compute the force constants matrix (see Sect. 6.1), and diagonalize it in the standard way. The latter is also called the "frozen phonon" approach.

7.3.1 Frozen-Phonon Calculations

Following (6.3) and (6.12), it emerges that the dynamical matrix $\mathbf{D}$ can be obtained from the knowledge of the derivatives, with respect of the independent $3N_{at}$ (cartesian) atomic displacements, of the *force* exerted on atoms. Such derivatives can be computed numerically, evaluating the change of the component β of the force generated on the atom m, when the atom n is displaced by a small amount in the direction α. This can be done by minimizing the DFT functional with respect to the electronic degrees of freedom and hence evaluating the forces on the ions (i.e., $\mathbf{F}_i$ in (7.11)), after having slightly displaced a given atom along a given cartesian direction. The process must be repeated for the three components x, y and z of every atom. Since the ionic displacements are small, the requested set of $3N_{at}$ electronic minimizations (i.e., solutions of (7.12)) is largely facilitated if one starts them from the electronic configuration which corresponds to the equilibrium ionic positions. Many practical applications of the above method can be found in the recent literature. A typical example from the field of carbon clusters is the case of C_{60}. Calculations [121] confirmed that good agreement with respect to the experimentally measured vibrational spectra can be obtained by using the Car–Parrinello method in the DFT–LD approximation (see Sect. 6.2.2 and in particular Table 6.2). Similarly, the cluster C_{28} has been studied at the same level of accuracy [160]. Its normal modes which couple to the LUMO electronic orbital (i.e., those with A_1, E and T_2 symmetry) are reported in Table 7.1.

The possibility to perform full ab initio calculations of the vibrational spectra of several different fullerene clusters allows one to make quantitative comparisons concerning, e.g., the quantities related to the electron–phonon coupling (see Chap. 8).

Table 7.1. A_1, E and T_2 normal modes of the cluster C_{28} calculated within DFT–LDA [160]

$1/\lambda$ [cm^{-1}]	irrep.
351	E
391	T_2
524	T_2
565	A_1
570	E
607	E
707	T_2
724	T_2
763	A_1
771	T_2
791	T_2
976	E
983	T_2
1093	T_2
1101	A_1
1116	E
1171	A_1
1191	T_2
1220	A_1
1260	T_2
1306	E
1381	T_2
1414	E

Application to small sodium clusters. Another interesting case where to apply the ab initio methods discussed above is that of metal aggregates. Among them, the simplest system is the closed-shell Na_8 cluster, already discussed in Sect. 5.4. Self-consistent ab initio DFT-LDA calculation for Na_8, based on norm-conserving pseudopotentials, and performed using a modified version of the CPMD code, yields the well-known geometrical and electronic structure of the ground state (see Fig. 5.14 and Table 5.8). In this state, the cluster has a $\mathcal{D}_{2d}$ symmetry, and 14 *distinct* nonzero vibrational eigenvalues, 4 of them being doubly degenerate[2]. The energies and eigenvectors of the normal modes have been obtained diagonalizing the dynamical matrix. The latter was built by a series of self-consistent calculations of the restoring forces induced by small cartesian displacements of single atoms of the cluster. At full convergence with respect to the energy cutoff (15 Ry) and to other computational parameters, the resulting energies and symmetries for the normal modes are those reported in Table 7.2. Figure 7.2 reproduces the atomic displacements associated to the highest totally symmetric mode (16.7 meV). The displacement field associated to specific modes is an important

---

[2] This corresponds to 18 nonzero vibrational modes, which added to the six (zero-frequency) rigid translations and rotations yields a total of $3\times8=24$ degrees of freedom.

physical quantity, to which ab initio calculations give direct access. The displacement fields (vibrational eigenvectors) are in fact hardly accessible from the experiment; they are, however, very important objects, since they govern the size of the coupling with the electronic states (electron–phonon coupling, see Chap. 8).

Table 7.2. Calculated frequencies for the Na_8 vibrational modes.

irrep	ω_ν [meV]	$1/\lambda$ [cm^{-1}]
B_1	5.84	47.07
B_2	6.85	55.26
A_1	7.93	63.92
E	8.77	70.74
E	9.36	75.52
A_1	9.48	76.45
A_2	10.2	82.15
B_2	12.3	99.50
E	14.1	113.75
B_1	14.9	119.83
A_1	15.9	128.57
A_1	16.7	134.83
E	17.5	140.95
B_2	18.6	150.00

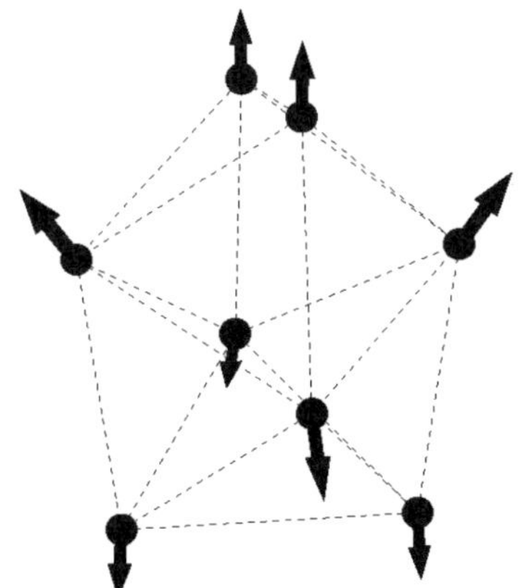

Fig. 7.2. Displacement associated with the A1 phonon in Na_8 at energy 16.7 meV

8. Coupling of Electrons to Phonons and to Plasmons

So far, electrons and ions have been considered as independent degrees of freedom. Ions have been taken either as fixed in their ground-state configuration or as moving under the action of a phonon displacement field. The electrons have been treated within the Born–Oppenheimer (BO) approximation (see Sect. 3.1 and Sect. 7.3). This means that it has been assumed that they are able to follow adiabatically the evolution of the ionic configuration. This approximation (based on the fact that typical frequencies obey $\omega_{\text{electrons}} \gg \omega_{\text{ions}}$) may break down in a number of cases in which the ionic motion induces correlations between electronic states, as it will be shown in the present chapter (see also Sect. 2.10).

Furthermore, many-body systems like bulk solids, metallic aggregates, fullerenes etc., display collective electronic vibrations (plasmons), with frequencies which are of the same order of magnitude of those associated with the single-electron motion (see Chaps. 4 and 5). Although the renormalization effects due to the coupling of electrons to plasmons are in principle included in the exact DFT, in practice they are incorporated only to some extent in the correlation potential of the LDA, where important effects of this coupling are left out. Moreover, the exchange of plasmons between electrons, which is the basic mechanism leading to the screening of the Coulomb field, gives also rise to an electron self-energy (see Sects. 2.4 and 2.7) whose frequency dependence is essential to obtain a proper description of this phenomenon (see Sects. 2.3-2.7).

The outline of the present chapter is the following: in Sect. 8.1 we define the coupling between the ionic normal modes (phonons) and the electrons and show how to perform the explicit evaluation of the corresponding matrix elements of this interaction. In Sect. 8.2 we show how the diagonal matrix elements are related to the change in the single-particle electronic energies induced by the electron–phonon coupling. In Sect. 8.3 we give an example of the electron–phonon coupling calculations, while in Sect. 8.4 we discuss a standard technique used to extract the electron–phonon coupling matrix elements from experiment. The effective interaction between electrons arising from the exchange of phonons is the subject of Sect. 8.5. In Sect. 8.6 we repeat the arguments developed in the previous sections, but for the case of the electron–plasmon coupling. Special attention is given to the effective electron–

electron interaction arising by the exchange of plasmons. This interaction (screening) is to be treated on equal footing to the interaction mediated by the exchange of phonons in describing phenomenona ranging from the optical response of atomic clusters (see Sect. 8.7 and 8.8) to superconductivity (see (2.72) as well as Chap. 9).

8.1 Definition of the Electron–Phonon Interaction

Let us start from the equations of Sect. 3.1. The total Hamiltonian H_{tot} of a system made up of ions and electrons was defined in (3.1) and split into an electronic Hamiltonian H_{elec} (see (3.10)) and a ionic Hamiltonian H_{ion} (see (3.13)). The solution of H_{elec} in terms of Kohn–Sham states was treated in Chap. 3 while that of H_{ion} was discussed in Chaps. 6 and 7 in terms of the normal modes of the system, that is, the phonons. In the present chapter we assume that both electron as well as phonon eigenvalues and eigenvectors are known. We adopt accordingly the notation H_{phon} instead of H_{ion} and write

$$H = H_{\text{elec}} + H_{\text{phon}},\tag{8.1}$$

where (see (4.1))

$$H_{\text{elec}} = -\frac{\hbar^2}{2m_{\text{e}}}\nabla^2 + v_{\text{eff}}[\varrho]\tag{8.2}$$

with (see (3.26))

$$v_{\text{eff}}[\varrho] \equiv v_{\text{eff}}(\boldsymbol{r}) = v_{\text{H}}(\boldsymbol{r}) + V_{\text{ext}}(\boldsymbol{r}) + v_{\text{xc}}(\boldsymbol{r}).$$

The definition of the different terms entering the above equation are given in (3.23), (3.4) and (3.24) plus (3.30) respectively, the corresponding Schrödinger equation being (see (3.42))

$$\left(-\frac{\hbar^2}{2m_{\text{e}}}\nabla^2 + v_{\text{eff}}(\boldsymbol{r})\right)\varphi_j(\boldsymbol{r}) = \lambda_j\varphi_j(\boldsymbol{r}).\tag{8.3}$$

The phonon Hamiltonian (see (6.24))

$$H_{\text{phon}} = \sum_{\nu}\hbar\omega_{\nu}\left(\Gamma_{\nu}^{\dagger}\Gamma_{\nu} + \frac{1}{2}\right),\tag{8.4}$$

second quantized version of (see (6.21))

$$H_{\text{phon}} = \frac{1}{2}\sum_{\nu}\omega_{\nu}(P_{\nu}^2 + Q_{\nu}^2),$$

is written in terms of the eigenfrequencies ω_{ν} determined by solving the relation

$$\omega^2\boldsymbol{w} = \boldsymbol{D}\boldsymbol{w},$$

where

$$D = M^{-1/2}\boldsymbol{\Phi}M^{1/2},$$

and

$$w = M^{1/2}u,$$

are defined in (6.14), (6.12) and (6.10) and controlled by the diagonal (mass) matrix $M_{n\alpha,m\beta} = \delta_{\alpha\beta}\delta_{nm}M_n$, the displacement operators u with elements $u_{n\alpha}$ and the force constant matrix $\boldsymbol{\Phi}$ with elements $\Phi_{n\alpha m\beta}$.

It is assumed that the ion displacements are small, so that one can use the approximation $|u|^2 \ll |u|$ (see also Sect. 2.10 and the discussion before (2.35)). Consequently, one can write (neglecting terms proportional to u^2 and using the notation $v(r, \{R\})$ for $v_{\mathrm{eff}}(r)$ to make evident the parametric dependence on the ion positions)

$$v(r, \{R\}) = v(r, \{R^{(0)}\}) + u \cdot \boldsymbol{\nabla}v(r, \{R^{(0)}\}), \tag{8.5}$$

where $u \cdot \boldsymbol{\nabla}$ is the scalar product between vectors in a $3 \times N_{\mathrm{at}}$-dimensional space. It represents the derivative of the electronic mean field along the direction of the ionic motion. The last term of the above equation is the *deformation potential*. It measures the coupling existing between electrons and phonons and defines the *electron–phonon Hamiltonian*,

$$H_{\mathrm{elec,phon}} \equiv \sum_{\nu\mu}(v_{\mathrm{def}}^{(\nu)})_\mu. \tag{8.6}$$

The sum runs over the phonons quantum numbers ν characterizing the phonon and μ labelling the degenerate modes within each group[1] (see e.g. Table 6.2, where ν stands for F_{1u}, A_g and H_g). The deformation potential is

$$(v_{\mathrm{def}}^{(\nu)})_\mu \equiv u_\mu^{(\nu)} \cdot \boldsymbol{\nabla}v(r, \{R^{(0)}\}). \tag{8.7}$$

It can be shown that the matrix elements of the electron–phonon Hamiltonian correspond to those of the total Hamiltonian which are neglected in the BO approximation [161, 162, 163] (see Appendix 8.9.1).

From (8.3) and (8.5) it is seen that the deformation potential is the directional derivative of the total electronic potential, in the direction of the normal modes. Once this direction is known (that is, once the phonon problem is solved and the matrix elements of u written as $\sqrt{\hbar/2M_\lambda\omega_\lambda}$ found as shown in Sect. 6.1), the ions can be moved along such directions by a unit distance namely, displaced by $u/|u|$. In this way, the matrix elements of the derivative of the effective potential can be calculated using, for instance, the basis of the Kohn–Sham orbitals.

---

[1] In the case of a spherical system ν stands for the multipolarity λ of the phonon and μ is the magnetic quantum number running over the $2\lambda + 1$ substates from $\mu = -\lambda$ to $\mu = \lambda$ (see e.g. Sects. 2.4 and 8.3).

8.2 Diagonal Matrix Elements

In what follows we provide a simple, but important example of relations which the matrix elements of the deformation potential have to fulfill. Starting from the ionic equilibrium configuration $\{R_i^{(0)}\}$, one displaces the ions in the direction of one particular eigenvector first by the amount $+u/2$ and then by $-u/2$. Calling $v^{(1)} = v(r, \{R^{(0)} + u/2\})$ and $v^{(2)} = v(r, \{R^{(0)} - u/2\})$ and $\varepsilon_j^{(1)}$, $\varepsilon_j^{(2)}$ and $\varphi_j^{(1)}$, $\varphi_j^{(2)}$ the associated (Kohn–Sham) eigenvalues and eigenfunctions solutions of (8.3) one can write

$$\varepsilon_j^{(1)} = \langle \varphi_j^{(1)} | H_{\text{elec}} | \varphi_j^{(1)} \rangle = \langle \varphi_j^{(1)} | - \frac{\hbar^2}{2m_{\text{e}}} | \varphi_j^{(1)} \rangle + \langle \varphi_j^{(1)} | v^{(1)} | \varphi_j^{(1)} \rangle$$

$$= t_j^{(1)} + \langle \varphi_j^{(1)} | v^{(1)} | \varphi_j^{(1)} \rangle, \tag{8.8}$$

where the diagonal matrix elements of the kinetic energy are labelled by t_j. Writing $\varphi_j^{(1)}$ as $\varphi_j^{(0)} + \delta\varphi_j^{(1)}$ (where $\varphi_j^{(0)}$ is the solution of (8.3) for $v(r, \{R_i^{(0)}\})$), and assuming $\delta\varphi_j^{(1)}$ to be small, one obtains to first order,

$$\varepsilon_j^{(1)} = t_j^{(1)} + \langle \varphi_j^{(0)} | v^{(1)} | \varphi_j^{(0)} \rangle + \langle \delta\varphi_j^{(1)} | v^{(0)} | \varphi_j^{(0)} \rangle + \text{h.c.}, \tag{8.9}$$

where h.c. denotes the Hermitian conjugate. A similar expression holds for $\varepsilon_j^{(2)}$. Thus,

$$\varepsilon_j^{(2)} - \varepsilon_j^{(1)} = t_j^{(2)} - t_j^{(1)} + \langle \varphi_j^{(0)} | v^{(2)} - v^{(1)} | \varphi_j^{(0)} \rangle$$

$$+ \langle \delta\varphi_j^{(2)} | v^{(2)} | \varphi_j^{(0)} \rangle + \text{h.c.} - \langle \delta\varphi_j^{(1)} | v^{(1)} | \varphi_j^{(0)} \rangle - \text{h.c.} \tag{8.10}$$

The last line of this equation can be approximated by the relation

$$\langle \delta\varphi_j^{(2)} | v^{(0)} + \delta v^{(2)} | \varphi_j^{(0)} \rangle - \langle \delta\varphi_j^{(1)} | v^{(0)} + \delta v^{(1)} | \varphi_j^{(0)} \rangle + \text{h.c.}$$

$$= \langle \delta\varphi_j^{(2)} - \delta\varphi_j^{(1)} | v^{(0)} | \varphi_j^{(0)} \rangle + \text{h.c.}, \tag{8.11}$$

while $t_j^{(2)} - t_j^{(1)}$ can be written as

$$\langle \delta\varphi_j^{(2)} - \delta\varphi_j^{(1)} | T | \varphi_j^{(0)} \rangle. \tag{8.12}$$

Summing up the expressions given in (8.11) and (8.12), one obtains

$$\langle \delta\varphi_j^{(2)} - \delta\varphi_j^{(1)} | H^{(0)} | \varphi_j^{(0)} \rangle = \varepsilon_j \langle \delta\varphi_j^{(2)} - \delta\varphi_j^{(1)} | \varphi_j^{(0)} \rangle, \tag{8.13}$$

a quantity which vanishes according to first-order perturbation theory. Consequently,

$$\varepsilon_j^{(2)} - \varepsilon_j^{(1)} = \langle \varphi_j^{(0)} | v^{(2)} - v^{(1)} | \varphi_j^{(0)} \rangle + O(|\delta\varphi|^2). \tag{8.14}$$

Because $v^{(2)} - v^{(1)}$ is by definition the deformation potential, the above equation states that the diagonal matrix elements of this potential are equal to the differences between Kohn–Sham energies. This fact provides an important numerical test for electron–phonon calculations. In particular, in the case of the matrix elements shown in Table 8.1 and Table 8.4, the equality (8.14) holds with an accuracy of 1% or better.

8.3 Electron–Phonon Coupling in C$_{60}^-$

We shall in this section work out the electron–phonon coupling Hamiltonian in the case of the fullerene ion C$_{60}^-$, that is, a molecule of C$_{60}$ with an electron moving in the LUMO (t_{1u}) level (see Fig. 5.1).

Let us assume $v(\boldsymbol{r}, \{\boldsymbol{R}^{(0)}\})$ to be the mean field LDA potential seen by an electron moving in the three-fold degenerate t_{1u} LUMO state (see Sect. 5.1.1). Because the nearest orbitals are well separated from it by energies of the order of the eV (see Fig. 5.1), much larger than typical phonon frequencies which are of the order of 1000 K ($\sim$100 meV, see Table 6.2), we shall only consider scattering processes among the degenerate substates of the t_{1u} orbital (see Fig. 8.1). Because of symmetry reasons (see Appendix 5.5.1), only couplings of the t_{1u} orbital to the eight H_g and the two A_g phonons are allowed. Thus,

$$
\begin{aligned}
H = {} & \varepsilon \sum_{i=1}^{3} \sum_{\sigma} a_{i\sigma}^{\dagger} a_{i\sigma} + \sum_{\nu=1}^{8} \hbar\omega_\nu \sum_{\mu=1}^{5} \left(\Gamma_{\nu\mu}^{\dagger} \Gamma_{\nu\mu} + \frac{1}{2} \right) \\
& + \sum_{\nu=9}^{10} \hbar\omega_\nu \left(\Gamma_{\nu}^{\dagger} \Gamma_{\nu} + \frac{1}{2} \right) \\
& + \sum_{\nu,\mu} \sum_{i,j,\sigma} \langle t_{1u}i, \nu\mu | (v_{def}^{(\nu)})_\mu | t_{1u}j \rangle \left(\Gamma_{\nu\mu}^{\dagger} + \Gamma_{\nu\mu} \right) a_{i\sigma}^{\dagger} a_{j\sigma},
\end{aligned}
\tag{8.15}
$$

where the first term describes the motion of the electrons, the second and third terms the H_g and the A_g phonons, while the last term, the coupling between the phonons and the electrons (see Fig. 8.1). The indices i and j label the substates of the t_{1u} orbital, while σ indicates the spin orientation. The indices ν and μ characterize the phonons.

In order to reduce the number of matrix elements of the deformation potential to be calculated one can make use of the symmetries of the icosahedral group and introduce the reduced matrix elements making use of the

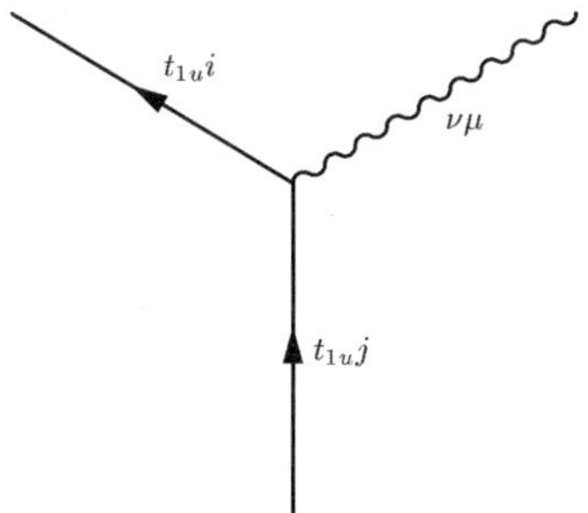

Fig. 8.1. Graphical representation of the electron–phonon coupling in C_{60}^-

Wigner–Eckart theorem. That is,

$$\langle t_{1u}, i; \nu\mu |(v_{def}^{(\nu)})_\mu| t_{1u}, j\rangle = \sqrt{\frac{\hbar}{2M\omega_\nu}} \langle t_{1u}, i| \frac{u_\mu^{(\nu)}}{|u_\mu^{(\nu)}|} \cdot \boldsymbol{\nabla}^{(\nu)} v(\boldsymbol{r}, \{\boldsymbol{R}^{(0)}\})| t_{1u}, j\rangle$$

$$= \sqrt{\frac{\hbar}{2M\omega_\nu}} \langle t_{1u}, i| \left(\boldsymbol{\nabla}^{(\nu)} v(\boldsymbol{r}, \{\boldsymbol{R}^{(0)}\})\right)_\mu \rangle| t_{1u}, j\rangle$$

$$= \frac{g_\nu}{2} C_{ij}^{(\mu)}, \tag{8.16}$$

where

$$\frac{g_\nu}{2} = \sqrt{\frac{\hbar}{2M\omega_\nu}} \langle t_{1u}|| \boldsymbol{\nabla}^{(\nu)} v(\boldsymbol{r}, \{\boldsymbol{R}^{(0)}\})|| t_{1u}\rangle. \tag{8.17}$$

The above reduced matrix element does not depend on the index μ nor on the indices i and j. This dependence is carried by the geometrical coefficients

$$C_{ij}^{(\mu)} = \begin{cases} V_{ij}^{(\mu)} & \nu \in H_g, \\ W_{ij} & \nu \in A_g, \end{cases} \tag{8.18}$$

which couples the t_{1u} electronic orbital to the H_g and A_g vibrational modes. These coupling coefficients are determined from symmetry considerations [164] and are given by

$$V^{(1)} = \begin{pmatrix} -1 & 0 & 0 \\ 0 & -1 & 0 \\ 0 & 0 & 2 \end{pmatrix}, \quad V^{(2)} = \begin{pmatrix} \sqrt{3} & 0 & 0 \\ 0 & -\sqrt{3} & 0 \\ 0 & 0 & 0 \end{pmatrix}, \quad V^{(3)} = \sqrt{3}\begin{pmatrix} 0 & 1 & 0 \\ 1 & 0 & 0 \\ 0 & 0 & 0 \end{pmatrix},$$

$$V^{(4)} = \sqrt{3}\begin{pmatrix} 0 & 0 & 1 \\ 0 & 0 & 0 \\ 1 & 0 & 0 \end{pmatrix}, \quad V^{(5)} = \sqrt{3}\begin{pmatrix} 0 & 0 & 0 \\ 0 & 0 & 1 \\ 0 & 1 & 0 \end{pmatrix},$$

$$\tag{8.19}$$

and by

$$W = \begin{pmatrix} \sqrt{2} & 0 & 0 \\ 0 & \sqrt{2} & 0 \\ 0 & 0 & \sqrt{2} \end{pmatrix}. \tag{8.20}$$

Consequently, the Hamiltonian coupling the t_{1u} electron to the H_g and A_g phonons (last term in (8.15)) can be written as,

$$H_{\text{elec,phon}} = \sum_{\nu=1}^{8} \frac{g_\nu}{2} \sum_{\mu=1}^{5} \sum_{i=1}^{3} \sum_{j=1}^{3} \sum_\sigma V_{ij}^{(\mu)}(\Gamma_{\nu\mu}^\dagger + \Gamma_{\nu\mu}) a_{i\sigma}^\dagger a_{j\sigma}$$

$$+ \sum_{\nu=9}^{10} \frac{g_\nu}{2} \sum_{i=1}^{3} \sum_{j=1}^{3} \sum_\sigma W_{ij}(\Gamma_\nu^\dagger + \Gamma_\nu) a_{i\sigma}^\dagger a_{j\sigma}. \tag{8.21}$$

In Table 8.1 the values of g_ν are shown as calculated making use of either the frozen phonon approximation described in Sect. 7.3.1 or by means of

Table 8.1. Diagonal electron-phonon (reduced) matrix elements g_ν measuring the strength with which the t_{1u} LUMO state of C_{60}^- couples to the different normal modes (H_g and A_g) resulting from two different calculations, one based on ab initio methods (frozen phonon approximation) and the other on the empirical Hybrid Orbital Model

	Frozen phonon[a]	HOM[b]
	$|g_\nu|$[meV] [c]	$|g_\nu|$[meV] [c]
$H_g(8)$	41	16
$H_g(7)$	43	55
$H_g(6)$	15	14
$H_g(5)$	20	28
$H_g(4)$	17	24
$H_g(3)$	22	17
$H_g(2)$	19	22
$H_g(1)$	9	17
$A_g(2)$	64	41
$A_g(1)$	0	30

[a] see [121] and the discussion in Sect. 7.3.1
[b] Hybrid Orbital Model (HOM), [165, 166]
[c] The matrix elements given e.g. in [165, 166] obey:
  matrix element [meV]$\times c_\nu = g_\nu$, where c_ν=2 for H_g and $\sqrt{2}$ for A_g

an empirical model introduced in [165, 166] and called the Hybrid Orbital Model (HOM). In this model, the C_{60} density is described as a superposition of standard s- and p-orbitals, whose parameters are adjusted for an optimal fit of the density resulting from a LDA calculation. This parametrization allows for an economic determination of the deformation potential.

A parallel can be traced between the formalism discussed in the beginning of this section and that associated with the particle–vibration coupling developed in connection with the surface vibrations of a finite system, where the fermions move in an average potential $U(r - R_0)$, R_0 being, for example, the radius of a spherical metal cluster or of an atomic nucleus [10].

As already discussed in Sect. 2.4 – discussion connected with the coupling of electrons to plasmons but equally applicable to the electron–phonon coupling (see also Sect. 2.10) – the surface of these systems can vibrate collectively with multipolarity L, and their radius can be parametrized as $R = R_0(1 + \sum_{n,LM} \alpha_{LM}(n) Y_{LM}^*(\hat{r}))$, $\alpha_{LM}(n)$ being the deformation parameters, coordinate of the associated harmonic oscillator, equivalent to the displacement vectors $\mathbf{u}$. In the present case L and M characterize the transformation properties (tensorial character) of the (dynamic) deformation, while n refers to the possibility of having more than one type of vibration with the same multipolarity L and magnetic quantum number M ($\nu \leftrightarrow (n, L)$).

Inserting R into U, and expanding to lowest order in α one obtains

$$U(r - R) = U(r - R_0) + \delta U, \tag{8.22}$$

where

$$\delta U = \sum_{n,LM} \delta U_{LM}(n), \tag{8.23}$$

and

$$\delta U_{LM} = \alpha_{LM}(n) f_{LM}(\boldsymbol{r}), \tag{8.24}$$

where

$$f_{LM}(\boldsymbol{r}) = -R_0 \frac{\partial U(r)}{\partial r} Y_{LM}^*(\hat{r}). \tag{8.25}$$

In second quantization

$$\alpha_{LM}(n) = \sqrt{\frac{\hbar}{2M\omega_{n,L}}} (\Gamma_{n,LM}^\dagger + \Gamma_{n,LM}). \tag{8.26}$$

The associated particle–vibration matrix elements diagonal in the total angular momentum of the single-particle orbital can be written as (see Fig. 8.2)

$$\langle jm', nLM | \delta U_{n,LM} | jm \rangle = \langle nLM | \alpha_{LM}(n) | 00 \rangle \, \langle jm' | f_{LM}(\boldsymbol{r}) | jm \rangle$$

$$= \sqrt{\frac{\hbar}{2M\omega_L(n)}} \, \langle jm' | f_{LM}(\boldsymbol{r}) | jm \rangle$$

$$= \frac{g_L(n)}{2} C_{mm'}^{(M)},$$

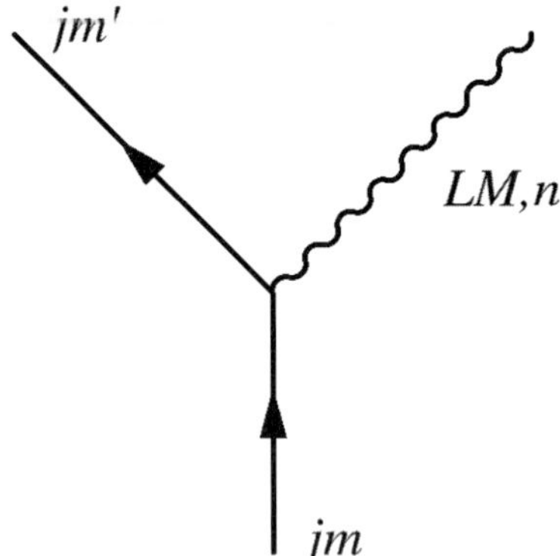

Fig. 8.2. Graphical representation of the particle–vibration coupling vertex between a fermion moving in a j-shell and the n-th collective vibration of multipolarity L and projection M

where

$$\frac{g_L(n)}{2} = \sqrt{\frac{\hbar}{2M\omega_L(n)}} \frac{\langle j||f_L||j\rangle}{\sqrt{2j+1}},$$

(8.27)

and

$$C_{mm'}^{(M)} = \frac{1}{\sqrt{2j+1}} \langle jmLM|jm'\rangle.$$

(8.28)

The quantity $\langle jmLM|jm'\rangle$ is the Clebsch–Gordan coefficient needed to couple the angular momenta j and L to j'. Because the magnetic quantum numbers are the eigenvalues of a scalar operator, that is, the third component of the total angular momentum, $m' - m = M$. Assuming, e.g., $L = 2$ (5-fold degeneracy like the H_g phonons) and $j = l = 1$ (3-fold degeneracy, as t_{1u}), one can write

$$V^{(0)} = \begin{pmatrix} \langle 1120|11\rangle & 0 & 0 \\ 0 & \langle 1020|10\rangle & 0 \\ 0 & 0 & \langle 1\text{-}120|1\text{-}1\rangle \end{pmatrix},$$

$$V^{(1)} = \begin{pmatrix} 0 & 0 & 0 \\ \langle 1021|11\rangle & 0 & 0 \\ 0 & \langle 1\text{-}121|10\rangle & 0 \end{pmatrix},$$

$$V^{(-1)} = \begin{pmatrix} 0 & \langle 112\text{-}1|10\rangle & 0 \\ 0 & 0 & \langle 102\text{-}1|1\text{-}1\rangle \\ 0 & 0 & 0 \end{pmatrix},$$

(8.29)

$$V^{(2)} = \begin{pmatrix} 0 & 0 & 0 \\ 0 & 0 & 0 \\ \langle 1\text{-}122|11\rangle & 0 & 0 \end{pmatrix},$$

$$V^{(-2)} = \begin{pmatrix} 0 & 0 & \langle 112\text{-}2|1\text{-}1\rangle \\ 0 & 0 & 0 \\ 0 & 0 & 0 \end{pmatrix}.$$

Consequently, in second quantization,

$$\delta U = \sum_n \frac{g_2(n)}{2} \sum_{M=-2}^{2} \sum_{\sigma} \sum_{m=-1}^{1} \sum_{m'=-1}^{1} V_{mm'}^{(M)}(\Gamma_{n,2M}^{\dagger} + \Gamma_{n,2M}) a_{m'\sigma}^{\dagger} a_{m\sigma}, \quad (8.30)$$

where the assumption was made that one is dealing with an uncoupled $l - s$ situation, that is $a_{m'\sigma}^{\dagger}|0\rangle = |lm'\sigma\rangle (l = 1, m' = \pm 1, 0, \sigma \pm 1/2)$. Consequently δU is equivalent, in the case of a spherical symmetric system in which the LUMO state is of $l=1$ character and the surface vibrations are of quadrupole type, to the first term of the electron–phonon coupling Hamiltonian defined in (8.21).

In order to assess the relevance the electron–phonon coupling has on the spectrum of finite systems like atomic nuclei and fullerenes, e.g. C$_{60}$, it is necessary to introduce some dimensionless quantity. That is, to provide an energy scale againts which to measure a typical value of g_ν. A useful parameter is provided by the ratio $g\Omega/D = x$, where $g = \sum_\nu g_\nu$, Ω is the degeneracy of the LUMO (for only one spin orientation), while D is the (average) distance between this single-particle orbital and its nearest neighbour level[2].

In the nuclear case $g \approx 2g_2 \approx$ -0.6 MeV in keeping with the fact that essentially only quadrupole and octupole vibrations couple effectively to the motion of the nucleons close to the Fermi energy. Making use of this value of g and of typical values of $\Omega=3$ and $D=2$ MeV for medium-heavy nuclei (see e.g. [45]), one obtains $x \approx 0.9$.

In the case of C_{60}^-, $g = \sum_\nu g_\nu \approx 250$ meV (see Table 8.1 (a)), $\Omega=3$ (for the t_{1u} orbital) and $D \approx 1$ eV (see Fig. 5.1), leading to $x \approx 0.8$. These results imply that the particle–vibration coupling associated with (8.21) and (8.30) are expected to be important. Some of the consequences of this couplings in the case of fullerenes and of atomic clusters are discussed in the remaining of the present chapter as well as in Chap. 9, while we refer to [45] concerning the nuclear case.

8.4 Signatures of the Electron–Phonon Interaction: Photoemission in C_{60}^-

Within the scheme of the LDA discussed in Chaps. 3 and 5, in the negatively charged C_{60}^- fullerene, the three-fold degenerate Kohn–Sham t_{1u} orbital (see Fig. 5.1) is singly occupied. Other orbitals are well separated from the t_{1u} orbital on the energy scale considered (room temperature $T \approx 25$ meV). This temperature is also lower than the energy of the lowest phonon mode of type H_g and A_g (see Table 8.2). Consequently, in the diagonalization of the Hamiltonian (8.15) one can assume that $T = 0$. The ground state wavefunction of (8.15) can be written as

$$|\Psi_{\text{init}}\rangle = \left(C_0 + \sum_{r=1}^{42} C_1(r)\Gamma_r^\dagger + \sum_{r=1}^{42}\sum_{s=1}^{42} C_2(r,s)\Gamma_r^\dagger \Gamma_s^\dagger + \ldots \right) a_{i\sigma}^\dagger |0\rangle, \quad (8.31)$$

where r (the same is true for s) stands for the labels ν and μ for $r \leq 40$ (H_g phonons) and for ν for $r = 41, 42$ (A_g phonons), while $|0\rangle$ is the product of the phonon and electron vacuum, which in this case corresponds to an antisymmetric wavefunction describing the state in which 240 electrons fully

---

[2] To be noted that $N(0) \approx \Omega/D$ is the density of levels at the Fermi energy for one spin orientation. Although one cannot produce femtometer materials made out of nuclei (aside from neutron stars, astronomical objects kept together by the gravitational force), one can create nanometer materials using e.g. C_{60} as building blocks. Because in such (van der Waals) materials the phonons remain essentially unchanged with respect to the isolated molecule, while the density of levels is that of the solid (see Fig. 9.1), the value of the parameter x is controlled by the strength of the isolated molecule and the density of states $N(0)$ of the solid.

occupy the orbitals of C_{60}^{-} up to and including the state h_u (HOMO state in C_{60}). The first term of (8.31) describes a state with no phonon, the second term a state with one phonon, and so on.

Typical calculations include states with up to 3–5 phonons [165, 166, 167]. Because of the very large dimensions of the associated matrices, and due to the fact that they contain many zeros, as a rule, the Lanczos method is used to calculate the lowest eigenvalue and eigenvector defined by the C coefficients of (8.31).

Information concerning the C coefficients can be obtained from the analysis of the photoemission reaction process

$$\text{C}_{60}^{-} + \hbar\omega \rightarrow \text{C}_{60} + e^{-}, \tag{8.32}$$

where a photon of energy $\hbar\omega$ impinging on the charged cluster excites one electron in the continuum, which is assumed not to interact with the system left behind (sudden approximation). For the final state of the neutral C_{60} molecule, there is no coupling between states with different number of phonons, since the electron has been emitted, and the eigenstates are therefore trivial. Neglecting furthermore the energy dependence of the dipole[3] matrix elements, the photoemission spectrum of C_{60}^{-} can be written in terms of the coefficients of (8.31) as

$$S(\omega) \sim C_0^2 \delta(\hbar\omega - E_\text{i} + E_\text{f}) + \sum_r C_1^2(r)\delta(\hbar\omega - E_\text{i} + E_\text{f} + E_r) + \ldots \tag{8.33}$$

In [167], the matrix elements of $H_\text{elec,phon}$ have been considered as free parameters and determined by means of a best fit to the experimental spectrum shown with the solid line in Fig. 8.3. In this fitting the phonon energies were taken from experiment (see Table 6.2). A Gaussian broadening of full width at half maximum (FWHM) of 41 meV was introduced to mimic the experimental resolution. Also shown in this figure, are the results of a number of ab initio and phenomenological calculations.

Instead of reporting directly the matrix elements g_ν, it is often costumary to report the (dimensionless) partial electron–phonon coupling constant (see Appendix 8.9.2)

$$\lambda_\nu = 2\frac{N(0)}{n_\text{deg}^2} \sum_\mu \frac{\sum_{ij} |\langle j, \nu\mu|(v_\text{def}^{(\nu)})_\mu|i\rangle|^2}{\hbar\omega_\nu}, \tag{8.34}$$

where $N(0)$ is the density of levels at the Fermi energy per spin orientation and per molecule, while n_deg is the degeneracy of the LUMO state $|i\rangle$ per spin orientation[4]. In the case of the C_{60}^{-} fullerite (see [164] and Sect. 8.3),

---

[3] The dipole approximation used to describe the interaction between the incident photon and the electron is justified by the fact that the wavelength associated with a photon of the energy of the order of the ionization energy (few eV) is much larger than the dimension of the cluster.

[4] To be noted that $|\langle i|(v_\text{def}^{(\nu)})_\mu|i\rangle|^2 \sim M^{-1}$ and $\hbar\omega_\nu \sim M^{-1/2}$. Thus λ_ν depends on the ion mass as $M^{-1/2}$. This fact will be of interest in Chap. 9.

$$\lambda_\nu = N(0)\frac{2}{9}\sum_\mu \frac{\sum_{ij}|\langle t_{1u},j,\nu\mu|(v_{\rm def}^{(\nu)})_\mu|t_{1u},i\rangle|^2}{\hbar\omega_\nu}$$

$$= N(0)\frac{2}{9}\frac{g_\nu^2}{4\hbar\omega_\nu}\times\begin{cases}\sum_\mu (V^{(\mu)})^2,\\ W^2.\end{cases} \tag{8.35}$$

Making use of (8.19) and (8.20) one obtains

$$\sum_\mu (V^{(\mu)})^2 = \begin{pmatrix} 10 & 0 & 0 \\ 0 & 10 & 0 \\ 0 & 0 & 10 \end{pmatrix},$$

and

$$W^2 = \begin{pmatrix} 2 & 0 & 0 \\ 0 & 2 & 0 \\ 0 & 0 & 2 \end{pmatrix}, \tag{8.36}$$

respectively. From this result and (8.35) one can write the total electron–phonon coupling constant, summed over the phonons, leading to

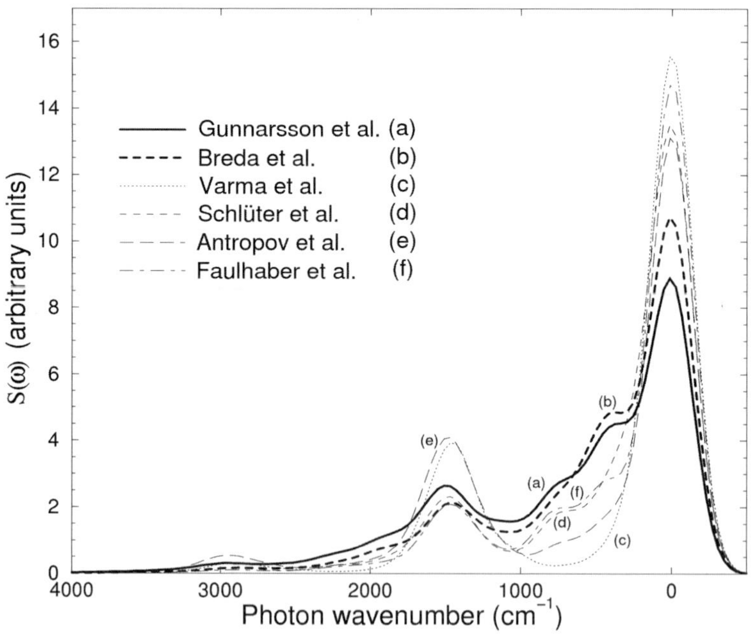

Fig. 8.3. Results for the photoemission spectrum $S(\omega)$ of C_{60}^-, obtained making use of the electron–phonon matrix elements calculated by means of different theoretical approaches. (a) corresponds to the fitting of the experimental results [167], (b) [165], (c) [168], (d) [169], (e) [170], and (f) [171]

$$\lambda = \sum_{\nu} \lambda_{\nu}, \tag{8.37}$$

where

$$\frac{\lambda_{\nu}}{N(0)} = \alpha \frac{g_{\nu}^2}{\hbar\omega_{\nu}}. \tag{8.38}$$

The quantity α is equal respectively to $\frac{5}{3}$ for the H_g modes and $\frac{1}{3}$ for the A_g modes.

In Table 8.2 we display the partial electron–phonon coupling constants $\lambda_{\nu}/N(0)$, as extracted from the phenomenological analysis of the photoemission spectra. Because of the experimental resolution, it is not possible to distinguish between the coupling of the A_g modes and to the H_g modes with similar energies. The couplings to the A_g modes were taken[5] from [170].

In Table 8.2 are also shown results of typical calculations of $\lambda_{\nu}/N(0)$ based on the semi-empirical MNDO method [168], on the empirical BCM for the phonons and the LDA for the electronic part [169], as well as ab initio LDA calculations both to generate the phonons and to calculate their interaction with the electrons [170, 171]. Furthermore, the results obtained making use of an ab initio calculation based on the Car–Parrinello methods (frozen-phonon approximation, see Sect. 7.3.1 [121]) and by means of an empirical model (HOM) [165, 166] (see also Table 8.1) are also displayed in Table 8.2.

These results testify to the fact that there are substantial differences between the different calculations. Let us start by discussing the results associated with the coupling to the eight H_g modes. In [168] it was found that most of the electron–phonon coupling strength is associated with the two highest modes. In [169, 170, 171] more coupling strength was found in the lower energy phonons, and this is even more so in the case of the results of [121]. In any case, all these calculations find the strongest coupling to one or two of the highest modes. This is quite different from the results obtained within the framework of the HOM [165, 166] as can be seen also from Fig. 8.4. In fact, in this case the largest coupling strength is associated with the second lowest phonon, although consistent strength is also found associated with the next to highest mode.

The large variation in the distribution of coupling between the different phonons observed from the results shown in Table 8.2 (and Fig. 8.4) is related to the sensitivity of the results to the phonon eigenvectors[6]. In

---

[5] To be noted that the coupling to $A_g(2)$ can, however, be varied between $\lambda_{A_g(2)}/N(0)=0$ and 3 meV without essentially worsening the fit, if the total coupling to $H_g(7)$ and $H_g(8)$ is changed correspondingly (between 7 meV and 0) [173].

[6] This situation finds similarities and differences with the case of other finite many-body systems like e.g. the atomic nucleus. In fact, in this case the system is self-bound, and the phonons are associated with vibrations of the fermion density, in a similar way as plasmons in atomic clusters (see Fig. 1.2 and Chap. 5). In the nuclear case low-energy quadrupole and octupole surface vibrational modes give

Table 8.2. Partial electron–phonon coupling constants $\lambda_\nu/N(0)$ in meV which measures the strength with which an electron moving in the t_{1u} LUMO state of C_{60}^- couples to the eight H_g and the two A_g phonons of C_{60} (see (8.16), also (1) of [172]). In column two and three the experimental energy of the phonons is given (see Table 6.2). In column four the values obtained from a phenomenological analysis of the photoemission electron spectra is given while in columns five-ten, the results of ab initio as well as of empirical models is displayed. The last two lines report the total values $\sum H_g = \sum_{i=1}^{8} H_g(i)$ and $\sum(H_g + A_g)$ with $\sum A_g = \sum_{i=1}^{2} A_g(i)$

	Frequency		$\lambda_\nu/N(0)$						
	Exp.		Exp.	Theory					
	λ^{-1} [b)	$\hbar\omega_\nu$ [c)	[d)	[a),e)	[a),f),g)	[h)	[i)	[j)	[k)
	$[\mathrm{cm}^{-1}]$	$[\mathrm{meV}]$							
$H_g(8)$	1575	195.3	23	14	8	22	11	9	9
$H_g(7)$	1428	177.1	17	17	16	20	34	13	15
$H_g(6)$	1250	155.0	5	2.5	3	8	0	3	2
$H_g(5)$	1099	136.3	12	4.8	6	3	6	1	2
$H_g(4)$	774	96.0	18	5	5	3	0	7	10
$H_g(3)$	710	88.0	13	9.5	11	3	1	4	1
$H_g(2)$	437	54.2	40	11	26	6	1	7	10
$H_g(1)$	273	33.9	19	4.3	10	3	3	8	1
$A_g(1)$	1470	182.3	11	7.5	6	11		5	
$A_g(2)$	496	61.5	0	0	1	0		0	
$\sum H_g$			147	68.1	85	68	56	52	49
$\sum(H_g + A_g)$			158	75.6	92	68			

[a) Obtained from Table 8.1 making use of the relation $\lambda_\nu/N(0) = (5/3)g_\nu^2/\hbar\omega_\nu$ for H_g and $\lambda_\nu/N(0) = (1/3)g_\nu^2/\hbar\omega_\nu$ for A_g (see (8.38)), and the experimental phonon energies displayed in column 3

[b) see Table 6.2, column labelled by Exp. b

[c) $\lambda = \frac{1240}{\hbar\omega}$ nm·eV $= \frac{0.124}{\hbar\omega}$ cm· meV, $\hbar\omega = (\frac{1}{\lambda})_{\mathrm{cm}}{}_{1} \times 0.124$ meV

[d) Extracted from photoemission spectra (PES) [167]

[e) see Colò et al. [121] and the discussion in Sect. 7.3.1

[f) Hybrid Orbital Model (HOM) [165, 166]

[g) The matrix elements given e.g. in [165, 166] obey:
Matrix element [meV]$\times c_\nu = g_\nu$, where $c_\nu=2$ for H_g and $\sqrt{2}$ for A_g

[h) Antropov et al. [170]

[i) Varma et al. [168]

[j) Schlüter et al. [169]

[k) Faulhaber et al. [171]

rise, as a rule, to the strongest couplings. Although different model calculations lead to quite similar phonon eigenvectors (transition densities, see Sect. 8.6), they can give rise to conspicous differences in the particle-vibration coupling strengths. This is because small variations in $\hbar\omega_\nu$ (E_n) and in $\varepsilon_p - \varepsilon_h$ can lead to significant variations in the amplitudes $X_{\mathrm{ph}}^{(n)}$ and $Y_{\mathrm{ph}}^{(n)}$ (see (4.40)) and in the zero-point fluctuations associated with the surface modes.

fact, calculations giving very similar results concerning the phonon energies, can give rise to quite different results concerning the phonon eigenvectors. Small admixtures of one normal mode into another can lead to important changes of the phonon energies and the associated zero-point amplitudes, and to considerable changes in the distribution of the values of $\lambda_\nu/N(0)$. These variations indicate that important uncertainities are to be ascribed to the results of the different calculations. On the other hand, the total value $\lambda/N(0) = \sum_\nu \lambda_\nu/N(0)$ displays less variation from calculation to calculation, than the individual $\lambda_\nu/N(0)$ values. This fact will be shown to be of relevance in the discussion of pairing in atomic aggregates carried out in the next chapter.

Concerning the coupling of the t_{1u} state to the two A_g modes, although one finds quantitative differences between the results of the different calculations, qualitatively all of them are similar.

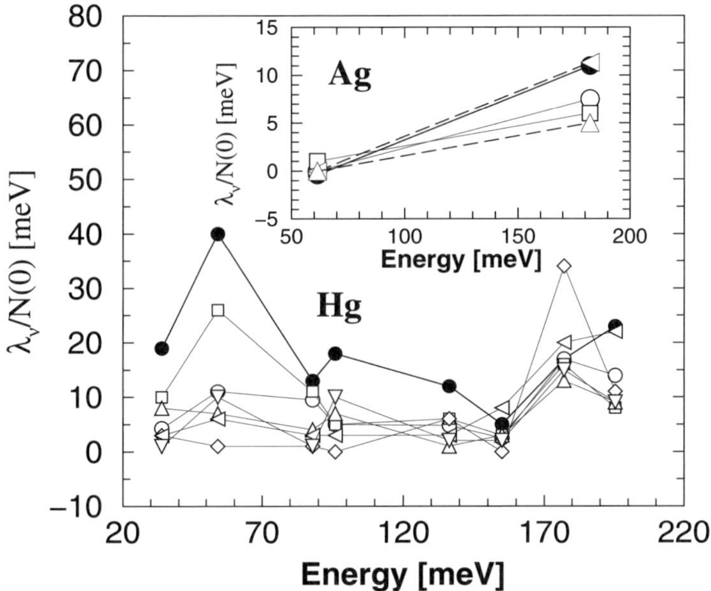

Fig. 8.4. Partial electron–phonon coupling constants $\lambda_\nu/N(0)$ vs. energy $\hbar\omega_\nu$ [meV] in Hg as extracted from a phenomenological analysis of the photoemission spectrum (PES) (*full circles*) [167], and as calculated in a variety of models: (*open circles*) [121], (*open squares*) [165, 166], (*open diamond*) [168], (*open up triangle*) [169], (*open left triangle*) [170] and (*open down triangle*) [171]. The inset shows, for Ag, the PES results (*full circles*) [167], (*open circles*) [121], (*open squares*) [165, 166], (*open left triangle – dashed line*) [170] and (*open up triangle – dashed line*) [169]

8.4.1 Electron–Phonon Coupling in Other Fullerenes

While C_{60} is the paradigm of the family of fullerenes, both smaller and larger C_n close cage clusters made out of carbon atoms have been produced (see Figs. 1.3 and 5.4).

Among these clusters C_{20} and C_{70} occupy a particular place. C_{20} because it is the smallest of fullerenes, while C_{70} can be viewed as two halves of a C_{60} fullerene joined by a belt of five benzoid rings. This belt adds ten carbon atoms to the molecule and provides the mechanism to produce carbon nanotubes. In what follows we shall discuss the case of C_{70} and leave to Chap. 9 the study of the electron–phonon coupling in the smallest fullerenes.

Nuclear magnetic resonance experiments [174] have shown the presence of five different atomic sites, indicating a $\mathcal{D}_{5h}$ symmetry. A number of calculations of the electronic structure of the rugby-ball-like C_{70} (see Figs. 1.3 and 5.4) have been reported in the literature [175, 176, 177, 178] which indicate this configuration leads to a binding energy per atom very similar to that of C_{60}. In Fig. 8.5 are shown the results of a LDA calculation [179] carried out on a spherical basis $|nlm\rangle$, characterized by the radial, orbital and magnetic quantum numbers. The calculated HOMO–LUMO gap is 1.86 eV, to be compared with the experimental value of 1.6±0.2 eV [180].

The effect of the deformation is evidenced by the structure of the single-particle wavefunctions of the states near the Fermi energy. In particular those

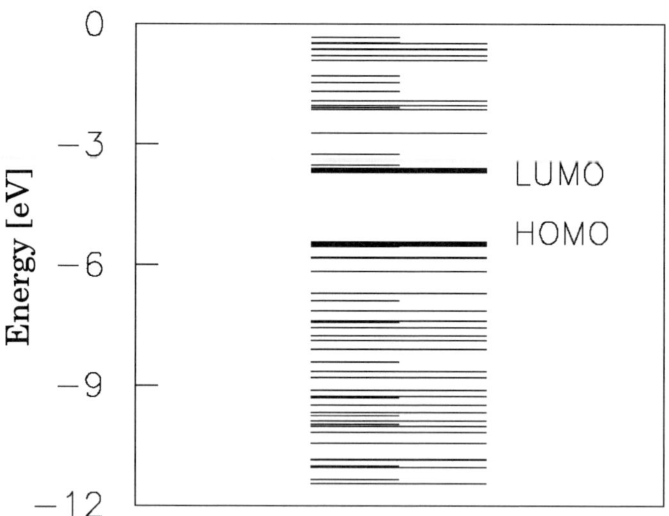

Fig. 8.5. The electronic energy levels around the HOMO–LUMO for the fullerene C_{70}. The energy levels are singly and doubly degenerate. Considering the spin, they admit two and four electrons, respectively

describing the $E1$ HOMO and LUMO levels, which in the $|nlm\rangle$ basis read[7]

$$|\text{HOMO}\rangle \approx 0.7(|154\rangle - |15\bar{4}\rangle) + 0.4(|161\rangle + |16\bar{1}\rangle)$$
$$+ 0.2(|074\rangle - |07\bar{4}\rangle) - 0.2(|094\rangle - |09\bar{4}\rangle), \tag{8.39}$$

and

$$|\text{LUMO}\rangle \approx 0.4(|154\rangle - |15\bar{4}\rangle) - 0.7(|161\rangle + |16\bar{1}\rangle)$$
$$+ 0.2(|176\rangle - |17\bar{6}\rangle) - 0.3(|081\rangle + |08\bar{1}\rangle). \tag{8.40}$$

While both states are to a large extent π-states, they are strongly mixed in the angular momentum quantum number. The importance of this mixing becomes more evident by comparing them with the corresponding HOMO, LUMO wavefunctions of the (almost spherical) C$_{60}$ fullerene (see (5.2) and (5.3)).

Making use of the phonons calculated in the BCM (see Sect. 6.2.2) and of the HOM, the reduced matrix elements g_ν and the associated partial electron–phonon coupling constants (see also (8.38))

$$\frac{\lambda_\nu}{N(0)} = C\frac{g_\nu^2}{\hbar\omega_\nu}, \tag{8.41}$$

have been calculated in [181]. In the above expression C is $d_\nu/2$, d_ν being the degeneracy of the phonon state ν. The results corresponding to the LUMO (e_1'') and to the A_1' and E_2' phonons, are collected in Table 8.3, where only those couplings leading to values of $\lambda_\nu/N(0) \geq 0.005$ meV are displayed.

Comparing the results displayed in column $^{f)}$ (HOM) of Table 8.2 and of Table 8.3 it is seen that the total electron–phonon coupling constant $\lambda/N(0)$ which is equal to 92 meV for C$_{60}$ reduces to ≈ 11 meV in the case of C$_{70}$. In other words, deformation measured, e.g., by the ratio of the shortest to the longest radius, see Fig. 5.4, and which changes from 1:1 (C$_{60}$) to 1:1.2 (C$_{70}$), i.e., from β_2=0 to β_2=0.2 (see footnote 6 of Chap. 2) plays an important role concerning the fermion–boson coupling[8], by breaking the symmetries (degeneracies) of electronic and vibrational levels, and by weakening the coupling. This effect can be understood by analyzing, for both clusters, the spatial localization of the electronic LUMO state and its overlap with the eigenvectors of the (low-frequency) phonons which couple to it [181].

---

[7] The bar on top of the number indicates the negative sign.

[8] Similar effects have been found in the case of spherical and quadrupole deformed nuclei (i.e., in systems for which β_2 changes from ≈ 0 to ≈ 0.2), where conspicous reductions of the mass parameter λ (see (8.85) and following equations) have been calculated (see e.g. [7] and references therein). In fact, while λ is found to be ≈ 0.6 for a spherical nucleus like ^{120}Sn, it becomes ≈ 0.3 in a typical deformed nucleus like ^{152}Dy [30, 182, 183, 184, 185, 186]. This reduction can be connected with the fact that a large fraction of the collectivity of the system becomes tied up to the static deformation of the system, aside from the fact that deformation implies breaking of degeneracies of not only the fermion but also of the boson states.

Table 8.3. Reduced electron–phonon matrix element g_ν and partial coupling constants $\lambda_\nu/N(0)$ for the LUMO electron state and selected phonons of C_{70} calculated in the HOM [181]. The first column give the experimental phonon frequencies while the subsequent columns report the calculated frequencies and symmetries obtained within a bond–charge model calculation (see [27] in [181])

Exp. [187]	Theory			
frequency	frequency	symm.	g_ν	$\lambda_\nu/N(0)$
$[\text{cm}^{-1}]$	$[\text{cm}^{-1}]$		[meV]	[meV]
226	230	$E_2^{'}$	4.77	0.798
256	271	$A_1^{'}$	21.9	7.136
—	308	$E_2^{'}$	6.03	0.951
394	425	$A_1^{'}$	3.59	0.122
408	442	$E_2^{'}$	3.76	0.258
454	448	$A_1^{'}$	4.95	0.220
—	508	$E_2^{'}$	3.26	0.169
—	590	$E_2^{'}$	3.09	0.130
567	625	$A_1^{'}$	1.95	0.025
702	701	$A_1^{'}$	1.71	0.017
712	736	$E_2^{'}$	6.98	0.533
737	764	$E_2^{'}$	2.48	0.065
—	1089	$E_2^{'}$	2.66	0.052
1181	1200	$A_1^{'}$	1.73	0.010
1228	1223	$A_1^{'}$	1.46	0.007
1444	1368	$A_1^{'}$	6.45	0.123
1459	1459	$A_1^{'}$	6.79	0.127
	Total		83.56	10.743

In the low-frequency region around 400-500 cm^{-1} (as illustrated in [159]), it is possible to recognize a correspondence between the $H_g(2)$ mode of C_{60}, to which one of the largest large electron–phonon coupling constant $\lambda_\nu/N(0)$ is associated, and a group of five (slightly splitted) vibrational modes of C_{70}. These five modes are the $A_1^{'}$, $E_1^{''}$ and $E_2^{'}$ modes of C_{70} at 425, 415 and 442 cm^{-1}, respectively. The correspondence can be established by classifying the modes of both clusters according to the largest common symmetry subgroup, C_{5v} (see also Sect. 6.2.2). The $E_1^{''}$ mode does not couple to the LUMO of C_{70}, hence about 2/5 of the coupling strength is lost. The remaining modes are associated with a higher $\hbar\omega$ with respect to the case of C_{60}, further reducing the coupling. Finally, a large reduction effect comes from the different spatial localization of the LUMO wavefunctions in C_{60} and C_{70}. In the case of C_{60} the LUMO wavefunction is concentrated in the neighborhood of the bonds which undergo the largest stretching (the darkest regions in Fig. 8.6). On the other hand, in C_{70}, the LUMO wave function is distributed over a larger region (see Fig. 8.7).

8.4.2 Electron–Phonon Coupling in Sodium Clusters

Small sodium clusters have been discussed in Sects. 5.4 and 7.3.1. In what follows we consider the electron–phonon coupling in Na_8 and Na_9^+. The electronic states and phonons have been worked out in the Car–Parrinello scheme (see Chap. 7), as well as the deformation potential which results from a frozen phonon type of calculation [108].

In the case of Na_8 the LUMO state has A_1 symmetry (see Sect. 7.3.1). The diagonal electron-phonon matrix elements $\langle A_1\ i, \nu | (v_{\mathrm{def}}^{(\nu)}) | A_1\ i \rangle$ are different from zero only for $\nu \equiv A_1$ (see Table 7.2). The associated reduced electron–phonon matrix elements g_ν are displayed in Table 8.4. Also shown are the partial coupling constants $\lambda_\nu/N(0)$. Comparing these results with those of Table 8.2, where the corresponding results for C_{60}^- are collected, it is seen that the partial coupling constants of Na_8 are, in average, larger than those associated with C_{60}^-.

Actually, the main reason for this difference lies in the fact that the lowest Na_8 phonons have energies which are almost one order of magnitude smaller than those of C_{60} (see Tables 8.4 and 8.1). Consequently, the square of the zero-point fluctuations $\sqrt{\hbar/2M_\lambda\omega_\lambda}$ entering the expression of $\lambda_\nu/N(0)$ (see Sect. 8.1) are larger also by one order of magnitude. In keeping with this result, the matrix elements of the gradient of the electronic potential, must

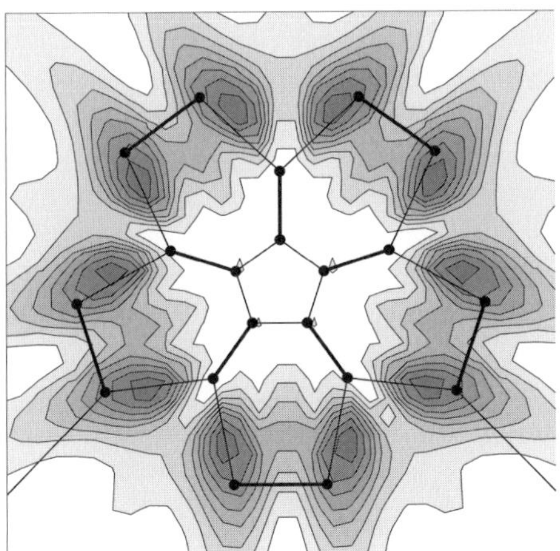

Fig. 8.6. Gnomonic projection on a plane of the values taken by the squared modulus of the wavefunction of the LUMO state in C_{60} on the upper half of the cluster surface (i.e., on a spherical hemisphere with radius 3.4 Å). The thick lines mark the bonds that vary more than 6% under the effect of the displacement associated with the $H_g(2)$ phonon. (After [181])

Table 8.4. Reduced electron–phonon matrix elements g_ν and partial coupling constants $\lambda_\nu/N(0)$ associated with the LUMO electron state and selected phonon states of Na_8

	$\hbar\omega$ [meV]	$\lambda_\nu/N(0)$ [meV]	g_ν [meV]
$A_1(1)$	7.93	36.3	12.0
$A_1(2)$	9.48	173.8	28.7
$A_1(3)$	15.94	56.9	21.3
$A_1(4)$	16.72	12.2	10.1
Total		279.2	72.1

be comparable or even smaller in Na_8 than in C_{60}^-. In this latter cluster, the delocalization of the 3p electrons is larger than that of the C_{60} electrons which tend to concentrate around the bonds (as seen in Fig. 8.6). The more the electrons are delocalized, and therefore far from the ion–ion bonds, the smaller are the matrix elements, as we have discussed in the previous section.

Fig. 8.7. Gnomonic projection on a plane of the values taken by the squared modulus of the wavefunction of the LUMO state in C_{70} on the upper part of an ellipsoidal surface with radii 3.55 Å and 4.20 Å. The thick lines mark the bonds that vary more than 6% under the effect of the displacement associated with the A_1' phonon. (After [181])

8.5 The Effective Electron–Electron Interaction

An electron can interact with the ions through the deformation potential and excite a phonon. This phonon can be absorbed by the same electron (self-energy process, see Chap. 2) or by a second electron (see Sect. 2.7). This second process gives rise to an effective electron–electron interaction which we shall denote $v_{\text{el}-\text{el}}^{(\text{ind})}$ and calculate below making use of the results discussed above.

We write the total Hamiltonian (see (8.15)-(8.18)) as

$$H = H_0 + \Lambda H_1 \tag{8.42}$$

where Λ is a perturbative parameter, and

$$H_0 = \sum_{i\sigma} \varepsilon_i \, a_{i\sigma}^\dagger a_{i\sigma} + \sum_{\nu,\mu} \hbar\omega_\nu \left(\Gamma_{\nu\mu}^\dagger \Gamma_{\nu\mu} + \frac{1}{2} \right) \tag{8.43}$$

while (see (8.16))

$$H_1 = \sum_{\nu\mu} \sum_{ij\sigma} \langle j, \nu\mu | (v_{\text{def}}^{(\nu)})_\mu | i \rangle (\Gamma_{\nu\mu}^\dagger + \Gamma_{\nu\mu}) \, a_{i\sigma}^\dagger a_{j\sigma}$$

$$= \sum_\nu \frac{g_\nu}{2} \sum_\mu (\Gamma_{\nu\mu}^\dagger + \Gamma_{\nu\mu}) \sum_{ij\sigma} C_{ij}^{(\mu)} \, a_{i\sigma}^\dagger a_{j\sigma}. \tag{8.44}$$

In order to derive an induced electron–electron interaction from (8.42) we start by defining an operator S given by

$$S = -\Lambda \sum_\nu \frac{g_\nu}{2} \sum_\mu \sum_{ij\sigma} a_{i\sigma}^\dagger a_{j\sigma} \left(\frac{C_{ij}^{(\mu)}}{\varepsilon_{ij} - \hbar\omega_\nu} \Gamma_{\nu\mu} + \frac{C_{ij}^{(\mu)}}{\varepsilon_{ij} + \hbar\omega_\nu} \Gamma_{\nu\mu}^\dagger \right), \tag{8.45}$$

where ε_{ij} is a shorthand notation for $\varepsilon_i - \varepsilon_j$ and where

$$\langle i\sigma | S | j\sigma, \nu\mu \rangle = \Lambda \frac{(g_\nu/2) C_{ij}^{(\mu)}}{\varepsilon_{ij} + \hbar\omega_\nu},$$

$$\langle i\sigma, \nu\mu | S | j\sigma \rangle = \Lambda \frac{(g_\nu/2) C_{ij}^{(\mu)}}{\varepsilon_{ij} - \hbar\omega_\nu}, \tag{8.46}$$

are the corresponding matrix elements. It can be shown that

$$[H_0, S] = -\Lambda H_1. \tag{8.47}$$

The calculation of this commutator is quite lenghty but straightforward (see Appendix 2.11.1).

We now use S to perform the canonical transformation

$$\tilde{H} = e^{-S} H e^{S} =$$

$$= H_0 + \Lambda H_1 + [H_0, S] + \Lambda [H_1, S] + \frac{1}{2} [[H_0 + \Lambda H_1, S], S] + \dots \tag{8.48}$$

Within the weak-coupling approximation, one can neglect terms beyond those proportional to Λ^2. Thus,

$$\tilde{H} = \Lambda[H_1, S] + \frac{1}{2}[[H_0, S], S],$$

where (8.47) has been used. Making once more use of (8.47), one obtains

$$\tilde{H} = -\frac{1}{2}\Lambda[H_1, S]. \tag{8.49}$$

Selecting from $\tilde{H}$ those terms which are proportional to $a^\dagger a^\dagger a a$ (see Appendix 8.9.3) leads to

$$\sum_\nu \frac{g_\nu^2}{4} \sum_\mu \sum_{ij\sigma,i'j'\sigma'} C_{i'j'}^{(\mu)} C_{ij}^{(\mu)} \frac{\hbar\omega_\nu}{(\varepsilon_i - \varepsilon_j)^2 - (\hbar\omega_\nu)^2} a_{i\sigma}^\dagger a_{i'\sigma'}^\dagger a_{j'\sigma'} a_{j\sigma}. \tag{8.50}$$

This is the effective electron–electron interaction induced by the exchange of phonons. From inspection of the diagonal terms

$$v_{\text{el-el}}^{(\text{ind})} = -\sum_\nu \frac{g_\nu^2}{4\hbar\omega_\nu} \sum_\mu \sum_{ij\sigma} (C_{ij}^{(\mu)})^2 a_{i\sigma}^\dagger a_{i\tilde{\sigma}}^\dagger a_{j\tilde{\sigma}} a_{j\sigma},$$

$$= -\sum_{\nu\mu} \sum_{ij\sigma} \frac{|\langle t_{1u}, i; \nu\mu|(v_{\text{def}}^{(\nu)})_\mu|t_{1u}, j\rangle|^2}{\hbar\omega_\nu} a_{i\sigma}^\dagger a_{i\tilde{\sigma}}^\dagger a_{j\tilde{\sigma}} a_{j\sigma}, \tag{8.51}$$

where use was made of (8.16) in writing the second expression. It is seen that the matrix elements can be attractive. In particular, in the case of C_{60}^- (see Sect. 8.3)

$$v_{\text{el-el}}^{(\text{ind})} = \sum_{i,j=1}^{3} \left(-\sum_{\nu=1}^{8} \frac{g_\nu^2}{2\hbar\omega_\nu} \sum_{\mu=1}^{5} (V_{ij}^{(\mu)})^2 - \sum_{\nu=9}^{10} \frac{g_\nu^2}{2\hbar\omega_\nu} (W_{ij})^2 \right) a_{i\sigma}^\dagger a_{i\tilde{\sigma}}^\dagger a_{j\tilde{\sigma}} a_{j\sigma}$$

$$= \left(-\sum_{\nu=1}^{8} \frac{\lambda_\nu}{N(0)} - \sum_{\nu=9}^{10} \frac{\lambda_\nu}{N(0)} \right) \sum_{ij} a_{i\sigma}^\dagger a_{i\tilde{\sigma}}^\dagger a_{j\tilde{\sigma}} a_{j\sigma}, \tag{8.52}$$

where (8.41) has been used in writing the last relation.

8.6 Coupling of Electrons to Plasmons

In this section we discuss the dynamical screening[9] arising from the coupling of electrons with plasmons. The electron–plasmon coupling which gives rise to the electron self-energy has been written in (2.39) using second-order perturbation theory,

---

[9] That is, the effect in which one electron penetrating into a cloud of particles possessing a charge of the same sign (the rest of the electrons), pushes away the remaining electrons and in the process gets dressed, acquiring among other things an effective charge.

$$\Sigma_j(\omega) = \sum_{j',n} \frac{|\langle n, j'|\delta U|j\rangle|^2}{\hbar\omega - (\varepsilon_{j'} + \hbar\omega_n)}$$

(see also Fig. 2.5 and [188, 189, 190]). The coupling potential δU is the mean field variation associated to a vibration of the system. In Sect. 2.4, the system under study has been considered to be spherical and its surface vibrations were written in terms of the multipole expansion parameters $\alpha_{\lambda\mu}$ (see (2.33)). Consequently (see (2.35)),

$$\delta U(\boldsymbol{r}) = U(\boldsymbol{r} - \boldsymbol{R}) - U(\boldsymbol{r} - \boldsymbol{R}_0) = -R_0 \frac{\partial U}{\partial r} \sum_{\lambda\mu} \alpha_{\lambda\mu} Y_{\lambda\mu}^*.$$

In the more general case we can use, for vibrations whose frequency is equal to ω, the expressions (4.5) and (4.6) for δU. These expressions correspond to the case in which the mean field U has been identified with the LDA effective potential v_{eff}. This is not associated to a single parameter $\boldsymbol{R}$ as it is above, but depends on the density[10] so that (see also (4.4))

$$\delta U(\boldsymbol{r}) = v_{\text{eff}}(\boldsymbol{r}; [\varrho]) - v_{\text{eff}}(\boldsymbol{r}; [\varrho_0]) = \int \mathrm{d}^3 r' \frac{\delta v_{\text{eff}}(\boldsymbol{r})}{\delta\varrho(\boldsymbol{r}')} \, \delta\varrho(\boldsymbol{r}', \omega).$$

We are now confronted with the evaluation of the matrix element $\langle n, j'|\delta U|j\rangle$. This can be done using the transition density of a given plasmon n, and again this can be understood in analogy with the discussion of Sect. 2.4. In that case $\alpha_{\lambda\mu}$ has been taken as a dynamical variable describing the collective vibration, its contribution to the matrix element $\langle n, j'|\delta U|j\rangle$ being $\langle n|\alpha|0\rangle$. Here, the corresponding role is played by the density variation operator $\delta\varrho$. Its matrix element between the ground state and the one boson (plasmon) state coincides with that of ϱ, that is (see (4.19)),

$$\langle n|\varrho(\boldsymbol{r}')|0\rangle \equiv \delta\varrho_n(\boldsymbol{r}'; \omega_n).$$

Then, using this property, one obtains

$$\langle n, j'|\delta U|j\rangle = \int \mathrm{d}^3 r \, \mathrm{d}^3 r' \, \varphi_{j'}^*(\boldsymbol{r}) \, \frac{\delta v_{\text{eff}}(\boldsymbol{r})}{\delta\varrho(\boldsymbol{r}')} \, \delta\varrho_n(\boldsymbol{r}'; \omega_n) \, \varphi_j(\boldsymbol{r}). \tag{8.53}$$

We remember also that if one works in the configuration space (i.e., with an explicit basis of particle–hole excitations), the plasmon state can be expanded according to (4.40),

$$|n\rangle \equiv \sum_{ph} \left[X_{\text{ph}}^{(n)} a_p^\dagger \, a_h - Y_{\text{ph}}^{(n)} a_h^\dagger \, a_p \right] |0\rangle.$$

In this case, we can express the transition density as in (4.42),

$$\delta\varrho_n(\boldsymbol{r}; \omega_n) = \sum_{ph} (X_{\text{ph}}^{(n)} - Y_{\text{ph}}^{(n)}) \, \varphi_p(\boldsymbol{r})\varphi_h^*(\boldsymbol{r}).$$

---

[10] At variance with the case of the phonons, here the ions do not vibrate.

Consequently, the matrix element (8.53) can be written as (see also (4.38)),

$$\langle j', n|\delta U|j\rangle = \sum_{ph}(X_{\mathrm{ph}}^{(n)} - Y_{\mathrm{ph}}^{(n)})$$

$$\int \mathrm{d}^3r\,\mathrm{d}^3r'\,\varphi_{j'}^*(\boldsymbol{r})\,\varphi_h^*(\boldsymbol{r}')\,\frac{\delta v_{\mathrm{eff}}(\boldsymbol{r})}{\delta\varrho(\boldsymbol{r}')}\,\varphi_p(\boldsymbol{r}')\,\varphi_j(\boldsymbol{r}). \tag{8.54}$$

It is seen that the electron–plasmon coupling corresponds to the interaction of an electron with a coherent superposition of the p–h electron excitations.

The electron–plasmon interaction with matrix elements $\langle i, n|\delta U|j\rangle$ is called the *transition potential*. The main contribution to the matrix element (8.54) comes from the Coulomb interaction, the (smaller) contribution arising from the second term of (4.16) being, as a rule, neglected.

If we wish to keep the parallel with the previous section and write the Hamiltonian which includes electrons, plasmons, and the electron–plasmon coupling in second quantization, we can express it as

$$H = \sum_i \varepsilon_i\, a_{i\sigma}^\dagger a_{i\sigma} + \sum_n \hbar\omega_n \left(\Gamma_n^\dagger\Gamma_n + \frac{1}{2}\right)$$

$$+ \sum_{ij\sigma,n} \langle i\sigma, n|\delta U|j\sigma\rangle\, a_{i\sigma}^\dagger a_{j\sigma} \left(\Gamma_n^\dagger + \Gamma_n\right), \tag{8.55}$$

where we stress the formal analogy with the electron–phonon Hamiltonian (8.44), and where the spin degrees of freedom are explicitly written. In fact, a simultaneous diagonalization of both the electron–phonon and the electron–plasmon interactions is called for, to account for the most important correlations of systems made out of electrons and ions. This amounts to solve the total Hamiltonian (see (8.42), (8.44) and (8.55),

$$H = H_{\mathrm{elec}} + H_{\mathrm{phon}} + H_{\mathrm{el,phon}} + H_{\mathrm{plas}} + H_{\mathrm{el,plas}}, \tag{8.56}$$

where

$$H_{\mathrm{elec}} = \sum_{i\sigma} \varepsilon_i a_{i\sigma}^\dagger a_{i\sigma}, \tag{8.57}$$

$$H_{\mathrm{phon}} = \sum_{\nu\mu} \hbar\omega_\nu \left(\Gamma_{\nu\mu}^\dagger\Gamma_{\nu\mu} + \frac{1}{2}\right), \tag{8.58}$$

$$H_{\mathrm{el,phon}} = \sum_{\nu\mu} \frac{g_\nu}{2} C_{ij}^{(\mu)} \sum_{ij\sigma} a_{i\sigma}^\dagger a_{j\sigma} \left(\Gamma_{\nu\mu}^\dagger + \Gamma_{\nu\mu}\right), \tag{8.59}$$

$$H_{\mathrm{plas}} = \sum_n \hbar\omega_n \left(\Gamma_n^\dagger\Gamma_n + \frac{1}{2}\right), \tag{8.60}$$

$$H_{\mathrm{el,plas}} = \sum_{ij\sigma,n} \langle n, i\sigma|\delta U|j\sigma\rangle\, a_{i\sigma}^\dagger a_{j\sigma} \left(\Gamma_n^\dagger + \Gamma_n\right). \tag{8.61}$$

Diagonalizing the total Hamiltonian in a basis of electrons (see Chap. 3), plasmons (see Chaps. 4 and 5) and phonons (see Chaps. 6 and 7), one obtains dressed (quasiparticle) states

$$|\tilde{i}\sigma\rangle = \alpha_i|i\sigma\rangle + \sum_{j'\sigma',n} \gamma^i_{j'\sigma',n}|j'\sigma',n\rangle + \sum_{i'\sigma',\nu\mu} \beta^i_{i'\sigma',\nu\mu}|i'\sigma',\nu\mu\rangle, \qquad (8.62)$$

linear combinations of bare electron states $|i\sigma\rangle$, of electrons $j'\sigma'$ coupled to plasmons n and of electrons $i'\sigma'$ coupled to phonons $\nu\mu$.

Density Functional Theory (DFT) discussed in previous chapters includes, in principle, all screening phenomena. In practice, whether the same is true in the case of the actual implementations of DFT which include a variety of appoximations (like the widely used LDA), is still an open question. In particular, the ω-dependence of the correlations induced in the system by the electron–plasmon coupling is approximated by the (static) correlation term of $v_{\mathrm{xc}}(r)$ (see e.g. (3.26), (3.30), also (2.65)).

To deal with (8.61) without introducing double counting of the interactions between fermions (electrons) and bosons (plasmons and phonons), one has to either use Hartree–Fock theory to describe electrons or to neglect v_{c} in the Kohn–Sham equations (the first implementation being arguably to prefer). In fact, in the case of small clusters, it is possible to perform Hartree–Fock calculations to build the term H_{elec} of (8.57), and include correlations explicitly by means of the terms H_{plas} and $H_{\mathrm{el,plas}}$.

We conclude this section by mentioning two calculations along this line. In [190], it has been shown, using the example of Na_{40}, that the coupling with surface vibration of the cluster, not only shift the Hartree-Fock states so that they become similar to the LDA ones (in agreement with the general discussion carried out in Chap. 2), but also produces partial occupancies of the states around the Fermi energy, the occupation probabilities Z deviating from unity by 30 to 40% (see Chap. 2, in particular Fig. 2.2 and (2.52), (2.55)). In [191], the explicit construction of the LDA correlations by means of the electron–plasmon coupling is shown to work well even for a smaller system (Na_8) by looking at the ground state energy and other global observables.

8.7 TDLDA with Electron–Plasmon and Electron–Phonon Couplings

Because normal vibrations of the delocalized electrons of a finite atomic aggregate (plasmons) can be viewed as a correlated particle–hole state, that is, a coherent linear combination of unperturbed particle–hole excitations (Chaps. 4 and 5), the dressing of the bare electronic states (8.62) give rise to important renormalization effects of the collective effects of the collective plasmon excitations as testified by the finite lifetime acquired by these states (damping) as well as by the associated changes in the energy centroids and lineshapes.

Labelling both plasmons and phonons by the index n and the states made up by a vibrational state $|n\rangle$ and an unperturbed particle–hole state by $|N\rangle$, that is,

$$|N\rangle \equiv |ph, n\rangle, \tag{8.63}$$

one can calculate the density–density correlation function (see (4.13)) going beyond the TDLDA, by diagonalizing in the basis $|ph\rangle$, $|hp\rangle$ and $|N\rangle$ the Hamiltonian

$$H = H_{\text{elec}} + H_{\text{phon}} + H_{\text{el,phon}} + H_{\text{plas}} + H_{\text{el,plas}}. \tag{8.64}$$

Here, H_{elec} stands for the TDLDA Hamiltonian (Chap. 4), while $H_{\text{el,phon}}$ and $H_{\text{el,plas}}$ have been defined in (4.38). The matrix elements of H_{elec} between the states $|ph\rangle$ and $|hp\rangle$ have already been defined in (4.38), while those between these states and the states $|N\rangle$ are zero. The spaces $\{|ph\rangle, |hp\rangle\}$ and $\{|N\rangle\}$ are connected by the matrix elements

$$(A_2)_{ph,p'h'n} = \langle p|V|p', n\rangle \delta_{pp'} - \langle h|V|h', n\rangle \delta_{hh'}$$
$$(A_2)_{hp,h'p'n} = -(A_2)^*_{ph,p'h'n}, \tag{8.65}$$

where V stands either for $H_{\text{el,phon}}$ or $H_{\text{el,plas}}$. Finally,

$$(A_3)_{phn,p'h'n} = (\varepsilon_p - \varepsilon_h + \hbar\omega_n)\delta_{pp'}\delta_{hh'}\delta_{nn'}. \tag{8.66}$$

Consequently, the eigenvalue equation associated with the Hamiltonian (8.64) describing the response of the system to a time-dependent external field (4.3) takes the form

$$\left(\begin{array}{cc|cc} A & B & A_2 & 0 \\ -B^* & -A^* & 0 & -A_2^* \\ \hline A_2^T & 0 & A_3 & 0 \\ 0 & -A_2^{T*} & 0 & -A_3^* \end{array}\right) \left(\begin{array}{c} X \\ Y \\ X' \\ Y' \end{array}\right) = \hbar\omega \left(\begin{array}{c} X \\ Y \\ X' \\ Y' \end{array}\right), \tag{8.67}$$

where the superscript T indicates matrix transposition.

The resulting (dressed) states are (see also (4.40))

$$|\bar{n}\rangle = \sum_{ph} [X_{ph}|ph\rangle - Y_{ph}|hp\rangle]$$

$$+ \sum_{N\equiv p'h',n'} \left[X'_{p'h'n'}|p'h'n'\rangle - Y'_{p'h'n}|h'p'n\rangle \right]. \tag{8.68}$$

It is possible to strongly reduce the dimensions of the matrix appearing in the l.h.s. of (8.67) by projecting this equation in the space of p-h excitations. This produces a RPA–like eigenvalue equation,

$$\left(\begin{array}{cc} A + \Sigma(\omega) & B \\ -B^* & -A^* - \Sigma^*(-\omega) \end{array}\right) \left(\begin{array}{c} \mathcal{X} \\ \mathcal{Y} \end{array}\right) = \hbar\omega \left(\begin{array}{c} \mathcal{X} \\ \mathcal{Y} \end{array}\right) \tag{8.69}$$

structurally similar to (4.39). The price paid for the (significant) dimensional reduction is that the Hamiltonian matrix is now energy dependent. The self-energy operator

$$\Sigma(\omega) = A_2 \frac{1}{\hbar\omega - A_3} A_2^T, \tag{8.70}$$

is defined in the p-h space by the elements

$$\Sigma_{ph,p'h'}(\omega) = \sum_N \frac{\langle ph|V|N\rangle\langle N|V|p'h'\rangle}{\hbar\omega - \hbar\omega_N}. \tag{8.71}$$

As above, V is either $H_{\text{el,phon}}$ or $H_{\text{el,plas}}$ according to whether N contains a phonon or a plasmon, while $\hbar\omega_N$ is a shorthand notation for $\varepsilon_{p''} - \varepsilon_{h''} + \hbar\omega_n$ (see (8.66)).

The model described here, has been initially developed within the context of nuclear physics [72, 192]. An application of the formalism to the case of small Na clusters [108] is the subject of the next section. In keeping with the fact that the phonon energies can be much smaller than the typical energies of the p-h pairs (see, e.g., Table 8.4), the coupling with states of type $|N\rangle$ but containing more than one phonon, must be envisaged (as in the case of photoemission, see Sect. 8.4).

Before concluding this section, let us note that the basis formed by the states $|n\rangle$ (where n identifies a plasmon state, see (4.40)), $|ph\rangle$ and $|hp\rangle$ is overcomplete. Consequently, the matrix elements discussed above are to be corrected to eliminate Pauli principle violating contributions. A consistent formalism, the nuclear field theory [188], has been developed to systematically deal with these violations, producing results which coincide with the exact results to all orders of perturbation theory (also infinite, equivalent to an exact diagonalization). In simple terms, the corresponding corrections essentially eliminate the contributions arising from non collective states $|N\rangle$, made out of a non collective state $|n\rangle \approx |ph\rangle$ and the unperturbed state $|ph\rangle$, while they reduce the contribution of a state $|N\rangle = \sum_{ph} [X_{ph}|ph\rangle - Y_{ph}|hp\rangle] |p'h'\rangle$ to that of the state (see e.g. [189]) $\sum_{ph\neq p'h'} [X_{ph}|ph\rangle - Y_{ph}|hp\rangle] |p'h'\rangle$.

8.8 An Example: Electron–Phonon and Electron–Plasmon Couplings in Na$_9^+$

In [108], a consistent calculation of the photoabsorbtion spectrum of two small metal clusters (Na$_8$ and Na$_9^+$ in their lowest-energy geometries, $\mathcal{D}_{2d}$ and $\mathcal{D}_{3h}$ respectively) at zero temperature was performed. Both the coupling of electrons to plasmons and to phonons were considered and calculated ab initio. After diagonalizing the Hamiltonian (8.64), the strength function (4.22) associated with the dipole operator was calculated, as well as the photoabsorbtion cross section (4.31).

The basic quantities entering the Hamiltonian were determined within the Kohn–Sham formalism. Since one takes explicitly into account the electron–plasmon coupling, the LDA calculations were carried out setting v_c equal to zero. To be noted, however, that zeroing the correlation part of v_{xc} leads to only a rigid shift of the electronic eigenvalues without essentially changing the unperturbed p–h energy differences which are the key ingredients which control the dipole spectrum. The phonon and plasmon states, as well as the matrix elements of their coupling with electrons, were also calculated ab initio using the frozen phonon method (Chap. 7) and the TDLDA (Chap. 4), respectively. In fact, the different quantities entering the calculation were obtained making use of a modified version of the Car–Parrinello code described in Chap. 7.

The Kohn–Sham electronic states were calculated using norm conserving pseudopotentials (see Sect. 3.3.2) with nonlinear core correction. The cutoff on the plane wave basis states and the dimension of the supercell where the calculations were carried out, are $E_{\mathrm{cut}} = 15$ Ry and $L_{\mathrm{cell}} = 40$ a_{B}, respectively. The TDLDA eigenvalue problem as described in Chap. 4, was solved in a discrete basis of particle–hole excitations exhausting a large fraction (i.e., larger than 90%) of the corresponding EWSR (see Sect. 4.2). The results for the TDLDA response to external fields of different multipolarity, has been already shown in Fig. 5.14.

The main result of the calculation is that the electron–plasmon coupling is essential to shift downwards the TDLDA spectrum. This is shown in Fig. 8.8 for the case of Na_9^+. The inset of the figure, with the TDLDA peaks displayed by dashed lines (see also Fig. 5.15) and those resulting from the full calculation by full lines, demonstrates that the red shift of the main collective peaks is systematic and of the order of 0.2 eV. This result indicates that the electron–plasmon coupling can be of importance in other contexts where calculations of the dynamical screening are in order. For instance, in

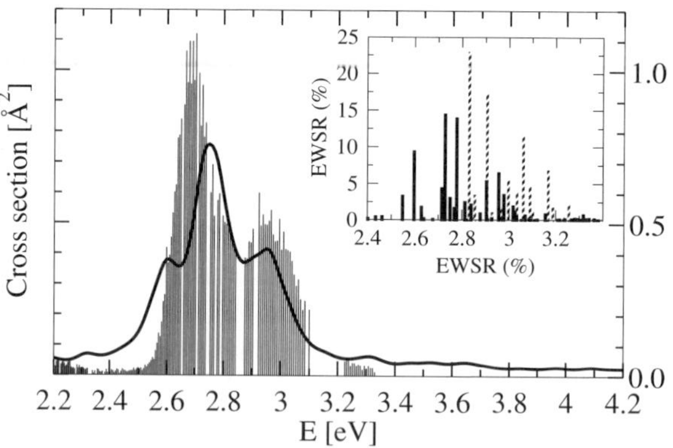

Fig. 8.8. Photoabsorption cross section of Na_9^+ versus energy E (*continuous heavy line*) calculated taking into account both the coupling to plasmons and to phonons. The experimental spectrum measured at $T \approx 39$ K [193] is displayed by the thin vertical lines. The inset shows the percentage of the EWSR calculated making use of the TDLDA results (*dashed lines*) and taking into account the coupling to plasmons (*continuous lines*)

the case of the screened Coulomb potential μ^* (see Chap. 9 and Appendix 9.5.1).

A second result of the full calculation, is that the electron–phonon coupling acts essentially as an averaging parameter of the order of 0.1 eV, leading to a smoothing of the spectrum. This has been checked also by including configurations with up to five phonons. Due to the large dimensions of the model space, this cannot be done with a full diagonalization. In fact, in [108] the strategy to overcome dimensionality problems has been the following: first, it has been verified that the one-, two- and three–phonon configurations provide simply an averaging of the spectrum, when added to a basis including the most collective TDLDA plasmons. After that, the large matrix including the p–h pairs and the states made up of p–h plus a plasmon, has been diagonalized. Finally, the resulting excitations have been coupled with configurations containing up to three phonons.

In the case of Na_8, no low temperature data are available. On the other hand, a measurement of the photoabsorbtion spectrum of Na_9^+ at $T \approx 39$ K has been performed in [193]. The experimental results are displayed in Fig. 8.8, as thin vertical lines.

8.9 Appendices

8.9.1 Terms Neglected in the Born–Oppenheimer Approximation

Using the notation of Sect. 3.1, the perturbation to the adiabatic Born-Oppenheimer Hamiltonian is given by (3.12),

$$C(m, m') = \sum_n \frac{-\hbar^2}{2M_n} \left[2\langle \Psi_m | \boldsymbol{\nabla}_n | \Psi_{m'} \rangle \cdot \boldsymbol{\nabla}_n + \langle \Psi_m | \nabla_n^2 | \Psi_{m'} \rangle \right] . \qquad (8.72)$$

This perturbing potential can move the system away from the stationary state described by the product wavefunction of an electronic and a ionic independent wavefunctions. The probability associated with the transition from the state m to the state m' is given, within the first order perurbation theory (see also Appendix 8.9.4), by

$$|a_{m'm}^{(1)}(t)|^2 = \frac{1}{\hbar^2} |C(m', m)|^2 \left(\frac{\sin(\Omega/2t)}{\Omega/2} \right)^2 , \qquad (8.73)$$

with

$$\hbar\Omega = E_{m'} - E_m. \qquad (8.74)$$

Let us write the Schrödinger equation (3.10) with the shorthand notation $V \equiv V_{\mathrm{ion,e}} + V_{\mathrm{e,e}}$, as well as the corresponding complex conjugate equation,

$$\left(-\frac{\hbar^2}{2m_\mathrm{e}}\sum_i \nabla_i^2 + V\right)\Psi_m = \varepsilon_m \Psi_m,$$

$$\left(-\frac{\hbar^2}{2m_\mathrm{e}}\sum_i \nabla_i^2 + V\right)\Psi_{m'}^* = \varepsilon_{m'}\Psi_{m'}^*. \tag{8.75}$$

We now apply the operator ∇_n to the first equation and multiply it from the left by $\Psi_{m'}^*$. After multiplying the second equation by $\nabla_n \Psi_m$ from the left and subtracting the second equation from the first, we integrate the resulting expression. Remembering that the electronic energies ε depend (implicitly) on the ionic coordinates, we obtain

$$-\frac{\hbar^2}{2m_\mathrm{e}}\int\left(\Psi_{m'}^*\sum_i \nabla_i^2 \nabla_n \Psi_m - \nabla_n \Psi_m \sum_i \nabla_i^2 \Psi_{m'}^*\right) + \int \Psi_{m'}^*(\nabla_n V)\Psi_m$$

$$= \nabla_n \varepsilon_m \int \Psi_{m'}^*\Psi_m + (\varepsilon_m - \varepsilon_{m'})\int \Psi_{m'}^*\nabla_n \Psi_m, \tag{8.76}$$

where the integrals are meant to be carried out over the electron coordinates. The first integral on the l.h.s. can be shown to vanish due to the asymptotic behaviour of the wavefunctions. For the same reason, also the first integral on the right-hand side vanishes. Thus

$$\int \Psi_{m'}^*\nabla_n \Psi_m = \frac{1}{\varepsilon_m - \varepsilon_{m'}}\int \Psi_{m'}^*(\nabla_n V)\Psi_m, \tag{8.77}$$

where in the last integral it is understood that the operator ∇_n acts only on V. We now come back to (8.72). We make the ansatz that, although the electronic quantities (potential and wavefunctions) vary with the ionic co-ordinates, this variation is small so that a first-order approximation holds (consistently with our statements in Sect. 8.1). In this way, the second term in (8.72) can be neglected. Within the same spirit, we replace the first term of (8.72) by using (8.77) and by assuming that the derivative $\nabla_n V$ is evaluated at the ionic equlibrium positions. Then,

$$C(m',m) = -\sum_n \frac{1}{\varepsilon_m - \varepsilon_{m'}}\int \Psi_{m'}^*\nabla_n V\Psi_m \cdot \frac{\hbar^2}{M_n}\langle m'|\nabla_n|m\rangle, \tag{8.78}$$

where the last matrix element is to be taken between the ionic wavefunctions. Using the commutator

$$[X,H] = \frac{\mathrm{i}\hbar}{M_n}P,$$

which holds since the potential part of the Hamiltonian commutes with X, we obtain

$$\frac{\hbar^2}{M_n}\langle m'|\nabla_n|m\rangle = \frac{\mathrm{i}\hbar}{M_n}\langle m'|P_n|m\rangle$$

$$= (E_m - E_{m'})\langle m'|X_n|m\rangle, \tag{8.79}$$

leading to

$$C(m', m) = \sum_n \frac{E_{m'} - E_m}{\varepsilon_m - \varepsilon_{m'}} \int \Psi_{m'}^* \nabla_n V \Psi_m \cdot \langle m' | X_n | m \rangle. \tag{8.80}$$

This is equal to the matrix element of the electron–phonon Hamiltonian discussed in this chapter, assuming that the pre-factor $(E_{m'} - E_m)/(\varepsilon_m - \varepsilon_{m'})$ is close to one. This last statement follows from the assumption of the (approximate) on-energy-shell condition, namely from the fact that if an electron has made a transition, the corresponding energy has been subtracted from the ionic degrees of freedom.

8.9.2 The Electron–Phonon Coupling Constant λ

In the first part of this Appendix we follow mainly the arguments of [194], to show that the electron–phonon coupling constant λ coincides with the mass enhancement factor introduced in Chap. 2. The expression for the electron self–energy Σ_{phon} due to the coupling to phonons can be written as (see (2.109))

$$\Sigma_j(\omega) = \sum_{j', \nu} |\langle j', \nu | H_{\mathrm{el-phon}} | j \rangle|^2$$

$$\times \left(\frac{1}{\hbar\omega - \varepsilon_{j'} - \hbar\omega_\nu} + \frac{1}{\hbar\omega + \varepsilon_{j'} + \hbar\omega_\nu} \right), \tag{8.81}$$

where now the index ν labels the phonons. We assume that the molecules are embedded in a solid and that the sums must be replaced by integrals over the appropriate energy bands. Equation (8.81) becomes

$$\Sigma(\omega) = 2 \int d\varepsilon \int d(\hbar\omega_q) \, \alpha^2(\omega_q) \, F(\omega_q)$$

$$\left(\frac{1}{\hbar\omega - \varepsilon - \hbar\omega_q} + \frac{1}{\hbar\omega + \varepsilon + \hbar\omega_q} \right), \tag{8.82}$$

where $\alpha^2(\omega_q)$ is an average of the squares of the electron–phonon coupling matrix elements for single-particle states lying around the Fermi surface (see (8.87)). $F(\omega_q)$ is the phonon density of states. The factor 2 takes into account the two possible spin orientations (the self-energy is the trace of a diagonal matrix in the spin space). If we wish to study the properties of the self-energy around the zero energy (see also Chap. 2 and Appendix 2.11.2), we can neglect $\hbar\omega$ in the energy denominators of (8.82). In this limit, and in the somewhat crude approximation that the upper integration limit for ε may be set at the phonon energy $\hbar\omega_q$, that is, if we can use $\int_0^{\omega_q} d\varepsilon \, (-2)\omega/(\varepsilon + \omega_q)^2 = -\omega/\omega_q$, we obtain

$$\Sigma(\omega) = -2 \int d\hbar\omega_q \, \frac{\alpha^2(\omega_q) F(\omega_q)}{\hbar\omega_q} \times \hbar\omega. \tag{8.83}$$

The dimensionless quantity

$$\lambda \equiv 2 \int \mathrm{d}\hbar\omega_q \, \frac{\alpha^2(\omega_q)F(\omega_q)}{\hbar\omega_q} \tag{8.84}$$

is defined as the electron–phonon coupling constant. Since

$$\Sigma(\omega) = -\lambda \cdot \hbar\omega, \tag{8.85}$$

according to (2.52) it is found that the renormalization of the electron mass is

$$m_\omega = m_{\mathrm{e}}(1 + \lambda) \tag{8.86}$$

(see Sect. 2.10).

In the second part of the Appendix, we show how to derive (8.34) from (8.84), by using the procedure discussed in [195]. We consider explicitly that around the Fermi energy there are different electronic states labelled by k and k' (their energies will be denoted by ε_k, $\varepsilon_{k'}$ while the phonon energy will be denoted by $\hbar\omega_{k-k'}$, in keeping with momentum conservation). The averaging of the squared electron-phonon matrix elements introduced in (8.82) is defined as

$$\alpha^2(\omega) = \frac{\sum_{kk'} |M_{kk'}|^2 \delta(\varepsilon_k)\delta(\varepsilon_{k'})}{\sum_k \delta(\varepsilon_k)}, \tag{8.87}$$

where M indicates the electron–phonon matrix element. Multiplying (8.87) by the function $F(\omega)$ introduces an additional factor $\delta(\omega - \omega_{k-k'})$. Inserting these results in (8.84) one obtains

$$\lambda = 2\sum_{kk'} |M_{kk'}|^2 \frac{1}{\hbar\omega_{k-k'}} \frac{1}{N(0)} \delta(\varepsilon_k)\delta(\varepsilon_{k'}),$$

where the electronic density of states at the Fermi surface for one spin orientation has been indicated by

$$N(0) = \sum_k \delta(\varepsilon_k). \tag{8.88}$$

Let us now make the approximation that the solid we wish to study, can be treated as if it were made up of molecules or clusters which are independent as far as the electron–phonon coupling is concerned. This approximation simplifies considerably the present problem, and can be considered a fairly good one in the case, e.g., of the van der Waals solid (fullerite) made out of C_{60} fullerenes. It has in fact been introduced in [164] for that system. In other cases it might prove necessary to calculate in full detail the properties of the solid, including all the dispersion effects in the electronic and vibrational spectra which are neglected in the following. This amounts to evaluating (8.34) numerically [12].

Let us assume that we have a single molecule or cluster per unit cell, and that its LUMO state has degeneracy n_{deg} (for each spin orientation). We also assume that the electronic states around the Fermi energy of the solid are built only from the LUMO molecular states and can be written in terms of Bloch wavefunctions as,

$$\sum_{l=1}^{n_{\mathrm{deg}}} c_i(l) \cdot \frac{1}{\sqrt{N}} \sum_R e^{ikR} \varphi_{l,R}. \tag{8.89}$$

Here R denotes the cell position and N is the number of cells. The phonon wavefunction will similarly include a Bloch factor $e^{iqR}/\sqrt{N}$ in addition to the molecular wavefunction, and the exponentials will be summed up using the relation

$$\frac{1}{N} \sum_R e^{i(k'-k+q)R} = \delta(k' - k + q).$$

This leads to

$$M_{kk'} = \sum_{ij} c_i^*(k) c_j(k-q) \langle i, \nu | H_{\mathrm{el,phon}} | j \rangle. \tag{8.90}$$

Inserting this expression in (8.34), neglecting the q-dependence of the phonon energy, and taking into account the fact that

$$\frac{1}{N} \sum_k c_i^*(k) c_j(k) \delta(\varepsilon_k) = N_{\nu\nu'}(0) \delta_{\nu\nu'}$$

is a (partial) density of states, one obtains

$$\lambda_\nu = 2 \frac{N(0)}{n_{\mathrm{deg}}^2} \sum_\mu \frac{\sum_{ij} |\langle j, \nu\mu | (v_{\mathrm{def}}^{(\nu)})_\mu | i \rangle|^2}{\hbar \omega_\nu}. \tag{8.91}$$

Here $N(0)$ is the density of states around the Fermi energy per spin orientation and per molecule, n_{deg} is the degeneracy of the LUMO state i also for a single spin orientation, while the factor of 2 arises from the two spin orientations.

8.9.3 Outline of the Derivation of (8.50)

We start from (8.49),

$$\tilde{H} = -\frac{1}{2} \lambda [H_1, S],$$

and we insert in this expression the explicit form of the operators, namely (8.44) and (8.45). This yields

$$-\frac{1}{2} \lambda [H_1, S] = -\frac{\lambda^2}{2} \sum_\nu \frac{g_\nu}{2} \sum_\mu \sum_{ij\sigma} C_{ij}^{(\mu)} \sum_{\nu'} \sum_{\mu'} \sum_{i'j'\sigma'}$$
$$\left[a_{i\sigma}^\dagger a_{j\sigma} (\Gamma_{\nu\mu} + \Gamma_{\nu\mu}^\dagger), a_{i'\sigma'}^\dagger a_{j'\sigma'} \left(\frac{C_{i'j'}^{(\mu')}}{\varepsilon_{i'j'} - \hbar\omega_{\nu'}} \Gamma_{\nu'\mu'} + \frac{C_{i'j'}^{(\mu')}}{\varepsilon_{i'j'} + \hbar\omega_{\nu'}} \Gamma_{\nu'\mu'}^\dagger \right) \right]. \tag{8.92}$$

Although the calculation of the commutator is quite lengthy, it does not involve more than the application of the basic rules of second quantization (see Appendix 2.11.1). The result is

$$
[a_{i\sigma}^{\dagger}a_{j'\sigma'}\delta(i'j)\delta(\sigma,\sigma')\Gamma_{\nu'\mu'}\Gamma_{\nu\mu} - a_{i'\sigma'}^{\dagger}a_{j\sigma}\delta(ij')\delta(\sigma,\sigma')\Gamma_{\nu'\mu'}\Gamma_{\nu\mu}]\frac{C_{i'j'}^{(\mu')}}{\varepsilon_{i'j'} - \hbar\omega_{\nu'}}
$$

$$
+[-a_{i\sigma}^{\dagger}a_{j\sigma}a_{i'\sigma'}^{\dagger}a_{j'\sigma'}\delta(\nu,\nu')\delta(\mu,\mu') + a_{i\sigma}^{\dagger}a_{j'\sigma'}\delta(i'j)\delta(\sigma,\sigma')\Gamma_{\nu'\mu'}\Gamma_{\nu\mu}^{\dagger}
$$

$$
- a_{i'\sigma'}^{\dagger}a_{j\sigma}\delta(ij')\delta(\sigma,\sigma')\Gamma_{\nu'\mu'}\Gamma_{\nu\mu}^{\dagger}]\frac{C_{i'j'}^{(\mu')}}{\varepsilon_{i'j'} - \hbar\omega_{\nu'}}
$$

$$
+[a_{i\sigma}^{\dagger}a_{j\sigma}a_{i'\sigma'}^{\dagger}a_{j'\sigma'}\delta(\nu,\nu')\delta(\mu,\mu') + a_{i\sigma}^{\dagger}a_{j'\sigma'}\delta(i'j)\delta(\sigma,\sigma')\Gamma_{\nu'\mu'}^{\dagger}\Gamma_{\nu\mu}
$$

$$
- a_{i'\sigma'}^{\dagger}a_{j\sigma}\delta(ij')\delta(\sigma,\sigma')\Gamma_{\nu'\mu'}^{\dagger}\Gamma_{\nu\mu}]\frac{C_{i'j'}^{(\mu')}}{\varepsilon_{i'j'} + \hbar\omega_{\nu'}}
$$

$$
+[a_{i\sigma}^{\dagger}a_{j'\sigma'}\delta(ij')\delta(\sigma,\sigma')\Gamma_{\nu'\mu'}^{\dagger}\Gamma_{\nu\mu} - a_{i'\sigma'}^{\dagger}a_{j\sigma}\delta(i'j)\delta(\sigma,\sigma')\Gamma_{\nu'\mu'}^{\dagger}\Gamma_{\nu\mu}^{\dagger}]\frac{C_{i'j'}^{(\mu')}}{\varepsilon_{i'j'} + \hbar\omega_{\nu'}}. \tag{8.93}
$$

One can recognize in this expression the presence of many second-order terms. As stated in Sect. 8.5, we are interested in the terms corresponding to an effective electron–electron interaction, namely terms proportional to $a^{\dagger}a^{\dagger}aa$. Isolating these terms and performing simple algebraic steps, leads to (8.50).

8.9.4 Equation (8.50) Using Second-Order Perturbation Theory

In what follows, we use standard first- and second-order perturbation theory (see, e.g., [57]) to derive the basic structure of (8.50). In the case under study, the perturbation on the electrons arizes from the harmonic motion of the ions and is therefore time-dependent. Let us write the total Hamiltonian as

$$
H = H_0 + \Lambda H_1(t), \tag{8.94}
$$

where H_0 describes the unperturbed electrons and ions (see (8.1)) while $H_1(t)$ is the electron–phonon Hamiltonian (the second term of (8.5)) where the time-dependence of the ionic motion has been set in evidence. We now expand the perturbed wavefunction Ψ of a generic state on the basis of the unperturbed wavefunctions Ψ_n eigenstates of H_0 (E_n being the corresponding eigenvalues). Thus

$$
\Psi = \sum_n a_n(t)\Psi_n e^{-\frac{i}{\hbar}E_n t}. \tag{8.95}
$$

Inserting this wavefunction in the time-dependent Schrödinger equation for the total Hamiltonian (8.94), one obtains, making the ansatz that the amplitudes a_n can be written as a power series in Λ

$$a_n = a_n^{(0)} + a_n^{(1)} + \dots, \tag{8.96}$$

a set of equations for $a_n^{(s)}$. In particular, neglecting terms of higher order than first, one obtains

$$\dot{a}_n^{(0)} = 0, \tag{8.97}$$

$$\dot{a}_k^{(1)} = -\frac{i}{\hbar} \sum_n \langle k|H_1(t)|n\rangle a_n^{(0)} e^{i\omega_{kn}t}, \tag{8.98}$$

where $\hbar\omega_{kn} \equiv E_k - E_n$. With the initial condition that at time $t = -\infty$ the state is pure (namely, $a_n^{(0)} = \delta_{nm}$), the equation for the first-order amplitudes becomes

$$\dot{a}_k^{(1)} = -\frac{i}{\hbar} \int_{-\infty}^{t} \langle k|H_1(t)|m\rangle e^{i\omega_{km}t'}\, dt'. \tag{8.99}$$

The time dependence of the perturbation H_1 is of the type

$$H_1(t) = \frac{H_1}{2} e^{i\omega_\lambda t} + \text{h. c..} \tag{8.100}$$

We assume that H_1 is real and that its action starts at $t = 0$. One can then find from (8.99) that

$$\dot{a}_k^{(1)} = -\frac{\langle k|H_1|m\rangle}{2\hbar} \left[\frac{e^{i(\omega_{km}+\omega_\lambda)t} - 1}{\omega_{km} + \omega_\lambda} + \frac{e^{i(\omega_{km}-\omega_\lambda)t} - 1}{\omega_{km} - \omega_\lambda} \right]. \tag{8.101}$$

where ω_λ are the phonon frequencies.

We look at transitions which occur between almost degenerate states at the second-order level. The equation analogous to (8.98) - and which can be derived in the same way - for $a_k^{(2)}$, reads

$$\dot{a}_k^{(2)} = -\frac{i}{\hbar} \sum_n \langle k|H_1(t)|n\rangle a_n^{(1)} e^{i\omega_{kn}t}. \tag{8.102}$$

If we replace (8.101) in (8.102) we obtain a rather complicated structure. We make the simplifying approximation that the time scale of the perturbation induced by the phonon is much longer than the electron time scale, so that the perturbation can be considered almost stationary. That is, $\omega_\lambda \approx 0$ and (8.101) simplifies to

$$\dot{a}_k^{(1)} = -\frac{\langle k|H_1|m\rangle}{\hbar} \frac{e^{i\omega_{km}t} - 1}{\omega_{km}}. \tag{8.103}$$

Introducing this expression in (8.102) we find

$$\dot{a}_k^{(2)} = \frac{i}{\hbar^2} \sum_n \langle k|H_1|n\rangle\langle n|H_1|m\rangle e^{i\omega_{kn}t} \frac{e^{i\omega_{nm}t} - 1}{\omega_{nm}}$$

$$= \frac{i}{\hbar^2} \sum_n \frac{\langle k|H_1|n\rangle\langle n|H_1|m\rangle}{\omega_{nm}} (e^{i\omega_{km}t} - e^{i\omega_{kn}t}). \tag{8.104}$$

If $a_k^{(2)}(t=0)=0$, then

$$a_k^{(2)} = \frac{1}{\hbar^2}\sum_n \frac{\langle k|H_1|n\rangle\langle n|H_1|m\rangle}{\omega_{nm}}\left[\frac{\mathrm{e}^{\mathrm{i}\omega_{km}t}-1}{\omega_{km}} - \frac{\mathrm{e}^{\mathrm{i}\omega_{kn}t}-1}{\omega_{kn}}\right]. \qquad (8.105)$$

We now choose the initial state m as made up of two (nearly) degenerate electrons in states j and j'. In the intermediate state n the electron j has made a transition to the state i and a phonon ν has been emitted. In the final state k the phonon has been absorbed by the electron in the state j' which has made a transition to the state i' (see Fig. 8.9). In this situation $\omega_{km}\sim 0$ while $\omega_{nm}\sim\omega_{kn}\neq 0$, consistently with what has been said after (8.101). We neglect therefore the second term in square brackets in (8.105) and we write the transiton probability to the state k as

$$P_k(t) = |a_k^{(2)}(t)|^2 = \frac{1}{\hbar^4}M^2\frac{|\mathrm{e}^{\mathrm{i}\omega_{km}t}-1|^2}{\omega_{km}^2} = \frac{1}{\hbar^2}M^2\frac{4\sin^2(\omega_{km}/2t)}{\hbar^2\omega_{km}^2}, (8.106)$$

where M is defined as $\langle k|H_1|n\rangle\langle n|H_1|m\rangle/\omega_{nm}$. We must sum (8.106) over the final states which fulfill $\omega_{km}\sim 0$ in order to obtain the total transition probability. Introducing the density of final states $\varrho(k)$ we write the transition probability per unit time w as

$$w = \frac{1}{t}\int |a_k^{(2)}(t)|^2\varrho(k)\mathrm{d}E_k. \qquad (8.107)$$

The density of states $\varrho(k)$ is here different from zero only in a small energy interval such that the approximation $\omega_{km}\sim 0$ is valid. However, for large values of t the function written at the right-hand side of (8.106) tends to a narrow peak, so that we can formally leave E_k as a continuum variable within that peak and assume $\varrho(k)$ as a constant, obtaining

$$w = \frac{1}{t}\frac{4M^2}{\hbar}\varrho(k)\int_{-\infty}^{\infty}\frac{\sin^2(\omega_{km}t/2)}{\omega_{km}^2}\mathrm{d}\omega_{km} = \frac{2\pi}{\hbar}\varrho(k)M^2. \qquad (8.108)$$

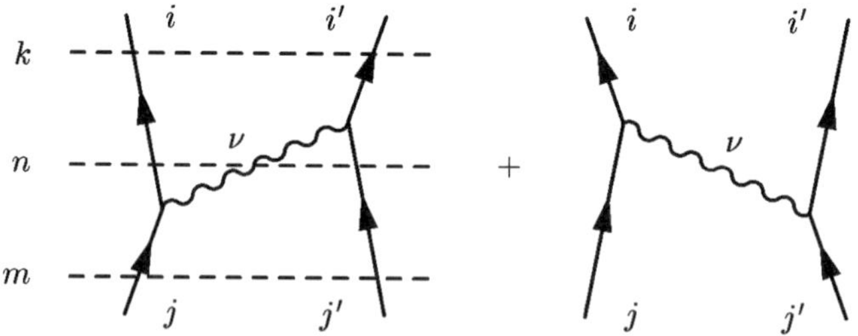

Fig. 8.9. Graphical representation of the effective electron–electron interaction induced by the exchange of a phonon. The calculation of the left-side diagram is discussed in detail in the text, using the second-order perturbation theory

This formula – which is nothing but the Golden Rule for the case at hand – shows that

$$M = \frac{(g_\lambda^2/4)C_{ij}^{(\mu)}C_{i'j'}^{(\mu')}}{(\varepsilon_i - \varepsilon_j) - \hbar\omega_\lambda}. \tag{8.109}$$

Adding to M the corresponding expression in the case in which the phonon is first emitted by the electron j' and then absorbed by the electron j, leads to (8.50).

9. Pairing in Atomic Aggregates

Electrons near the Fermi surface in a superconductor, that is in a system which conducts with zero resistivity, interact to form correlated pairs. This idea was first suggested by Cooper [196] and since then the pairs are often called "Cooper pairs". They are constructed from states in which the two electrons have zero total spin and equal and opposite linear momentum $\boldsymbol{k}$ and $-\boldsymbol{k}$. As discussed in the previous chapter, the effective interaction between electrons is a result of the virtual emission of a phonon by one electron and its absorption by another. This interaction causes scattering of an electron pair from states $(\boldsymbol{k}, -\boldsymbol{k})$ to states $(\boldsymbol{k}', -\boldsymbol{k}')$ with an amplitude $V_{\boldsymbol{k}'\boldsymbol{k}}$ which depends on the electron–phonon coupling and on the phonon spectrum.

Each Cooper pair has a binding energy 2Δ which is much smaller than the Fermi energy ε_{F}. The main components of the pair wave function come from electron states with energies ε within Δ of the Fermi energy,

$$\varepsilon_{\mathrm{F}} - \Delta < \varepsilon < \varepsilon_{\mathrm{F}} + \Delta . \tag{9.1}$$

The energy spread $\delta\varepsilon \approx 2\Delta$ corresponds to a momentum range $\delta p \approx 2\Delta/v_{\mathrm{F}}$ where v_{F} is the Fermi velocity. The uncertainty relation $\delta x \approx \hbar/\delta p$, that is

$$\xi_{\circ} = \frac{\hbar v_{\mathrm{F}}}{2\Delta}, \tag{9.2}$$

provides an estimate of the size of the Cooper pair, the quantity $\xi_{\circ} = \delta x$ being called the coherence length of the superconductor.

In 1957 Bardeen, Cooper and Schrieffer [197] developed a microscopic theory of superconductivity (called the BCS theory) which incorporated the idea of Cooper pairs and gave a consistent treatment of the Pauli principle. According to the BCS theory, electrons near the Fermi surface in the ground state of a superconductor form correlated Cooper pairs. Excited states are formed by breaking pairs and there is an energy gap 2Δ between the ground state and the lower excited states. It is this energy gap which stabilizes the superconducting state. Thermal effects can break pairs, and in BCS theory the presence of unpaired electrons reduces the binding of those pairs which remain. Thus the gap parameter 2Δ is temperature dependent and decreases as T increases. At a critical temperature T_{c} the energy gap becomes zero, all pairs are broken and there is a phase transition from the superconducting phase and the normal phase. BCS theory predicts a definite relation between the transition temperature T_c and the energy gap $\Delta(0)$ at $T = 0$ namely, $2\Delta(0)/K_B T_c = 3.51$ (see 9.26). This relation can be checked experimentally because both $\Delta(0)$ and T_c can be measured. For most normal superconductors the ratio lies in the range 3.2–4.6 and is close to the BCS value (see Table 9.1).

Leggett [198] points out that the Cooper pairs in the BCS theory of the superconducting state must all behave in exactly the same way, not only as regards their internal structure but also as regards their center of mass motion. Each Cooper pair is made up of two fermions and therefore the pairs behave like bosons. From this point of view superconductivity is due to Bose condensation of the pairs. The analogy is not complete. In a Bose liquid like ^{4}He (i.e. a system made out of two protons, two neutrons and two electrons) the bosons exist even when they are not condensed, while in the superconducting state either the Cooper pairs are condensed or they do not exist.

There is another important fact to be considered in connection with the description of superconductors in terms of electron pairs. As pointed out by Schrieffer [44] the pairs could be treated as independent if they were well separated and Cooper's discussion would be appropriate. However, actual superconductors differ in a fundamental manner from a bound pair model in which the pairs are well separated in space and weakly interacting. The pairs overlap strongly and there are on the average one million bound pairs which have their centers of mass falling within the region occupied by a given pair wave function. This is a consequence of the fact that for metals like Pb or Sn, the coherence length ξ_o is much larger than the crystal lattice spacing (~ 5 Å). In fact, because the Fermi velocity of electrons in these materials is normally large ($v_F \approx 10^6$ m/s) and the energy gap is small (few °K, see Table 9.2), one obtains $\xi_o \approx 2300$ Å for Sn and $\xi_o \approx 830$ Å for Pb.

9.1 The BCS Solution

Let us start with the Hamiltonian

$$H = H_{sp} + H_p \,,$$

sum of single-particle Hamiltonian

$$H_{sp} = \sum_{\nu>0}(\varepsilon_\nu - \lambda_F)(a_\nu^\dagger a_\nu + a_{\bar\nu}^\dagger a_{\bar\nu}) \,,$$

where the single-particle energies ε_ν are measured from the Fermi energy λ_F, and a pairing interaction with constant matrix elements (i.e. $\sum_{\nu\mu}(g_\nu^2/4\hbar\omega_\nu)(C_{ij}^{(\mu)})^2 = G$, see (8.51))

$$H_p = -G \sum_{\nu>0,\nu'>0} a_\nu^\dagger a_{\bar\nu}^\dagger a_{\bar\nu'} a_{\nu'}$$

where $a_\nu^\dagger$ and a_ν are creation and annihilation fermion operators (second quantization). The state $\bar\nu$ is obtained from the state ν by the operation of time reversal. Consequently, $H_{\mathrm p}$ scatters pairs of fermions moving in time reversal states.

In what follows we shall solve H in mean field approximation. For this purpose we introduce the pair creation operator,

$$P^\dagger = \sum_{\nu>0} a_\nu^\dagger a_{\bar\nu}^\dagger = \alpha_0 + (P^\dagger - \alpha_0),$$

and add and subtract from it the mean field value

$$\alpha_0 = \langle \mathrm{BCS}|P^\dagger|\mathrm{BCS}\rangle = \langle \mathrm{BCS}|P|\mathrm{BCS}\rangle,$$

of the pair transfer operator in the (still unknown) mean field ground state. This state is called the $|\mathrm{BCS}\rangle$ state, in keeping with the fact that this solution was first proposed by Bardeen, Cooper and Schrieffer [197]. Note that $\langle \mathrm{BCS}|\mathrm{BCS}\rangle = 1$. We can now write

$$H_{\mathrm p} = -G(\alpha_0 + (P^\dagger - \alpha_0))(\alpha_0 + (P - \alpha_0))$$

$$= -G(\alpha_0^2 + \alpha_0(P^\dagger + P - 2\alpha_0) + (P^\dagger - \alpha_0)(P - \alpha_0)) \ .$$

Assuming that the matrix elements of the operators $(P^\dagger - \alpha_0)$ and $(P - \alpha_0)$ in the states near to the ground state are much smaller than α_0, one obtains the pairing field

$$V_{\mathrm p} = -\Delta(P^\dagger + P) + \frac{\Delta^2}{G} \ ,$$

where

$$\Delta = G\alpha_0 \ . \tag{9.3}$$

The mean field Hamiltonian then becomes

$$H_{\mathrm{MF}} = H_{\mathrm{sp}} + V_{\mathrm p}$$

$$= \sum_{\nu>0}(\varepsilon_\nu - \lambda_{\mathrm F})(a_\nu^\dagger a_\nu + a_{\bar\nu}^\dagger a_{\bar\nu}) - \Delta\sum_{\nu>0}(a_\nu^\dagger a_{\bar\nu}^\dagger + a_{\bar\nu}a_\nu) + \frac{\Delta^2}{G} \ .$$

This is a bilinear expression in the creation and annihilation operators. Consequently, it can be diagonalized by a rotation in the $(a^\dagger, a)$-space. This can be accomplished through the linear transformation

$$\alpha_\nu^\dagger = U_\nu a_\nu^\dagger - V_\nu a_{\bar\nu}.$$

The creation operator of a quasiparticle $\alpha_\nu^\dagger$ creates a particle in the single-particle state ν with probability U_ν^2, while it creates a hole (annihilates a particle) with probability V_ν^2. Because to be able to create a particle, the state ν should be empty, while to create a hole it has to be filled, U_ν^2 (V_ν^2) is the probability that the state ν is empty (occupied).

Expressing the creation and annihilation operators $(a_\nu^\dagger, a_\nu)$ in terms of the quasiparticle operators $(\alpha_\nu^\dagger, \alpha_\nu)$, and writing H_{MF} in terms of quasiparticles, one has two parameters U_ν and V_ν for each level ν to make diagonal the Hamiltonian.

Making use of the anticommutation relation

$$\{a_\nu, a_{\nu'}^\dagger\} = \delta(\nu, \nu')$$
$$\{a_\nu, a_{\nu'}\} = \{a_\nu^\dagger, a_{\nu'}^\dagger\} = 0 \, ,$$

one obtains

$$\{\alpha_\nu, \alpha_{\nu'}^\dagger\} = \{(U_\nu a_\nu - V_\nu a_{\bar\nu}^\dagger), (U_{\nu'} a_{\nu'}^\dagger - V_{\nu'} a_{\bar{\nu'}})\}$$
$$= (U_\nu U_{\nu'} + V_\nu V_{\nu'})\, \delta(\nu, \nu') \, .$$

That is, for the quasiparticle transformation to be unitary, the U_ν, V_ν occupation factors have to fulfill the relation

$$U_\nu^2 + V_\nu^2 = 1 \, , \tag{9.4}$$

implying also that the one-quasiparticle states are normalized. In particular

$$\langle \nu | \nu \rangle = 1 = \langle \mathrm{BCS} | \alpha_\nu \alpha_\nu^\dagger | \mathrm{BCS} \rangle = \langle \mathrm{BCS} | \{\alpha_\nu, \alpha_\nu^\dagger\} | \mathrm{BCS} \rangle$$
$$= U_\nu^2 + V_\nu^2 \, ,$$

where

$$|\nu\rangle = \alpha_\nu^\dagger |\mathrm{BCS}\rangle.$$

Note that $|\mathrm{BCS}\rangle$ is also the quasiparticle vacuum, that is,

$$\alpha_\nu |\mathrm{BCS}\rangle = 0.$$

Let us now invert the quasiparticle transformation, that is express $a^\dagger$ in terms of $\alpha^\dagger$ and α. Multiplying $\alpha_\nu^\dagger$ by U_ν and $\alpha_{\bar\nu}$ by V_ν and adding the resulting expressions

$$U_\nu \alpha_\nu^\dagger = U_\nu^2 a_\nu^\dagger - U_\nu V_\nu a_{\bar\nu},$$
$$V_\nu \alpha_{\bar\nu} = U_\nu V_\nu a_{\bar\nu} + V_\nu^2 a_\nu^\dagger,$$

one obtains

$$a_\nu^\dagger = U_\nu \alpha_\nu^\dagger + V_\nu \alpha_{\bar\nu}.$$

We shall now express $a^\dagger a$ in terms of quasiparticles, that is,

$$a_\nu^\dagger a_\nu = (U_\nu \alpha_\nu^\dagger + V_\nu \alpha_{\bar\nu})(U_\nu \alpha_\nu + V_\nu \alpha_{\bar\nu}^\dagger)$$
$$= U_\nu^2 \alpha_\nu^\dagger \alpha_\nu + U_\nu V_\nu (\alpha_\nu^\dagger \alpha_{\bar\nu}^\dagger + \alpha_{\bar\nu} \alpha_\nu) - V_\nu^2 \alpha_{\bar\nu}^\dagger \alpha_{\bar\nu} + V_\nu^2. \tag{9.5}$$

Taking the time reversed of this expression one obtains

$$a_{\bar\nu}^\dagger a_{\bar\nu} = U_\nu^2 \alpha_{\bar\nu}^\dagger \alpha_{\bar\nu} + U_\nu V_\nu (\alpha_\nu^\dagger \alpha_{\bar\nu}^\dagger + + \alpha_{\bar\nu} \alpha_\nu) - V_\nu^2 \alpha_\nu^\dagger \alpha_\nu + V_\nu^2,$$

where the phase relation $|\bar{\bar\nu}\rangle = -|\nu\rangle$ (with $a_{\bar{\bar\nu}}^\dagger = -a_\nu$, $a_{\bar\nu} = -a_\nu$) has been used. One can then write

$$(a_\nu^\dagger a_\nu + a_{\bar\nu}^\dagger a_{\bar\nu}) = (U_\nu^2 - V_\nu^2)(\alpha_\nu^\dagger \alpha_\nu + \alpha_{\bar\nu}^\dagger \alpha_{\bar\nu})$$
$$+ 2U_\nu V_\nu(\alpha_\nu^\dagger \alpha_{\bar\nu}^\dagger + \alpha_{\bar\nu}\alpha_\nu) + 2V_\nu^2.$$

Note that

$$N = \langle\mathrm{BCS}|\hat{N}|\mathrm{BCS}\rangle = \langle\mathrm{BCS}|\sum_{\nu>0}(a_\nu^\dagger a_\nu + a_{\bar\nu}^\dagger a_{\bar\nu})|\mathrm{BCS}\rangle = 2\sum_{\nu>0} V_\nu^2, \quad (9.6)$$

where N is the average number of particles in the pairing mean field ground state (BCS state).

Let us now express the pair creation field $a_\nu^\dagger a_{\bar\nu}^\dagger$ in terms of quasiparticles,

$$a_\nu^\dagger a_{\bar\nu}^\dagger = (U_\nu\alpha_\nu^\dagger + V_\nu\alpha_{\bar\nu})(U_\nu\alpha_{\bar\nu}^\dagger - V_\nu\alpha_\nu)$$
$$= U_\nu^2\alpha_\nu^\dagger \alpha_{\bar\nu}^\dagger - U_\nu V_\nu(\alpha_\nu^\dagger \alpha_\nu + \alpha_{\bar\nu}^\dagger \alpha_{\bar\nu}) - V_\nu^2\alpha_{\bar\nu}\alpha_\nu + U_\nu V_\nu. \quad (9.7)$$

The hermitian conjugate of this operator is

$$a_{\bar\nu}a_\nu = U_\nu^2\alpha_{\bar\nu}\alpha_\nu - U_\nu V_\nu(\alpha_\nu^\dagger \alpha_\nu + \alpha_{\bar\nu}^\dagger \alpha_{\bar\nu}) - V_\nu^2\alpha_\nu^\dagger \alpha_{\bar\nu}^\dagger + U_\nu V_\nu.$$

Summing up the two expressions above leads to,

$$(a_\nu^\dagger a_{\bar\nu}^\dagger + a_{\bar\nu}a_\nu) = (U_\nu^2 - V_\nu^2)(\alpha_\nu^\dagger \alpha_{\bar\nu}^\dagger + \alpha_{\bar\nu}\alpha_\nu)$$
$$- 2U_\nu V_\nu(\alpha_\nu^\dagger \alpha_{\bar\nu}^\dagger) + 2U_\nu V_\nu .$$

Note that

$$\alpha_0 = \langle\mathrm{BCS}|P^\dagger|\mathrm{BCS}\rangle = \sum_{\nu>0}\langle\mathrm{BCS}|a_\nu^\dagger a_{\bar\nu}|\mathrm{BCS}\rangle = \sum_{\nu>0} U_\nu V_\nu,$$

and, according to (9.3),

$$\Delta = G\alpha_0 = G\sum_{\nu>0} U_\nu V_\nu. \quad (9.8)$$

Making use of the relations written above one can express H_{MF} in terms of quasiparticles, that is,

$$H_{\mathrm{MF}} = U + H_{11} + H_{20},$$

where

$$U = 2\sum_{\nu>0}(\varepsilon_\nu - \lambda_{\mathrm{F}})V_\nu^2 - \Delta\sum_{\nu>0} 2U_\nu V_\nu + \frac{\Delta^2}{G},$$

$$H_{11} = \sum_{\nu>0}\{(\varepsilon_\nu - \lambda_{\mathrm{F}})(U_\nu^2 - V_\nu^2) + \Delta 2U_\nu V_\nu\}(\alpha_\nu^\dagger \alpha_\nu + \alpha_{\bar\nu}^\dagger \alpha_{\bar\nu}),$$

$$H_{20} = \sum_{\nu>0}\{(\varepsilon_\nu - \lambda_{\mathrm{F}})2U_\nu V_\nu - \Delta(U_\nu^2 - V_\nu^2)\}(\alpha_\nu^\dagger \alpha_{\bar\nu}^\dagger + \alpha_{\bar\nu}\alpha_\nu).$$

In other words, one can separate the mean field pairing Hamiltonian in three terms: one which is a constant (ground state energy), a second which is diagonal in the quasiparticle basis, and a third one, which, although bilinear in the operators $\alpha^\dagger$ and α, is not diagonal. Setting $H_{20} = 0$, that is,

$$(\varepsilon_\nu - \lambda)2U_\nu V_\nu = \Delta(U_\nu^2 - V_\nu^2) \tag{9.9}$$

diagonalizes the pairing Hamiltonian in the mean field approximation.

Let us next use this relation, together with the normalization condition (9.4), to calculate the coefficients U_ν and V_ν. From the normalization relation one can write

$$(U_\nu^2 + V_\nu^2)^2 = 1 = U_\nu^4 + V_\nu^4 + 2U_\nu^2 V_\nu^2 \ ,$$

and

$$U_\nu^4 + V_\nu^4 = 1 - 2U_\nu^2 V_\nu^2 \ ,$$

leading to

$$(U_\nu^2 - V_\nu^2)^2 = U_\nu^4 + U_\nu^4 - 2U_\nu^2 V_\nu^2 = 1 - 4U_\nu^2 V_\nu^2 \ .$$

Inserting this relation in the square of the relation given in (9.9) leads to

$$4U_\nu^2 V_\nu^2((\varepsilon_\nu - \lambda_{\rm F})^2 + \Delta^2) = \Delta^2 \ ,$$

a relation which can be rewritten as

$$2U_\nu V_\nu = \frac{\Delta}{E_\nu} \ , \tag{9.10}$$

where the $+$ sign of the square operation implies minimization of the ground state energy U, and where

$$E_\nu = \sqrt{(\varepsilon_\nu - \lambda_{\rm F})^2 + \Delta^2} \ . \tag{9.11}$$

Making use again of the condition $H_{20} = 0$ (9.9) one obtains

$$(\varepsilon_\nu - \lambda_{\rm F})\frac{1}{E_\nu} = (U_\nu^2 - V_\nu^2). \tag{9.12}$$

Consequently,

$$U_\nu^2 - V_\nu^2 = 1 - 2V_\nu^2 = \frac{\varepsilon_\nu - \lambda_{\rm F}}{E_\nu} ,$$

and

$$V_\nu^2 = \frac{1}{2}\left(1 - \frac{\varepsilon_\nu - \lambda_{\rm F}}{E_\nu}\right),$$

leading to

$$V_\nu = \frac{1}{\sqrt{2}}\left(1 - \frac{\varepsilon_\nu - \lambda_{\rm F}}{E_\nu}\right)^{1/2} ,$$

$$U_\nu = \frac{1}{\sqrt{2}}\left(1 + \frac{\varepsilon_\nu - \lambda_{\rm F}}{E_\nu}\right)^{1/2} . \tag{9.13}$$

Let us now substitute these expressions in the relation $\Delta = G\alpha_0$ (9.8). One obtains

$$\Delta = \frac{G}{2} \sum_{\nu > 0} \left(1 - \frac{(\varepsilon_\nu - \lambda_{\mathrm{F}})^2}{E_\nu^2}\right)^{1/2},$$

$$= \frac{G}{2} \sum_{\nu > 0} \frac{\Delta}{E_\nu}.$$

This relation, together with (9.6) are the BCS equations namely,

$$N = 2 \sum_{\nu > 0} V_\nu^2, \qquad \text{(number equation)} \qquad (9.14)$$

$$\frac{1}{G} = \sum_{\nu > 0} \frac{1}{2E_\nu}, \qquad \text{(gap equation)}. \qquad (9.15)$$

One can now write U in terms of the parameters λ and Δ, that is,

$$U = 2 \sum_{\nu > 0} (\varepsilon_\nu - \lambda_{\mathrm{F}}) V_\nu^2 - 2\frac{\Delta^2}{G} + \frac{\Delta^2}{G}$$

$$= 2 \sum_{\nu > 0} (\varepsilon_\nu - \lambda_{\mathrm{F}}) V_\nu^2 - \frac{\Delta^2}{G}.$$

Making use of (9.7)–(9.9), one can write H_{11} as

$$H_{11} = \sum_{\nu > 0} \left\{ \frac{(\varepsilon_\nu - \lambda_{\mathrm{F}})^2}{E_\nu} + \frac{\Delta^2}{E_\nu} \right\} (\alpha_\nu^\dagger \alpha_\nu + \alpha_{\bar\nu}^\dagger \alpha_{\bar\nu})$$

$$= \sum_{\nu > 0} E_\nu (\alpha_\nu^\dagger \alpha_\nu + \alpha_{\bar\nu}^\dagger \alpha_{\bar\nu}) = \sum_\nu E_\nu \alpha_\nu^\dagger \alpha_\nu$$

$$= \sum_\nu E_\nu N_\nu, \qquad (9.16)$$

where

$$N_\nu = \alpha_\nu^\dagger \alpha_\nu,$$

is the operator counting the number of quasiparticles in the state ν. The lowest energy state of the system is a two-quasiparticle excitation

$$\alpha_{\nu_1}^\dagger \alpha_{\nu_2}^\dagger |\mathrm{BCS}\rangle = |\nu_1 \nu_2\rangle,$$

with energy

$$H_{11} |\nu_1 \nu_2\rangle = \sum_\nu E_\nu \big(\delta(\nu, \nu_2) \alpha_{\nu_1}^\dagger \alpha_{\nu_2}^\dagger + \delta(\nu, \nu_1) \alpha_\nu^\dagger \alpha_{\nu_2}^\dagger \big) |\mathrm{BCS}\rangle$$

$$= (E_{\nu_1} + E_{\nu_2}) \alpha_{\nu_1}^\dagger \alpha_{\nu_2}^\dagger |\mathrm{BCS}\rangle.$$

Because $E_{\nu_1} + E_{\nu_2} \geq 2\Delta$, the lowest excitation in the pairing correlated system lies at an energy $\geq 2\Delta$, that is, the energy which takes to break a pair. In fact, in the paired system, the only excitations possible are those associated with the breaking of pairs of particles moving in time reversal states, an operation which takes an energy of the order of 2Δ.

9.2 Superconductivity in Metals

Measurements of the critical temperature T_c as a function of the isotopic mass M of the ions revealed that $T_c \approx M^{-1/2}$. This result strongly suggested lattice phonons as the carriers of pairing in metals (see e.g. (8.16), (8.41), (9.18) and (9.19) below), although not all superconducting metals displayed this behaviour, in particular not Ru, Os, Ir and Rh (transition metals).

In elements which exhibit an isotope effect the attractive interactions responsible for superconductivity can be assumed to arise from the exchange of phonons between electrons near the Fermi energy moving in time-reversal states. Below, we shall see that also for transition metals this mechanism is the one which is at the basis of superconductivity (see Table 9.2).

Let us for simplicity neglect the frequency dependence (retardation) of the interaction and use instead a pairing force with constant matrix elements, as done in the previous section. In addition let us consider a set of continuous, or closely spaced levels. In this case, the sum $\sum_{\nu>0} = \frac{1}{2}\sum_{\nu}$ can be replaced by the integral $(1/2)\int_A^B d\epsilon' N'(\epsilon')$. The limits of integration A, B depend on the mechanism which is at the basis of the pairing interaction, while $N'(\epsilon')$ is the density of electronic levels. In the case where the pairing interaction is due to the exchange of phonons between electrons with energies lying in the interval $A = -\hbar\omega_D$, $B = \hbar\omega_D$, where ω_D is the Debye frequency. In this case, the gap equation (9.15) reads

$$1 = \frac{G}{2}N'(0)\int_{-\hbar\omega_D}^{\hbar\omega_D} \frac{de}{\sqrt{e^2 + \Delta^2}} = GN(0)\sinh^{-1}\left(\frac{\hbar\omega_D}{\Delta}\right) \tag{9.17}$$

where $N(0)$ is the density of levels of one spin orientation at the Fermi energy. In deriving the above expression, the assumption has been made that the $N(\epsilon)$ changes little within the energy range $2\omega_D$ around ϵ_F.

In the weak coupling limit ($\lambda = GN(0) \ll 1$),

$$\Delta(T = 0) = 2\,\hbar\omega_D\,e^{-1/\lambda}, \tag{9.18}$$

while in the strong coupling limit ($\lambda \approx 1$)

$$\Delta(T = 0) = 2\,\hbar\omega_D\,\lambda, \tag{9.19}$$

where λ was defined in the previous Chapter (see also Chap. 2).

9.2.1 Finite Temperature

For $T \neq 0$, states with a number of quasiparticles are thermally excited. When $T \ll \Delta$, only few of these excitations are independent but for $T \approx \Delta$, the number of excitations is large. In what follows we study the average effect of the interaction between them.

At temperature T, the probability of finding a quasiparticle in the state ν (or $\tilde{\nu}$) is

$$\langle a_\nu^\dagger a_\nu \rangle_T = \langle a_{\bar\nu}^\dagger a_{\bar\nu} \rangle_T = F_\nu, \tag{9.20}$$

where the symbol $\langle\ \rangle_T$ denotes a thermal average. Minimizing the free energy $F = \langle H_{\mathrm{sp}} + H_{\mathrm{p}} \rangle_T - TS$ with respect to V_ν and F_ν (see Appendices 9.5.2 and 9.5.3) one obtains

$$F_\nu = \frac{1}{1 + \exp\left(E_\nu/T\right)}, \tag{9.21}$$

$$\Delta = G \sum_{\nu>0} U_\nu V_\nu (1 - 2F_\nu), \tag{9.22}$$

and

$$\frac{1}{G} = \sum_{\nu>0} \frac{1}{2E_\nu}(1 - 2F_\nu). \tag{9.23}$$

Making use of the fact that

$$1 - 2F_\nu = \tanh\left(\frac{E_\nu}{T}\right), \tag{9.24}$$

and that at the temperature required to break Cooper pairs, $T = T_{\mathrm{c}}$, the gap $\Delta = 0$, one can write

$$\frac{2}{GN(0)} = \int_{-\omega_{\mathrm{D}}}^{\omega_{\mathrm{D}}} \frac{\mathrm{d}e}{e} \tanh\left(\frac{e}{T_{\mathrm{c}}}\right).$$

Introducing the change of variables $x = e/2T_{\mathrm{c}}$, one can write

$$\frac{2}{\lambda} = 2 \int_0^{\omega_{\mathrm{D}}/2T_{\mathrm{c}}} \mathrm{d}x \frac{\tanh x}{x}.$$

In the weak coupling limit ($\lambda \ll 1$), one obtains by numerical integration,

$$T_{\mathrm{c}} = 1.14\, \omega_{\mathrm{D}}\, \mathrm{e}^{-1/\lambda}. \tag{9.25}$$

From this relation and (9.18), one can write

$$\frac{2\Delta(0)}{T_{\mathrm{c}}} = 3.5\ . \tag{9.26}$$

This prediction works reasonably well for metals, with some conspicuous exceptions, like e.g. Pb and Hg(α) (see Table 9.1).

To better understand this result we need to look in more detail at the basic assumptions used to derive both (9.18) and (9.25), in particular $\lambda \ll 1$ and $\omega_{\mathrm{D}} \gg T_{\mathrm{c}}$. For this purpose we turn to experiments to obtain the values of λ and of ω_{D}. Specific heat measurements provide information on both quantities (see also Sect. 2.10).

In fact, following standard theory, we write the total heat capacity of a metal (non magnetic), at low temperatures, as

$$C = C_{\mathrm{e}} + C_{\mathrm{ph}} = \gamma\, T + AT^3,$$

Table 9.1. Ratio of twice the pairing gap at zero temperature and the critical temperature for a number of metals (after [35], Table 34.3)

Element	$2\Delta(0)/T_c$
Al	3.4
Cd	3.2
Hg(α)	4.6
In	3.6
Nb	3.8
Pb	4.3
Sn	3.5
Ta	3.6
Tl	3.6
V	3.4
Zn	3.2

where C is the heat capacity per mole. For the free electron gas, the electronic heat capacity is $C_e = \gamma\, T$ with

$$\gamma = \frac{\pi^2 k_B^2 N_0 m_e}{\hbar^2 (3\pi^2 n)^{3/2}}.$$

It is convenient to express γ for a real metal as proportional to the average electronic band mass at the Fermi level,

$$\gamma = \gamma_b = \frac{N(0)}{N^{(0)}(0)}\gamma^{(0)} = \frac{m_b}{m}\gamma^{(0)},$$

where the superscript (0) is the free-electron value.

In the phonon part AT^3 of the heat capacity,

$$A = \frac{12\pi^4}{5} k_B N_0 \Theta_D^{-3}.$$

This relation follows from the limit $T \ll \Theta_D$ of the lattice heat capacity C_D in a Debye model with Debye temperature Θ_D determined from the relation

$$C_D = 9k_F N_0 \left(\frac{T}{\Theta_D}\right)^3 \int_0^{\Theta_D/T} dx \frac{e^x x^u}{(e^x - 1)^2}.$$

To be noted that $k_B\Theta_D = \omega_D$. Values of ω_D are shown in Table 9.2. Also shown in this table are the observed values of T_c. In all cases $T_c \ll \omega_D$.

From the ratio between the experimental and the free-electron values of γ, one can define a thermal effective electron mass m_{th} as

$$m_{th} = \frac{\gamma_{exp}}{\gamma^{(0)}} m_e.$$

To be noted that the ratio of the thermal mass m_{th} obtained from heat capacity measurements to the average band mass at the Fermi level m_b obtained from electron band calculations, that is, m_{th}/m_b deviates, in the case of metals like Li, Na, Pb, Rb, etc, considerably from the expected value of one.

Table 9.2. Debye temperature ω_D (known also as Θ_D), electron–phonon coupling constant λ, and critical temperature T_c for a number of metals. In the last column are displayed the results obtained making use of (9.28) and the experimental values for λ and ω_D listed in columns 2 and 3 and $\mu^* = 0.15$

	ω_D (K) [e]	λ [f]	T_c(K) exp.	T_c(K) th. [g]
Mg [a]	405	0.35 ± 0.05	-	7.7×10^{-2}
Al [b]	430	0.43 ± 0.05	1.196	0.73
Cd [b]	209	0.40 ± 0.05	0.56	0.19
Ru [c]	600	0.40 ± 0.10	0.49	0.53
Os [c]	500	0.40 ± 0.10	0.655	0.44
Ir [c]	420	0.40 ± 0.10	0.14	0.37
Pb [d]	105	1.55 ± 0.05	7.19	10.59
α-Hg [d]	72	1.60 ± 0.1	4.15	7.51

[a] Simple metals - not superconductors
[b] Simple metals - superconductors (weak coupling)
[c] Transition metals
[d] Simple metals - strong coupling superconductors
[e] Table 3.1 [43]
[f] Table 11.1 [43]
[g] Equation (9.28), with $\mu^* = 0.15$

The discrepancy is due to the neglect, in the calculations of m_b of the renormalization effects arizing from the electron–phonon coupling. In fact, a correct theory requires that the band mass at the Fermi level is enhanced by an electron–phonon mass enhancement-factor λ, that is (see Sect. 2.10),

$$m_{\text{th}} = (1 + \lambda)m_b, \tag{9.27}$$

as well as

$$\gamma = (1 + \lambda)\gamma^{(0)}.$$

Typical values of λ are reported in Table 9.2. From these results it is seen that Pb and Hg(α) are not weak-coupling superconductors and that (9.18) and (9.25) are hardly applicable to these systems. In fact these metals are strong-coupled superconductors. Consequently, to neglect the retardation (ω-dependence) of the interaction associated with the exchange of phonons between pairs of electrons (constant G), as well as to use bare, instead of phonon dressed electrons, is less justified than in the case of $\lambda \ll 1$.

On the other hand, the values of λ do not explain why some transition metals do not display the isotope effect. This result is closely connected with the fact that very weak coupled superconductors, with $T_c \ll \Theta_D$, should indeed have isotope effects below $\frac{1}{2}$, because of the sharp competition between phonon interactions (see Sect. 9.2 and (8.16) and (8.26)), which do scale with

$M^{-1/2}$, and the Coulomb interactions and electronic screening (electron–phonon couplings, see Sect. 8.6), which do not.

To take into account all these effects, Eliashberg derived a pair of coupled integral equations which relate a complex energy gap function $\Delta(\omega)$ and a complex renormalization parameter Z_ω (see (2.55)) for the superconducting state to the electron–phonon and the electron–electron interactions in the normal state [199]. A succesful parametrization of the solution of these equations for the critical temperature is due to McMillan [194]

$$T_c = \frac{\hbar\omega_{\ln}}{1.2} \exp\left[-\frac{1.04(1+\lambda)}{\lambda - \mu^*(1+0.62\lambda)}\right], \tag{9.28}$$

where $\omega_{\ln}$ is a typical phonon frequency (logarithmic average), λ is the electron–phonon coupling, and

$$\mu^* = \frac{\mu}{1 + \mu\ln(B/\omega_{\mathrm{ph}})}, \tag{9.29}$$

is the Coulomb pseudopotential (see also (2.72)), while

$$\mu = N(0)U_c \tag{9.30}$$

with U_c some typical (screened) Coulomb interaction. The quantities B and ω_{ph} are a typical electron energy (half the band width) and a representative phonon energy.

Typical values of μ^* for metals are estimated to lie in the interval $\mu^* \approx$ 0.1-0.2 (see Appendix 9.5.1). Making use of this result and of the values of ω_D and λ obtained from experimental data, (9.28) predicts the values of T_c reported in Table 9.2. These values are in overall agreement with the experimental findings.

9.3 Superconductivity in Fullerides

Fullerenes like, e.g., C_{60}, have attracted much interest since their discovery in 1985 [200], not least because they are the only finite, and most symmetric, allotropic forms of carbon. This interest increased drastically when methods were found to produce some of these fullerenes in large quantities [201], so it was possible to make solids (fullerites). Intercalating alkali metal atoms in solid C_{60} (a van der Waals solid), leads to metallic behaviour displaying a transition temperature of the order of 30 K [173].

Although there are still a number of open questions, several experimental findings testify to the fact that superconductivity in doped fullerides is likely to be driven by the coupling of electrons to intramolecular vibrations (phonons) of the fullerene molecules. In what follows we shall discuss some of these findings and ellaborate on the expectations opened by the possibilities of superconducting materials based on smaller fullerenes than C_{60} like C_{36}, C_{28} and C_{20}.

9.3.1 C_{60} Fullerene Based Superconductors

As already pointed out in previous chapters, the C_{60} molecule is the most symmetric of the fullerenes in the sense that its point group (icosahedral) with 120 symmetry operations is the largest point group of the known molecules. It has the shape of a soccer ball (see Fig. 1.3). The 60 carbon atoms are all equivalent and form 12 pentagons and 20 hexagons. It has 174 normal modes of vibrations (phonons, $3 \times 60 - 6 = 174$, 6= three rigid rotations and translations) and 240 valence (delocalized) electrons. The HOMO, of h_u symmetry (π electrons), is five-fold orbitally degenerate. The LUMO is a three-fold degenerate t_{1u} state (see Fig. 5.1). The energy gap between the HOMO and the LUMO state is predicted in LDA to be ≈ 1.9 eV (see, e.g., [59]) as compared to the value of 1.86 ± 0.1 eV measured experimentally [27] (see Chap. 5).

The high stability of C_{60} is associated with the fact that its HOMO state is fully occupied (closed shell system). These molecules condense into a solid of weakly bound molecules (Van der Waals solid). While the shortest separation between two atoms on the same molecule is about 1.4 Å, the shortest separation between two atoms of different molecules is about 3.1 Å, the fullerite displays a lattice constant of 14.15 Å (to be compared with the radius of C_{60} equal to 3.53 Å). The fullerites are therefore molecular solids, in which many of the molecular properties essentially survive. The discrete levels of a free C_{60} molecule are only weakly broadened in the solid, which leads to a set of essentially non overlapping bands with a band width of about 0.5 eV (see Fig. 9.1). For an undoped C_{60} solid, the h_u band is filled and the t_{1u} band is empty, and this system is therefore a band insulator. When solid C_{60} is doped by alkali atoms, the alkali atoms donate about one electron each to the t_{1u} band.

The phonons of alkali fullerides can be divided into subgroups that reflect the molecular nature of the solid. The highest are due to intramolecular vibrations and have energies in the range 273-1575 cm^{-1} (0.034-0.115 eV) for an undoped C_{60} solid (see Sect. 6.2, in particular Table 6.2). The intermolecular phonons occur at substantially lower energies in the range up to about 140 cm^{-1} (17 meV). It is believed that the intramolecular modes play the main role in the superconductivity of C_{60}-based materials. For the intramolecular modes, the effect of the interaction between the C_{60} molecules is small. These phonons can therefore be approximately classified by the icosahedral point group. In alkali fullerides, the t_{1u} band is (partly) populated. The coupling of the phonons to this band is therefore of particular interest. As discussed in the previous chapter, because of symmetry arguments only the two A_g phonons and the eight H_g phonons can couple to the t_{1u} electrons. The corresponding parameters λ_ν, obtained from the analysis of a photoemission experiment, are shown in Table 9.3 (see also Tables 8.1 and 8.2). Also given are the frequencies of the phonons.

Making use of the density of states $N(0) \approx 8.1$ eV^{-1} deduced in [170] for Rb$_3$C$_{60}$ as well as of the results of Table 9.3 ($\lambda/N(0)$=158 meV), one obtains for the total electron–phonon coupling

$$\lambda = 1.3. \tag{9.31}$$

Aside from this quantity and the logarithmic average

$$\hbar\omega_{\text{ln}} = 102 \text{ meV} = 1184 \text{ K}, \tag{9.32}$$

one needs the Coulomb pseudopotential μ^* (see (9.29) and (9.30)) to calculate T_c from (9.28). The best available estimates for the C$_{60}$ molecule are $0.3 \leq \mu^* \leq 0.4$ [173]. Making use of the results given in (9.31) and (9.32), and of the fact that for Rb$_3$C$_{60}$, T_c =29.5 K [172, 203, 204], one obtains from (9.28) $\mu^* \approx 0.34$.

Summing up, within the picture developed in this Section (see also [173]), superconductivity in materials constructed making use of atomic clusters as building blocks is controlled essentially by three parameters: (a) the electron–phonon coupling λ, (b) the density of levels $N(0)$, (c) the Coulomb pseu-

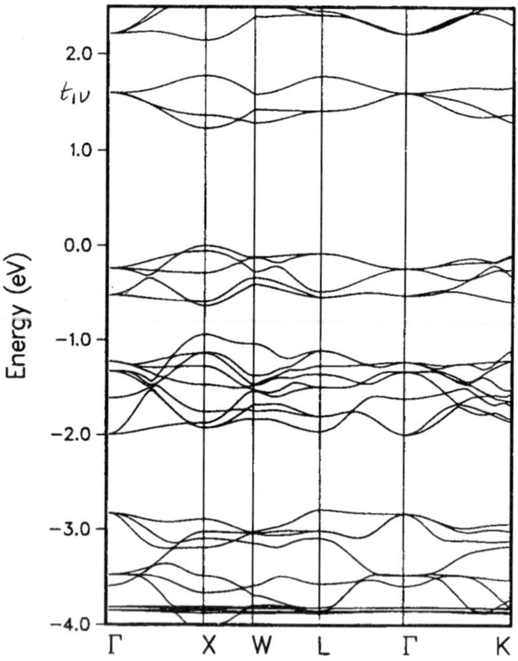

Fig. 9.1. Some of the subbands around the Fermi energy for solid C$_{60}$ in the $Fm\bar{3}$ structure. The bands at about -0.5 eV are the h_u bands, which are occupied in solid C$_{60}$, and the bands at about 1.5 eV are the t_{1u} bands, which become populated in A_nC_{60}. The bands around -1.5 eV result from the overlapping h_g and g_g bands. From [202]

Table 9.3. The partial electron–phonon coupling constants $\lambda_\nu/N(0)$ in C_{60-} according to photoemission data. Because it is not possible in the analysis of these data to distinguish between the coupling to the A_g modes and the H_g modes with similar energies, the coupling to the A_g modes were taken from the calculation of [170]. With this assumption the couplings to the H_g modes can be determined uniquely. Also given, at the bottom of columns 2 and 3 respectively, are the energies of the different phonons, as well as their logarithmic average $\hbar\omega_{ln}$ and the total electron–phonon coupling constant $\lambda = \sum_\nu \lambda_\nu/N(0)$ (see Chap. 8, Table 8.2)

	$\hbar\omega$ [meV]	$\lambda_\nu/N(0)$ [meV]
$H_g(8)$	195.3	23
$H_g(7)$	177.1	17
$H_g(6)$	155.0	5
$H_g(5)$	136.3	12
$H_g(4)$	95.0	18
$H_g(3)$	88.0	13
$H_g(2)$	54.2	40
$H_g(1)$	33.9	19
$A_g(2)$	182.2	11
$A_g(1)$	61.5	0
	102	158

dopotential μ^*. The first quantity depends on the properties of the (isolated) building blocks. The second and, to a large extent, the third, on the properties of the solid. This is because ions are anchored to the clusters, while delocalized electrons feel the whole system.

9.3.2 Nanometer Superconductors Based on Small Fullerenes

The interest for using smaller fullerenes than C_{60} to produce nanometer superconducting materials can be simply understood by remembering that the critical temperature below which graphite compounds are superconducting is $T_c=5$ K, while the same temperature has reached ≈ 30 K for C_{60}. Because C_{60} can be viewed as made out of a wrapped carbon lattice plane by introducing defects in the honeycomb lattice so as to create 12 pentagons (C_{60} is made out of these 12 pentagons and of 20 hexagons), curvature effects and electron spill out seems to be at the basis of the increased electron–phonon coupling and screening of the Coulomb field in going from graphite to C_{60} based materials. It is thus expected that materials made out of fullerenes with higher curvature and electron spill out than C_{60}, like C_{36}, C_{28} and C_{20} will display higher values of T_c.

This situation is very similar to that found in the case of other finite systems, in particular of the atomic nucleus. In this case, the strong force provides the basic glue of Cooper pairs in nuclei. Nonetheless, the contribution to this glue arizing from the exchange of low-lying surface collective vibrations between pairs of nucleons moving in time reversal states close to the Fermi energy cannot be neglected in a quantitative description of pairing in nuclei. In fact, it is found that in the case of spherical stable nuclei lying along the stability valley, about half of the pairing correlations arise from this exchange [45, 182, 183, 184, 185]. When e.g. neutrons are progressively added to these nuclei, in particular in the case of light nuclei like $_3$Li and $_4$Be (i.e. nuclei with 3 and 4 protons respectively), the Pauli principle forces them into states of higher momentum which in the case of $^{11}_{3}$Li$_8$ and $^{12}_{4}$Be$_8$ essentially drip out from the corresponding nuclei being quite loosely bound and giving rise to halo nuclei. In halo nuclei, some of the constituents neutrons venture beyond the "core" surface and form a very low-density misty cloud or halo. Not surprisingly, these extended nuclei behave very differently from ("normal") nuclei lying along the stability valley in the chart of nuclides. For example, ^{11}Li is twice as large as one of these nuclei of the same mass number and essentially half the size the lead nucleus ^{208}Pb, which holds 197 more particles. In the case of ^{11}Li, the last two neutrons are very weakly bound. Consequently, these neutrons need very little energy to move away from the nucleus. If one neutron is taken away from ^{11}Li, a second neutron comes out immediately, leaving behind the core of the system, the normal nucleus ^{9}Li (lying along the stability valley). This result indicates that pairing plays an important role in the stability of ^{11}Li. In fact, the properties of the exotic nucleus ^{11}Li can be understood, even quantitatively, in terms of the simplest scenario imaginable: the formation of a single Cooper pair which is held together to the ^{9}Li core essentially by the exchange of surface vibrations [184]. The importance of these vibrations is due to the high polarizability displayed by ^{11}Li, a consequence of the high curvature and large nucleon spill out of this system, as compared to "normal" nuclei.

Fullerenes with higher curvature than C_{60} can be obtained by reducing the number of hexagonal faces of the cage as one cannot reduce the number of pentagons. This is because according to Euler's theorem, 12 is the minimum number of pentagons a polyhedron can have. Therefore C_{20}, which is made out of exactly 12 pentagons (see Fig. 1.3), is the smallest member of the family of fullerenes, displaying the largest surface to volume ratio. Consequently, displaying also the largest curvature and, potentially, the largest electron spill out. The associated phonons are thus likely to provide an important glue between pairs of electrons in fullerene based materials. The C_{20} molecule has been synthesized in the gas phase[1] [206]. Experimental evidence about the synthesis of a crystalline solid based on C_{20} fullerenes has been provided in [207]. A face-centered-cubic symmetry has been inferred from Transmission Electron Microscopy data, and the presence of C_{20} clusters as building blocks of the new crystal has been determined by mass spectroscopy of laser-desorbed fragments. Based on this information a C_{20}-based solid was proposed [12]. The structure comprises C_{20} cages in an fcc lattice intercon-

---

[1] The synthesis of the fullerene C_{20} was achieved from the fully hydrogenated $C_{20}H_{20}$ molecule, dodecahedrone, which turns out to be stable [205].

nected by two bridging atoms per unit cell in the tetrahedral interstitial sites, the unit cell containing 22 atoms. On the basis of (9.28), the calculated $\lambda = 1.12$ for this fcc-C_{22} solid, and a μ^* assumed in the range 0.3–0.1, a T_c in the range 15–55 K is predicted.

In supersonic cluster beams obtained from laser vaporization, C_{28} is the smallest even-numbered cluster, and thus the fullerene displaying the largest curvature which is produced with special abundance. In fact, under suitable conditions, C_{28} is almost as abundant as C_{60} [208]. At variance to C_{60}, C_{28} is expected to form a covalent crystal (like C_{36}, [209, 210]) and not a van der Waals solid [3]. However, similar to C_{60}, C_{28} maintains most of its intrinsic characteristics when placed inside an infinite crystalline lattice [211]. The transport properties of the associated doped fullerides, in particular super-conductivity, can thus be calculated in terms of the electron–phonon coupling strength λ of the isolated molecule, and of the density of states of the solid [173, 211]. Calculations of $\lambda/N(0)$ have been carried out making use of ab initio methods for C_{36} and C_{28} and estimates of T_c based on (9.28) worked out (see also [212]).

Table 9.4. Electron–phonon coupling constants (2nd column) for the fullerenes displayed in the first column calculated making use of the model discussed in the text (ab initio calculations). The numbers displayed in the third column correspond to those shown in column 2, but multiplied by 158/84.6 so as to obtain for C_{60} the value of λ extracted from PES (see Table 9.3)

	$\lambda/N(0)$ [meV]	$\times$ 158/84.6	$\lambda^{a)}$	T_c [°K]$^{b)}$	$T_c/T_c(C_{60})$
$C_{70}^{c)}$	10.7	20	0.2	~ 0	0
$C_{60}^{d)}$	84.6	158	1.3	30	1
$C_{36}^{e)}$	186	347.4	2.8	120	4
$C_{28}^{f)}$	214	400.4	3.2	134	4.4

a) Obtained from the values of $\lambda/N(0)$ reported in the third column, and $N(0){=}8.1$ eV^{-1} typical of Rb_3C_{60}
b) Calculated making use of (9.28) and of the λ values of the preceding column, as well as $\mu^* \approx 0.34$ (in fact 0.3425) and $\hbar\omega_{ln} = 1184$ K
c) [179]
d) [165]
e) [214]
f) [160]

The first-principles investigation of the electronic and vibrational properties of C_{28}, as well as of the corresponding electron–phonon coupling strength within the framework of DFT in the local-spin-density approximation (LSDA) has been carried out in [160]. The equilibrium geometry of C_{28} obtained in this calculation has a full T_d point–group symmetry (see also [213]). All atoms are three fold coordinated, arranged in 12 pentagons and 4 hexagons. The resulting estimates for $\lambda/N(0)$ and of T_c are shown in Table 9.4, together with those associated with C_{70} and C_{36} fullerenes, all referred to the empirically obtained values of C_{60}. Although the estimates displayed in this Table depend on a number of assumptions they underscore the potentiality of fullerene based superconductors.

9.3.3 Thin Carbon Nanotubes

Closed carbon nanotubes of different diameters and thus having caps each containing six pentagons and an appropriate number of hexagons that are selected to fit perfectly to the cylindrical section have been produced. Within this content, while in principle there is no limitation on how large the diameter of a nanotube can be, the thinnest possible closed nanotube is that capped by two halves of C_{20} [14, 15, 215]. Also in this case each belt added to the two zigzag halves of C_{20} is made out of 5 benzoid rings (with a common edge), each containing 10 carbon atoms. Ab initio calculations of the electronic and ionic degrees of freedom as well as of their coupling of a (incipient) C_{20}-capped nanotube containing 4 belts (and thus 60 carbon atoms, see Fig. 9.2), have been carried out by one of the authors (H.E.R.). The results are reported in Fig. 9.3 and Fig. 9.4.

The reduced electron–phonon coupling constants $g_\nu^2/\hbar\omega_\nu$ have been calculated for each phonon of energy $\hbar\omega_\nu$. The mean coupling per phonon is found to be $\langle g_\nu^2/\hbar\omega_\nu\rangle \approx 0.33$ meV, yielding $\lambda/N(0) = 2 \times 174 \times 0.33\text{meV} = 0.114$ eV. Assuming $N(0) = 8$ eV^{-1} for a nanotube of same diameter but of a length of tens of nanometers, an ansatz which is not inconsistent with the states available per unit of energy around the Fermi energy (see Fig. 9.3), we find $\lambda = 0.9$. Using the McMillan relation (9.28) with $\hbar\omega_{\ln} = 1302$ ^{o}K and $\mu^* = 0.2$, we obtain $T_c \approx 38$ ^{o}K.

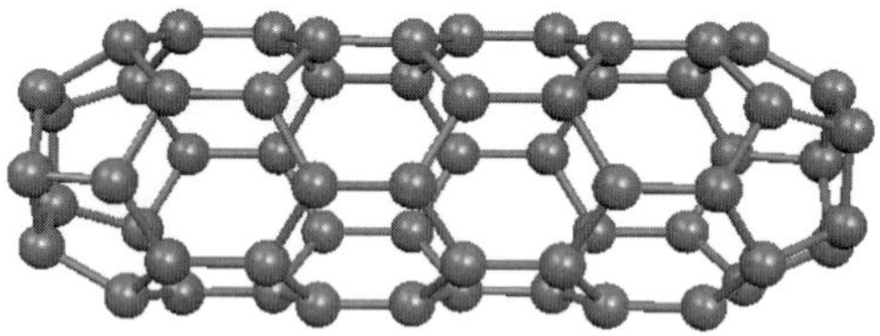

Fig. 9.2. The thinnest carbon nanotube closed with C_{20} caps, illustrated here for a 60 atoms molecule

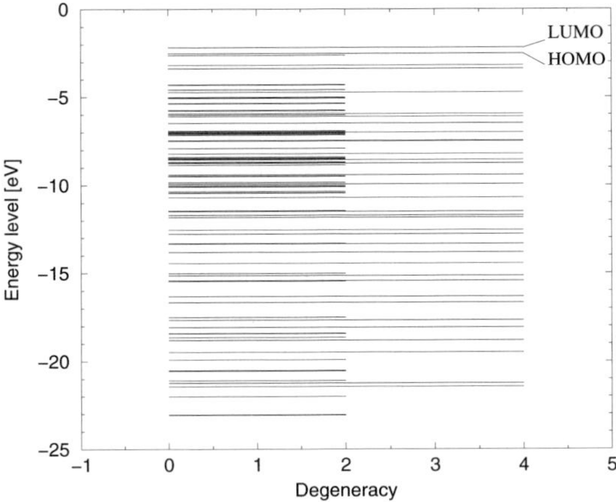

Fig. 9.3. The occupied Khon–Sham orbitals of the incipient nanotube shown in Fig. 9.2. The degeneracy of the electronic states can be either 2 or 4, including the spin

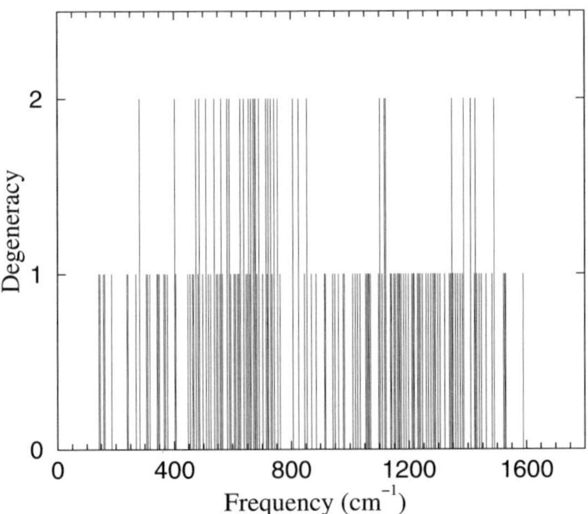

Fig. 9.4. Phonon spectrum of the incipient nanotube shown in Fig. 9.2. The vibrational states can have degeneracies one or two

9.4 Selected Open Questions

The finding that some alkali-doped C_{60} compounds (fullerides) are superconducting (with T_c that is only surpassed by that of the cuprates) provides an important example of the statement made in the preface that finite manybody systems (FMBS) can be used as building blocks to construct new materials (in particular superconductors), displaying novel properties directly related to those of the FMBS.

Although based on circumstancial evidence, neutron stars provide another examples of non-conventional superconducting and superfluid systems based on FMBS. In fact, the inner crust of neutron stars are thought to be made of a Coulomb lattice of neutron rich atomic nuclei, inmersed in a see of free neutrons (see [216] and refs. therein). In spite of the fact that pairing in atomic nuclei have been studied now for over fourty years, many open questions await an answer. In particular, the role the exchange of collective modes between pairs of nucleons moving close to the Fermi energy play in the formation of Cooper pairs, and in the resulting value of the pairing gap [45]. It will thus not come as a surprise that superconductors made out of atomic aggregates, discovered only a decade ago, have not yet been fully understood.

9.4.1 Hole Doped Fullerites

Although the so-called single j-shell (with degeneracy $2j + 1$) model is only an approximation to the mean field (independent-particle) description of the fermion motion in finite many-body systems, it has proved a useful model. In particular in the case of atomic nuclei [45]. Because the HOMO and LUMO states of the variety of fullerenes (C_{60}, C_{36}, C_{28}) are quite isolated from the other electronic states, these systems can be viewed as a concrete realization of the j-shell model. Within this model, the solution of the gap equation is quite simple. In fact, the probability that the different substates of such a system containing N particles are occupied is

$$V^2 = \frac{N}{2\Omega},\tag{9.33}$$

where

$$\Omega = \frac{2j + 1}{2},\tag{9.34}$$

is half the degeneracy of the orbital (degeneracy associated with a single spin orientation, or number of pairs of fermions with spin up and spin down which can be accomodated in the orbital. For example, in the case of the LUMO of C_{60}, where $j \equiv t_{1u}$, $\Omega = 3$).

Making use of the normalization condition (9.4) and of (9.33) one obtains

$$U^2 = 1 - \frac{N}{2\Omega},\tag{9.35}$$

and (see (9.8))

$$\Delta = \frac{G}{2}\sqrt{N(2\Omega - N)},\tag{9.36}$$

where the fact that $\sum_{\nu>0} = \Omega$ has been used. Consequently, Δ acquires its maximum value

$$\Delta_{\text{max}} = \frac{G\Omega}{2},\tag{9.37}$$

for $N = \Omega$ (half filled shell, see Fig. 9.5), in keeping with the fact that the two contradictory conditions have to be fulfilled to obtain a large pairing gap (for fixed N), namely: (a) large number of valence (conduction) fermions (note that $\Delta = 0$ for $N = 0$), (b) large number of (empty) final states (note that $\Delta = 0$ for $N = 2\Omega$). The situation $N = \Omega$ (number of particles and of holes equal) provides the best solution of this frustrated system. This result is likely to be at the basis of the fact that one needs to dope C_{60} fulleride with three electrons per unit cell to obtain a superconducting material, in keeping with the fact that t_{1u} (LUMO state of C_{60}) is three fold degenerate for each spin orientation.

From (9.37) and from the fact that in BCS theory $\Delta(0)$ is proportional to $T_{\rm c}$ (see (9.26)), we find that

$$\frac{\Delta(T = 0, C_{60}^{+++++})}{\Delta(T = 0, C_{60}^{---})} \approx \frac{T_{\rm c}(C_{60}^{5+})}{T_{\rm c}(C_{60}^{3-})} \approx \frac{5}{3}, \tag{9.38}$$

where we have indicated with C_{60}^{+} a fullerene doped with a hole in the HOMO. The ratio $5/3$ is in keeping with the corresponding degeneracies of the HOMO(h_u) and the LUMO(t_{1u}) of C_{60} (see Fig. 5.1).

Similar reasoning suggests that (see Fig. 1 in [160])

$$T_{\rm c}(C_{28}^{20+}) = \frac{20}{14} T_{\rm c}(C_{28}^{14-}). \tag{9.39}$$

Thus, making use of the values reported in Table 9.4 one can write,

$$T_{\rm c}(C_{28}^{20+}) \approx \frac{20}{14} \times 4.4 \; T_{\rm c}(C_{60}^{3-}). \tag{9.40}$$

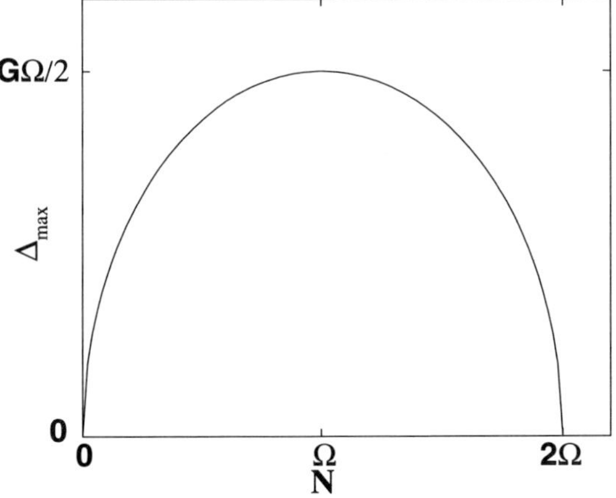

Fig. 9.5. Pairing gap in the single j-shell model (see (9.36)) as a function of the number of particles

These results are suggestive[2] and worth investigating both theoretically and experimentally.

9.4.2 Screening Effects

The fact that the screening of the Coulomb field arising from the electron–phonon coupling (see Sect. 8.6) in Na_8 and Na_9^+ leads to an important softening (red shift) of the dipole spectrum (see Sect. 8.8) solving a long standing problem [5, 6], and bringing theory essentially into agreement with experiment, indicates that a proper description of such coupling can be essential to achieve a quantitative description of the detailed properties of these systems. In particular, of the total effective interaction acting between pairs of electrons and resulting from the interplay of the Coulomb repulsion and of the exchange of phonons and of plasmons.

This emerges, for example, from (9.28). In this formula, the relevant quantity, namely the exponent

$$\frac{\lambda}{1+\lambda} - \mu^* \frac{1+0.62\lambda}{1+\lambda} \approx \frac{\lambda}{1+\lambda} - \mu^* \tag{9.41}$$

indicates that μ^* is as important as λ in determining T_c (see also [173] and [219]). Now, a proper calculation of the screened Coulomb field requires the detailed calculation of the effective interaction arising from the exchange of plasmons between electrons, in the same way to what is done in connection with the electron–phonon coupling (see Chap. 8).

9.5 Appendices

9.5.1 Simple Estimate of μ

The screened Coulomb potential (within the volume V) (see Fig. 9.6)

$$U_c(q,\omega) = \frac{4\pi e^2}{V q^2 \varepsilon(q,\omega)}, \tag{9.42}$$

displays an ω-dependence (retardation) arising from the exchange of plasmons between electrons. In the limit of $\omega = 0$ one can write (Thomas–Fermi approximation)

$$\epsilon(q,0) = 1 + \frac{k_s^2}{q^2}, \tag{9.43}$$

where the screening wave vector is given by

$$k_s^2 = \frac{4}{\pi}\frac{k_F}{a_0} = \frac{6\pi n e^2}{\epsilon_F}, \tag{9.44}$$

---

[2] This in spite of refs. [217] and of their corresponding retractions [218].

leading to

$$U_{\mathrm{c}} = \frac{4\pi e^2}{V(q^2 + k_{\mathrm{s}}^2)}. \tag{9.45}$$

In (9.44), k_{F} is the Fermi momentum, while n is the density of the system.

We now define U_{c} as an average over all q vectors connecting two points on the Fermi surface and introduce a dimensionless parameter $\mu = N(0)U_{\mathrm{c}}$. One obtains

$$\begin{aligned}
\mu = N(0)U_{\mathrm{c}} &= \frac{N(0)}{V 2k_{\mathrm{F}}^2} \int_0^{2k_{\mathrm{F}}} \mathrm{d}q \, \frac{4\pi e^2}{q^2 + k_{\mathrm{s}}^2} \\
&= \frac{N(0)}{V 2k_{\mathrm{F}}^2} 2\pi e^2 \ln\left(\frac{4k_{\mathrm{F}}^2 + k_{\mathrm{s}}^2}{k_{\mathrm{s}}^2}\right).
\end{aligned} \tag{9.46}$$

Making use of the relation given in (9.44) and of

$$N(0) = \frac{V}{2\pi^2} \frac{m k_{\mathrm{F}}}{\hbar^2}, \tag{9.47}$$

one obtains

$$\mu = \frac{k_{\mathrm{s}}^2}{8k_{\mathrm{F}}^2} \ln\left(\frac{4k_{\mathrm{F}}^2 + k_{\mathrm{s}}^2}{k_{\mathrm{s}}^2}\right). \tag{9.48}$$

To approximately take into account the ω-dependence of the Coulomb pseudopotential in a way which is consistent with the ω-dependence kept in the electron–phonon induced interaction in (9.28), one can change μ to a new effective interaction μ^* obtained as [44]

$$\mu^* = \frac{\mu}{1 + \mu \ln(B/\omega_{\mathrm{ph}})}, \tag{9.49}$$

where B is a typical electron energy (half the bandwidth) and ω_{ph} is a typical phonon energy. If $B/\omega_{\mathrm{ph}} \gg 1$, μ^* can be strongly reduced relative to μ. In the limit $\mu \ln(B/\omega_{\mathrm{ph}}) \gg 1$, (9.49) simplifies to $\mu^* \approx 1/\ln(B/\omega_{\mathrm{ph}})$, which may be of the order $0.1 - 0.2$.

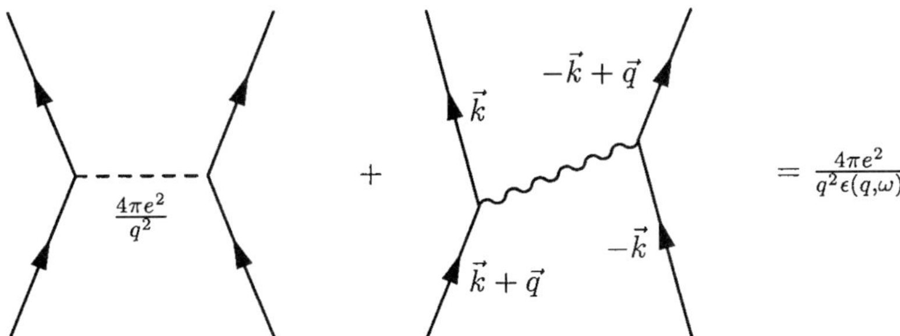

Fig. 9.6. Scattering process of two electrons (*arrowed lines*) interacting via the bare Coulomb interaction (*dotted line*) and by exchanging a plasmon (*wavy line*)

9.5.2 Solution of the Pairing Hamiltonian

We start by defining the pairing Hamiltonian (see Sect. 9.1)

$$H = H_{\mathrm{sp}} + H_{\mathrm{p}} \tag{9.50}$$

with

$$H_{\mathrm{sp}} = \sum_{\nu>0}(\varepsilon_\nu - \lambda)(a_\nu^+ a_\nu + a_{\bar\nu} a_{\bar\nu}), \tag{9.51}$$

and

$$H_{\mathrm{p}} = -G \sum_{\nu,\omega>0} a_\nu^+ a_{\bar\nu}^+ a_{\bar\omega} a_\omega. \tag{9.52}$$

Then, this interaction means that each state ν is correlated with its time reversal state $\bar\nu$. In order to take into account this effect we introduce, instead of a_ν and $a_{\bar\nu}$, the new Fermi operators

$$\alpha_\nu = U_\nu a_\nu - V_\nu a_{\bar\nu}^+,$$

$$\alpha_{\bar\nu} = U_\nu a_{\bar\nu} + V_\nu a_\nu^+, \tag{9.53}$$

where U_ν and V_ν are real numbers which obey the condition

$$U_\nu^2 + V_\nu^2 = 1, \tag{9.54}$$

and where $a_{\bar{\bar\nu}}^+ = -a_\nu^+$, in keeping with the fact that the time reversal operator is antiunitary.

In order to write the Hamiltonian in the new variables use is made of the transformation inverse to (9.53), that is,

$$a_\nu^+ = U_\nu \alpha_\nu^+ + V_\nu \alpha_{\bar\nu}. \tag{9.55}$$

Single-particle Hamiltonian. Making use of the relation (9.55) and the corresponding relations obtained from it by hermitian conjugation and time reversal operation one can write

$$a_\nu^+ a_\omega = (U_\nu \alpha_\nu^+ + V_\nu \alpha_{\bar\nu})(U_\omega \alpha_\omega + V_\omega \alpha_{\bar\omega}^+)$$

$$= U_\nu U_\omega \alpha_\nu^+ \alpha_\omega + U_\nu V_\omega \alpha_\nu^+ \alpha_{\bar\omega}^+ + V_\nu U_\omega \alpha_{\bar\nu} \alpha_\omega + V_\nu V_\omega \alpha_{\bar\nu} \alpha_{\bar\omega}^+ ,$$

and

$$a_{\bar\nu}^+ a_{\bar\nu} = (U_\nu \alpha_{\bar\nu}^+ - V_\nu \alpha_\nu)(U_\omega \alpha_{\bar\omega} - V_\omega \alpha_\omega^+)$$

$$= U_\nu U_\omega \alpha_{\bar\nu}^+ \alpha_{\bar\omega} - U_\nu V_\omega \alpha_{\bar\nu}^+ \alpha_\omega^+ - V_\nu U_\omega \alpha_\nu \alpha_{\bar\omega} + V_\nu V_\omega \alpha_\nu \alpha_\omega^+ ,$$

leading to

$$(a_\nu^+ a_\omega + a_{\bar\nu} a_{\bar\omega})$$

$$= U_\nu U_\omega (\alpha_\nu^+ \alpha_\omega + \alpha_{\bar\nu}^+ \alpha_{\bar\omega}) + U_\nu V_\omega (\alpha_\nu^+ \alpha_{\bar\omega}^+ - \alpha_{\bar\nu}^+ \alpha_\omega^+)$$

$$+ V_\nu U_\omega (\alpha_{\bar\nu} \alpha_\omega - \alpha_\nu \alpha_{\bar\omega}) + V_\nu V_\omega (\alpha_{\bar\nu} \alpha_{\bar\omega}^+ + \alpha_\nu \alpha_\omega^+). \tag{9.56}$$

We now introduce the definitions

$$\nu_\nu = \alpha_\nu^+ \alpha_\nu + \alpha_{\bar\nu}^+ \alpha_{\bar\nu},$$

$$z_\nu = 1 - \nu_\nu,$$

$$h_\nu = 2 U_\nu V_\nu,$$

$$f_\nu = U_\nu^2 - V_\nu^2. \tag{9.57}$$

In the single particle Hamiltonian, all terms have $\nu = \omega$, that is $(a_\nu^+ a_\omega + a_{\bar\nu}^+ a_{\bar\omega})\delta(\nu,\omega)$. Consequently the first and fourth terms of (9.56) become

$$U_\nu^2 \, (\alpha_\nu^+ \alpha_\nu + \alpha_{\bar\nu}^+ \alpha_{\bar\nu})\delta(\nu,\omega),$$

$$= U_\nu^2 \nu_\nu \delta(\nu,\omega) \,,$$

and

$$V_\nu^2 \, (\alpha_{\bar\nu} \alpha_{\bar\nu}^+ + \alpha_\nu \alpha_\nu^+)\delta(\nu,\omega)$$

$$= V_\nu^2 (2 - \nu_\nu)\delta(\nu,\omega),$$

their sum being

$$\left[(U_\nu^2 - V_\nu^2)\nu_\nu + 2V_\nu^2\right] \delta(\nu,\omega). \tag{9.58}$$

Making use of the relation

$$2V_\nu^2 = 2V_\nu^2 - U_\nu^2 + U_\nu^2$$

$$= -(U_\nu^2 - V_\nu^2) + 1,$$

expression (9.58) becomes

$$(1 - f_\nu z_\nu)\delta(\nu,\omega). \tag{9.59}$$

One now rewrites the second and third terms of (9.56) as

$$2U_\nu V_\nu \alpha_\nu^+ \alpha_{\bar\nu}^+ \delta(\nu,\omega) \,,$$

and

$$2V_\nu U_\nu \alpha_{\bar\nu} \alpha_\nu \delta(\nu,\omega) \,, \tag{9.60}$$

their sum being

$$2U_\nu V_\nu (\alpha_\nu^+ \alpha_{\bar\nu}^+ + \alpha_{\bar\nu} \alpha_\nu)\delta(\nu,\omega). \tag{9.61}$$

Introducing the definitions

$$X_\nu = \frac{1}{\sqrt{2}}(\alpha_\nu^+ \alpha_{\bar\nu}^+ + \alpha_{\bar\nu}\alpha_\nu)\,,$$

$$P_\nu = \frac{i}{\sqrt{2}}(\alpha_\nu^+ \alpha_{\bar\nu}^+ - \alpha_{\bar\nu}\alpha_\nu)\,, \tag{9.62}$$

one can finally write

$$H_{sp} = \sum_{\nu>0}(\varepsilon_\nu - \lambda) - \sum_{\nu>0}(\varepsilon_\nu - \lambda)f_\nu z_\nu + \sqrt{2}\sum_{\nu>0}(\varepsilon_\nu - \lambda)h_\nu X_\nu\,. \tag{9.63}$$

Quasi-particle Hamiltonian. We start by expressing $a_\nu^+ a_{\bar\nu}^+$ in terms of quasiparticles,

$$\begin{aligned}
a_\nu^+ a_{\bar\nu}^+ &= (U_\nu\alpha_\nu^+ + V_\nu\alpha_{\bar\nu})(U_\nu\alpha_{\bar\nu}^+ - V_\nu\alpha_\nu) \\
&= U_\nu^2\alpha_\nu^+\alpha_{\bar\nu}^+ - U_\nu V_\nu\alpha_\nu^+\alpha_\nu + V_\nu U_\nu\alpha_{\bar\nu}\alpha_{\bar\nu}^+ - V_\nu^2\alpha_{\bar\nu}\alpha_\nu \\
&= U_\nu^2\alpha_\nu^+\alpha_{\bar\nu}^+ - V_\nu^2\alpha_{\bar\nu}\alpha_\nu + U_\nu V_\nu z_\nu\,.
\end{aligned} \tag{9.64}$$

Inverting the relations (9.62) one can write,

$$\alpha_\nu^+\alpha_{\bar\nu}^+ = \frac{1}{\sqrt{2}}X_\nu - \frac{i}{\sqrt{2}}P_\nu,$$

$$\alpha_{\bar\nu}\alpha_\nu = \frac{1}{\sqrt{2}}X_\nu + \frac{i}{\sqrt{2}}P_\nu. \tag{9.65}$$

Replacing (9.65) into (9.64) leads to

$$a_\nu^+ a_{\bar\nu}^+ = \frac{1}{2}h_\nu z_\nu + \frac{1}{\sqrt{2}}f_\nu X_\nu - \frac{i}{\sqrt{2}}P_\nu\,. \tag{9.66}$$

From the relation $(a_\nu^+ a_{\bar\nu}^+)^+ = a_{\bar\nu}a_\nu$ one can also write

$$a_{\bar\omega}a_\omega = \frac{1}{2}h_\omega z_\omega + \frac{1}{\sqrt{2}}f_\omega X_\omega + \frac{i}{\sqrt{2}}P_\omega. \tag{9.67}$$

Making use of (9.66) and (9.67) the pairing Hamiltonian can be written as

$$H_{\mathrm{p}} = -\frac{G}{4}\left(\sum_{\nu>0}h_\nu z_\nu\right)^2 - \frac{G}{2\sqrt{2}}\left(\sum_{\nu>0}h_\nu z_\nu\right)\left(\sum_{\omega>0}f_\omega X_\omega\right) \tag{9.68}$$

$$- \frac{iG}{2\sqrt{2}}\left(\sum_{\nu>0}h_\nu z_\nu\right)\left(\sum_{\omega>0}P_\omega\right) - \frac{G}{2\sqrt{2}}\left(\sum_{\nu>0}f_\nu X_\nu\right)\left(\sum_{\omega>0}h_\omega z_\omega\right)$$

$$- \frac{G}{2}\left(\sum_{\nu>0}f_\nu X_\nu\right)^2 - \frac{iG}{2}\left(\sum_{\nu>0}f_\nu X_\nu\right)\left(\sum_{\omega>0}P_\omega\right)$$

$$+ \frac{iG}{2\sqrt{2}} \left(\sum_{\nu>0} P_\nu \right) \left(\sum_{\omega>0} h_\omega z_\omega \right) + \frac{iG}{2} \left(\sum_{\nu>0} P_\nu \right) \left(\sum_{\omega>0} f_\omega X_\omega \right)$$

$$- \frac{G}{2} \left(\sum_{\nu>0} P_\nu \right)^2 .$$

It can be shown that

$$[z_\nu, P_\omega] = -2iX_\nu \delta(\nu, \omega). \tag{9.69}$$

Thus

$$z_\nu P_\omega = P_\omega z_\nu - 2iX_\nu \delta(\nu, \omega). \tag{9.70}$$

Making use of this relation, the third term of (9.68) can be written as

$$- \frac{iG}{2\sqrt{2}} \left(\sum_{\nu,\omega>0} h_\nu z_\nu P_\omega \right)$$

$$= - \frac{iG}{2\sqrt{2}} \left(\sum_{\omega>0} P_\omega \right) \left(\sum_{\nu>0} h_\nu z_\nu \right) - \frac{G}{\sqrt{2}} \sum_{\nu>0} h_\nu X_\nu.$$

Consequently, the sum of the third and the seventh terms of (9.68) leads to

$$- \frac{G}{\sqrt{2}} \sum_{\nu>0} h_\nu X_\nu. \tag{9.71}$$

We now consider the sixth term of (9.68)

$$- \frac{iG}{2} \left(\sum_{\nu>0} f_\nu X_\nu \right) \left(\sum_{\omega>0} P_\omega \right)$$

$$= - \frac{iG}{2} \left\{ \left(\sum_{\omega>0} P_\omega \right) \left(\sum_{\nu>0} f_\nu X_\nu \right) + \sum_{\nu,\omega>0} f_\nu [X_\nu, P_\omega] \right\}$$

$$= - \frac{iG}{2} \left(\sum_{\omega>0} P_\omega \right) \left(\sum_{\nu>0} f_\nu X_\nu \right) + \frac{G}{2} \sum_{\nu>0} f_\nu z_\nu,$$

where use was made of

$$[X_\nu, P_\omega] = iz_\nu \delta(\nu, \omega). \tag{9.72}$$

Consequently, the sum of the sixth and eighth terms of (9.68) is equal to

$$\frac{G}{2} \sum_{\nu>0} f_\nu z_\nu. \tag{9.73}$$

Thus, H_p can be written as

$$H_\mathrm{p} = -\frac{G}{4}\left(\sum_{\nu>0} h_\nu z_\nu\right)^2 \tag{9.74}$$

$$-\frac{G}{2\sqrt{2}}\left\{\left(\sum_{\nu>0} h_\nu z_\nu\right)\left(\sum_{\omega>0} f_\omega X_\omega\right) + \left(\sum_{\nu>0} f_\nu X_\nu\right)\left(\sum_{\omega>0} h_\omega z_\omega\right)\right\}$$

$$-\frac{G}{2}\left(\sum_{\nu>0} f_\nu X_\nu\right)^2 - \frac{G}{2}\left(\sum_{\nu>0} P_\nu\right)^2$$

$$-\frac{G}{\sqrt{2}}\sum_{\nu>0} h_\nu X_\nu + \frac{G}{2}\sum_{\nu>0} f_\nu z_\nu.$$

The fifth and sixth terms are one-body terms, as can be seen from the relation

$$\sum_{\nu>0}(a_\nu^+ a_\nu + a_{\bar\nu}^+ a_{\bar\nu}) = \sum_{\nu>0}(1 - f_\nu z_\nu + \sqrt{2}h_\nu X_\nu) = \Omega - \sum_{\nu>0}(f_\nu z_\nu - \sqrt{2}h_\nu X_\nu).$$

They are the self-consistent contributions of the pairing force to the single-particle field. Also the term $-\frac{G}{4}(\sum_{\nu>0} h_\nu z_\nu)^2$ gives rise to a one-body and a constant contributions, besides a two-body term. In fact,

$$\left(\sum_{\nu>0} h_\nu z_\nu\right)^2 = \left(\sum_{\nu>0} h_\nu(1 - \nu_\nu)\right)^2 = \left\{\left(\sum_{\nu>0} h_\nu\right) - \left(\sum_{\nu>0} h_\nu \nu_\nu\right)\right\}^2$$

$$= \left(\sum_{\nu>0} h_\nu\right)^2 - 2\left(\sum_{\nu>0} h_\nu\right)\left(\sum_{\nu>0} h_\nu \nu_\nu\right) + \left(\sum_{\nu>0} h_\nu \nu_\nu\right)^2.$$

Making use of the fact that

$$\left(\sum_{\nu>0} h_\nu\right)\left(\sum_{\nu>0} h_\nu \nu_\nu\right) = \left(\sum_{\nu>0} h_\nu\right)\left(\sum_{\nu>0} h_\nu(-z_\nu + 1)\right)$$

$$= -\left(\sum_{\nu>0} h_\nu\right)\left(\sum_{\nu>0} h_\nu z_\nu\right) + \left(\sum_{\nu>0} h_\nu\right)^2,$$

one can write

$$\left(\sum_{\nu>0} h_\nu z_\nu\right)^2 = \left(\sum_{\nu>0} h_\nu\right)^2 + 2\left(\sum_{\nu>0} h_\nu\right)\left(\sum_{\nu>0} h_\nu z_\nu\right) - 2\left(\sum_{\nu>0} h_\nu\right)^2$$

$$+ \left(\sum_{\nu>0} h_\nu \nu_\nu\right)^2,$$

and thus

$$-\frac{G}{4}\left(\sum_{\nu>0}h_\nu z_\nu\right)^2 = -\frac{G}{2}\left(\sum_{\nu>0}h_\nu\right)\left(\sum_{\nu>0}h_\nu z_\nu\right) + \frac{G}{4}\left(\sum_{\nu>0}h_\nu\right)^2$$
$$-\frac{G}{4}\left(\sum_{\nu>0}h_\nu\nu_\nu\right)^2. \tag{9.75}$$

Making use of this result, (9.74) becomes

$$H_{\mathrm{p}} = -\frac{G}{4}\left(\sum_{\nu>0}h_\nu\nu_\nu\right)^2 \tag{9.76}$$
$$-\frac{G}{2\sqrt{2}}\left\{\left(\sum_{\nu>0}h_\nu z_\nu\right)\left(\sum_{\omega>0}f_\omega X_\omega\right) + \left(\sum_{\nu>0}f_\nu X_\nu\right)\left(\sum_{\omega>0}h_\omega z_\omega\right)\right\}$$
$$-\frac{G}{2}\left(\sum_{\nu>0}f_\nu X_\nu\right)^2 - \frac{G}{2}\left(\sum_{\nu>0}P_\nu\right)^2$$
$$-\frac{G}{\sqrt{2}}\sum_{\nu>0}h_\nu X_\nu + \frac{G}{2}\sum_{\nu>0}f_\nu z_\nu$$
$$-\frac{G}{2}\left(\sum_{\nu>0}h_\nu\right)\left(\sum_{\nu>0}h_\nu z_\nu\right) + \frac{G}{4}\left(\sum_{\nu>0}h_\nu\right)^2.$$

The usual quasi-particle (BCS) approximation consists in : a) Not consider the first, third and fourth terms in (9.76), b) to replace z_ν with 1 in the second term, c) to disregard the contribution to the single-particle self-consistent field given by the fifth term.

Within this approximation

$$H = H_{\mathrm{p}} + H_{\mathrm{sp}}$$
$$= \sum_{\nu>0}(\varepsilon_\nu - \lambda) - \sum_{\nu>0}(\varepsilon_\nu - \lambda)f_\nu z_\nu + \sqrt{2}\sum_{\nu>0}(\varepsilon_\nu - \lambda)h_\nu X_\nu$$
$$-\frac{G}{\sqrt{2}}\left(\sum_{\nu>0}h_\nu\right)\left(\sum_{\omega>0}f_\omega X_\omega\right) - \frac{G}{2}\left(\sum_{\nu>0}h_\nu\right)\left(\sum_{\nu>0}h_\nu z_\nu\right) + \frac{G}{4}\left(\sum_{\nu>0}h_\nu\right)^2$$
$$= \sum_{\nu>0}(\varepsilon_\nu - \lambda) - \sum_{\nu>0}\left\{(\varepsilon_\nu - \lambda)f_\nu + \frac{G}{2}\left(\sum_{\omega>0}h_\omega\right)h_\nu\right\}z_\nu + \frac{G}{4}\left(\sum_{\nu>0}h_\nu\right)^2$$
$$+ \sqrt{2}\sum_{\nu>0}(\varepsilon_\nu - \lambda)h_\nu X_\nu - \frac{G}{\sqrt{2}}\left(\sum_{\nu>0}h_\nu\right)\left(\sum_{\nu>0}f_\nu X_\nu\right).$$

Making use of the definition $z_\nu = 1 - \nu_\nu$, one can write

$$H = \sum_{\nu>0}(\varepsilon_\nu - \lambda) - \sum_{\nu>0}\left\{(\varepsilon_\nu - \lambda)f_\nu + \frac{G}{2}\left(\sum_{\omega>0}h_\omega\right)h_\nu\right\} + \frac{G}{4}\left(\sum_{\nu>0}h_\nu\right)^2$$

$$+ \sum_{\nu>0}\left\{(\varepsilon_\nu - \lambda)f_\nu + \frac{G}{2}\left(\sum_\omega h_\omega\right)h_\nu\right\}\nu_\nu$$

$$+ \sqrt{2}\sum_{\nu>0}\left\{(\varepsilon_\nu - \lambda)h_\nu - \frac{G}{\sqrt{2}}\left(\sum_{\omega>0}h_\omega\right)g_\nu\right\}X_\nu.$$

Calling

$$U = \sum_{\nu>0}(\varepsilon_\nu - \lambda) - \sum_{\nu>0}\left\{(\varepsilon_\nu - \lambda)f_\nu + \frac{G}{2}\left(\sum_{\omega>0}h_\omega\right)h_\nu\right\} + \frac{G}{4}\left(\sum_{\nu>0}h_\nu\right)^2,$$

$$H_{11} = \sum_{\nu>0}\left\{(\varepsilon_\nu - \lambda)f_\nu + \frac{G}{2}\left(\sum_{\omega>0}h_\omega\right)h_\nu\right\}\nu_\nu,$$

$$H_{20} = \sqrt{2}\sum_{\nu>0}\left\{(\varepsilon_\nu - \lambda)h_\nu - \frac{G}{2}\left(\sum_{\omega>0}h_\omega\right)f_\nu\right\}X_\nu.$$

and making use of the definition

$$\Delta = G\sum_{\nu>0}U_\nu V_\nu, \qquad\qquad (9.77)$$

as well as (9.52) and (9.62) one can write

$$U = \sum_{\nu>0}(\varepsilon_\nu - \lambda) - \sum_{\nu>0}\left\{(\varepsilon_\nu - \lambda)(U_\nu^2 - V_\nu^2) + \Delta 2U_\nu V_\nu\right\} + \frac{\Delta^2}{G},$$

$$H_{11} = \sum_{\nu>0}\left\{(\varepsilon_\nu - \lambda)(U_\nu^2 - V_\nu^2) + \Delta 2U_\nu V_\nu\right\}(\alpha_\nu^+\alpha_\nu + \alpha_{\bar\nu}^+\alpha_{\bar\nu}),$$

$$H_{20} = \sum_{\nu>0}\left\{(\varepsilon_\nu - \lambda)2U_\nu V_\nu - (U_\nu^2 - V_\nu^2)\Delta\right\}(\alpha_\nu^+\alpha_{\bar\nu}^+ + \alpha_{\bar\nu}\alpha_\nu).$$

We can now choose the coefficients U_ν, V_ν of the canonical transformations (9.55) so as to make H correspond to an independent-quasiparticle system. This is possible only if $H_{20} = 0$. The condition $H_{20} = 0$ may be shown to be equivalent to the requirement of a minimum "vacuum" energy U. Therefore, the ground state of the system in terms of the new particles is a "vacuum" state.

If one is interested in learning what type of collective excitations are associated with the pairing Hamiltonian one has to diagonalize the Hamiltonian $H = U + H_{11} + H_{\mathrm{int}}$ where $H_{\mathrm{int}} = -\frac{G}{2}(\sum_{\nu>0}f_\nu X_\nu)^2 - \frac{G}{2}(\sum_{\nu>0}P_\nu)^2$, in the quasi-particle basis. This can be done within the framework of two equivalent approximations: the quasi-boson approximation and the linearization method.

9.5.3 Temperature Dependent BCS Solution

We start this appendix by working out the thermal average of the different terms entering in the pairing Hamiltonian (9.50) :

$$\langle \alpha_\nu^+ \alpha_\nu \rangle_T = \langle \alpha_{\bar\nu}^+ \alpha_{\bar\nu} \rangle_T = F_\nu$$

where (see next section, (9.83))

$$F_\nu = \frac{1}{1 + e^{E_\nu/T}} \ .$$

Thus

$$\langle \alpha_\nu \alpha_\nu^+ \rangle_T = \langle \alpha_{\bar\nu} \alpha_{\bar\nu}^+ \rangle_T = 1 - F_\nu \ ,$$

$$\langle z_\nu \rangle_T = 1 - 2F_\nu \ ,$$

$$\langle \nu_\nu \rangle_T = 2F_\nu \ ,$$

$$\langle (\alpha_\nu^+ \alpha_{\bar\nu}^+ + \alpha_{\bar\nu} \alpha_\nu)(\alpha_{\nu'}^+ \alpha_{\bar\nu'}^+ + \alpha_{\bar\nu'} \alpha_{\nu'}) \rangle_T$$

$$= \langle \alpha_\nu^+ \alpha_{\bar\nu}^+ \alpha_{\bar\nu'} \alpha_{\nu'} \rangle_T + \langle \alpha_{\bar\nu} \alpha_\nu \alpha_{\nu'}^+ \alpha_{\bar\nu'}^+ \rangle_T$$

$$= \delta(\nu,\nu') F_\nu^2 + \delta(\nu,\nu')(1 - F_\nu^2)^2 = \delta(\nu,\nu')(1 - 2F_\nu + 2F_\nu^2) \ ,$$

$$\langle (\alpha_\nu^+ \alpha_{\bar\nu}^+ - \alpha_{\bar\nu} \alpha_\nu)(\alpha_{\nu'}^+ \alpha_{\bar\nu'}^+ - \alpha_{\bar\nu'} \alpha_{\nu'}) \rangle_T$$

$$= -\delta(\nu,\nu')(1 - 2F_\nu + 2F_\nu^2) \ ,$$

$$\langle (\alpha_\nu^+ \alpha_\nu + \alpha_{\bar\nu}^+ \alpha_{\bar\nu})(\alpha_{\nu'}^+ \alpha_{\nu'} + \alpha_{\bar\nu'}^+ \alpha_{\bar\nu'}) \rangle_T$$

$$= \langle \alpha_\nu^+ \alpha_\nu \alpha_{\nu'}^+ \alpha_{\nu'} \rangle_T + \langle \alpha_\nu^+ \alpha_\nu \alpha_{\bar\nu'}^+ \alpha_{\bar\nu'} \rangle_T$$

$$+ \langle \alpha_{\bar\nu}^+ \alpha_{\bar\nu} \alpha_{\nu'}^+ \alpha_{\nu'} \rangle_T + \langle \alpha_{\bar\nu}^+ \alpha_{\bar\nu} \alpha_{\bar\nu'}^+ \alpha_{\bar\nu'} \rangle_T$$

$$= 2F_\nu(1 - F_\nu)\delta(\nu,\nu') + 4F_\nu F_{\nu'} \ .$$

Making use of these results one can calculate the thermal average of the pairing Hamiltonian obtaining

$$\langle H \rangle_T = -\frac{G}{2} \left(\sum_\nu h_\nu \right) \left(\sum_\nu h_\nu (1 - 2F_\nu) \right) + \frac{G}{4} \left(\sum_\nu h_\nu \right)^2$$

$$+ \frac{G}{2} \left(\sum_\nu f_\nu (1 - 2F_\nu) \right)$$

$$- \frac{G}{4} \sum_{\nu,\nu'} f_\nu f_{\nu'} \delta(\nu,\nu')(1 - 2F_\nu + 2F_\nu^2)$$

$$+ \frac{G}{4} \sum_{\nu,\nu'} (-\delta(\nu,\nu')(1 - 2F_\nu + 2F_\nu^2)$$

$$- \frac{G}{4} \sum_{\nu,\nu'} h_\nu h_{\nu'} (4F_\nu F_{\nu'} + 2\delta(\nu,\nu')F_\nu(1 - F_\nu))$$

$$= -\frac{G}{4} \left(\sum_\nu h_\nu \right)^2 + G \left(\sum_\nu h_\nu \right) \left(\sum_\nu h_\nu F_\nu \right)$$

$$- \frac{G}{4} \sum_\nu \left[(-2f_\nu + f_\nu^2 + 1)(1 - 2F_\nu) + (f_\nu^2 + 1)2F_\nu^2 \right]$$

$$- G \sum_{\nu,\nu'} h_\nu h_{\nu'} F_\nu F_{\nu'} - \frac{G}{2} \sum_\nu h_\nu^2 F_\nu (1 - F_\nu)$$

$$= G \left(\sum_\nu h_\nu \right) \left(\sum_\nu F_\nu h_\nu \right) - \frac{G}{4} \left(\sum_\nu h_\nu \right)^2$$

$$- \frac{G}{4} \sum_\nu \left[(f_\nu - 1)^2 (1 - 2F_\nu) + (f_\nu^2 + 1)2F_\nu^2 \right]$$

$$- G \sum_{\nu,\nu'} h_\nu h_{\nu'} F_\nu F_{\nu'} - \frac{G}{2} \sum_\nu h_\nu^2 F_\nu (1 - F_\nu)$$

$$= G \left(\sum_\nu h_\nu (1 - F_\nu) \right) \left(\sum_\nu h_\nu F_\nu \right) - \frac{G}{4} \left(\sum_\nu h_\nu \right)^2$$

$$- \frac{G}{4} \sum_\nu \left[(f_\nu - 1)^2 (1 - 2F_\nu) + 2(f_\nu^2 + 1)F_\nu^2 + 2h_\nu^2 F_\nu (1 - F_\nu) \right] \; .$$

Making use of the relations

$$(f_\nu - 1)^2 = \left(U_\nu^2 - V_\nu^2 - (U_\nu^2 + V_\nu^2) \right)^2 = 4V_\nu^4$$

$$f_\nu^2 + 1 = (U_\nu^2 - V_\nu^2)^2 + (U_\nu^2 + V_\nu^2)^2 = 2(U_\nu^4 + V_\nu^4)$$

one can write

$$\langle H\rangle_T = 4\,G\left(\sum_\nu U_\nu V_\nu(1-F_\nu)\right)\left(\sum_\nu U_\nu V_\nu F_\nu\right) - G\left(\sum_\nu U_\nu V_\nu\right)^2$$

$$-\frac{G}{4}\sum_\nu\left[4V_\nu^4(1-2F_\nu)+4(U_\nu^4+V_\nu^4)F_\nu^2+8U_\nu^2V_\nu^2F_\nu(1-F_\nu)\right]$$

$$= 4\,G\left(\sum_\nu U_\nu V_\nu(1-F_\nu)\right)\left(\sum_\nu U_\nu V_\nu F_\nu\right) - G\left(\sum_\nu U_\nu V_\nu\right)^2$$

$$- G\sum_\nu\left[V_\nu^2(1-F_\nu)+U_\nu^2F_\nu\right]^2\ .$$

Let us now calculate the thermal average of

$$H_{\rm sp} = \sum_\nu e_\nu - \sum_\nu e_\nu f_\nu z_\nu + \sum_\nu e_\nu h_\nu(\alpha_\nu^+\alpha_{\bar\nu}^+ + \alpha_{\bar\nu}\alpha_\nu)$$

where

$$e_\nu = \varepsilon_\nu - \lambda\ .$$

one obtains

$$\langle H_{\rm sp}\rangle_T = 2\sum_\nu e_\nu\left[V_\nu^2(1-F_\nu)+U_\nu^2F_\nu\right]\ .$$

Thus

$$\langle H_{\rm sp}+H_{\rm p}\rangle_T = 2\,\sum_\nu e_\nu\left[V_\nu^2(1-F_\nu)+U_\nu^2F_\nu\right]\tag{9.78}$$

$$+ 4G\left(\sum_\nu U_\nu V_\nu(1-F_\nu)\right)\left(\sum_\nu U_\nu V_\nu F_\nu\right) - G\left(\sum_\nu U_\nu V_\nu\right)^2$$

$$- G\sum_\nu\left[V_\nu^2(1-F_\nu)+U_\nu^2F_\nu\right]^2\ .$$

We now minimize the ground state energy with respect to V_ν obtaining

$$\frac{\partial}{\partial V_\nu}\langle H_{\rm sp}+H_{\rm p}\rangle_T = 2e_\nu\left[2V_\nu(1-F_\nu)+2U_\nu\frac{\partial U_\nu}{\partial V_\nu}F_\nu\right]$$

$$-2G\left[V_\nu^2(1-F_\nu)+U_\nu^2F_\nu\right]\left[2V_\nu(1-F_\nu)+2U_\nu\frac{\partial U_\nu}{\partial V_\nu}F_\nu\right]$$

$$+4G\left(\sum_\nu U_\nu V_\nu(1-F_\nu)\right)\left[U_\nu F_\nu+V_\nu F_\nu\frac{\partial U_\nu}{\partial V_\nu}\right]$$

$$+4G\left[U_\nu(1-F_\nu)+V_\nu\frac{\partial U_\nu}{\partial V_\nu}(1-F_\nu)\right]\left(\sum_\nu U_\nu V_\nu F_\nu\right)$$

$$-2G\left(\sum_\nu U_\nu V_\nu\right)\left[U_\nu+V_\nu\frac{\partial U_\nu}{\partial V_\nu}\right]$$

$$=\{2e_\nu-2G[V_\nu^2(1-F_\nu)+U_\nu^2 F_\nu]\}\left[2V_\nu(1-F_\nu)+2U_\nu\frac{\partial U_\nu}{\partial V_\nu}F_\nu\right]$$

$$+4GF_\nu\left[U_\nu+V_\nu\frac{\partial U_\nu}{\partial V_\nu}\right]\left(\sum_\nu U_\nu V_\nu(1-F_\nu)\right)$$

$$-4GF_\nu\left[U_\nu+V_\nu\frac{\partial U_\nu}{\partial V_\nu}\right]\left(\sum_\nu U_\nu V_\nu F_\nu\right)$$

$$+4G\left[U_\nu+V_\nu\frac{\partial U_\nu}{\partial V_\nu}\right]\left(\sum_\nu U_\nu V_\nu F_\nu\right)$$

$$-2G\left[U_\nu+V_\nu\frac{\partial U_\nu}{\partial V_\nu}\right]\left(\sum_\nu U_\nu V_\nu\right)=0\ .$$

From the relations

$$\frac{\partial U_\nu}{\partial V_\nu}=\frac{-V_\nu}{(1-V_\nu^2)^{1/2}}=-\frac{V_\nu}{U_\nu}\ ,$$

$$U_\nu+V_\nu\frac{\partial U_\nu}{\partial V_\nu}=U_\nu-\frac{V_\nu^2}{U_\nu}=\frac{1}{U_\nu}[U_\nu^2-V_\nu^2]\ ,$$

one obtains,

$$2\left[e_\nu-G\left[V_\nu^2(1-F_\nu)+U_\nu^2 F_\nu\right]\right]U_\nu V_\nu$$

$$-G(U_\nu^2-V_\nu^2)\sum_\nu U_\nu V_\nu(1-2F_\nu)=0.$$

Introducing the definitions

$$\tilde{e}_\nu=e_\nu-G\left[V_\nu^2(1-2F_\nu)+F_\nu\right]$$

and

$$\Delta=G\sum_\nu U_\nu V_\nu(1-2F_\nu) \tag{9.79}$$

one obtains

$$\tilde{e}_\nu 2U_\nu V_\nu=(U_\nu^2-V_\nu^2)\Delta\ . \tag{9.80}$$

Taking the square of this expression and making use of

$$U_\nu^4 + 2U_\nu^2 V_\nu^2 + V_\nu^4 = 1 \; ,$$

leads to

$$4\tilde{e}_\nu{}^2 U_\nu^2 V_\nu^2 + 2\Delta^2 U_\nu^2 V_\nu^2 = \Delta^2(U_\nu^4 + V_\nu^4) \; .$$

With the help of

$$1 - 2U_\nu^2 V_\nu^2 = U_\nu^4 + V_\nu^4 \; ,$$

one arrives to

$$U_\nu V_\nu = \frac{\Delta}{2E_\nu} \; , \tag{9.81}$$

where

$$E_\nu = \sqrt{\tilde{e}_\nu{}^2 + \Delta^2} \; . \tag{9.82}$$

Making use of this relation and (9.80) leads to

$$U_\nu^2 - V_\nu^2 = 1 - 2V_\nu^2 = \frac{\tilde{e}_\nu}{E_\nu} \; .$$

Thus

$$V_\nu = \frac{1}{\sqrt{2}} \left(1 - \frac{\tilde{e}_\nu}{E_\nu} \right)^{1/2} \; ,$$

and

$$U_\nu = \frac{1}{\sqrt{2}} \left(1 + \frac{\tilde{e}_\nu}{E_\nu} \right)^{1/2} \; .$$

The BCS gap equation is obtained by inserting (9.81) into (9.79),

$$\frac{1}{G} = \sum_{\nu>0} \frac{1 - 2F_\nu}{2E_\nu} \; .$$

Calculation of the Occupation Factor F_ν. We start by calculating the free energy

$$F = \langle H_{\mathrm{sp}} + H_{\mathrm{p}} \rangle_T - TS,$$

where the entropy is given by

$$S = -2 \sum_\nu \left[F_\nu \ln F_\nu + (1 - F_\nu) \ln(1 - F_\nu) \right].$$

Making use of (9.79) one can write

$$F = 2 \sum_\nu e_\nu \left[U_\nu^2 F_\nu + V_\nu^2 (1 - F_\nu) \right]$$

$$-G \sum_\nu \left[U_\nu^2 F_\nu + V_\nu^2 F_\nu (1 - F_\nu) \right]^2$$

$$+4G \left(\sum_\nu U_\nu V_\nu (1 - F_\nu) \right) \left(\sum_\nu U_\nu V_\nu F_\nu \right) - G \left(\sum_\nu U_\nu V_\nu \right)^2$$

$$+2T \sum_\nu \left[F_\nu \ln F_\nu + (1 - F_\nu) \ln(1 - F_\nu) \right].$$

Let us now minimize this quantity with respect to F_ν,

$$\frac{\partial F}{\partial F_\nu} = 2e_\nu \left[U_\nu^2 - V_\nu^2 \right] - 2G \left[U_\nu^2 F_\nu + V_\nu^2 (1 - F_\nu) \right] (U_\nu^2 - V_\nu^2)$$

$$+2k_{\mathrm{B}} \left(\ln F_\nu + \frac{F_\nu}{F_\nu} - \ln(1 - F_\nu) - \frac{1 - F_\nu}{1 - F_\nu} \right)$$

$$= 2 \left\{ e_\nu - G[V_\nu^2 (1 - 2F_\nu) + F_\nu] \right\} (U_\nu^2 - V_\nu^2)$$

$$+4G U_\nu V_\nu \sum_\nu U_\nu V_\nu (1 - 2F_\nu)$$

$$+2T \left(\ln \frac{F_\nu}{1 - F_\nu} \right)$$

$$= 2\tilde{e}_\nu (U_\nu^2 - V_\nu^2) + 4 U_\nu V_\nu \Delta + 2T \left(\ln \frac{F_\nu}{1 - F_\nu} \right) = 0.$$

Making use of (9.81) and (9.82) leads to

$$E_\nu^2 + E_\nu T \ln \frac{F_\nu}{1 - F_\nu} = 0 ,$$

which can be rewritten as

$$\ln \frac{F_\nu}{1 - F_\nu} = -\frac{E_\nu}{T} ,$$

leading to

$$F_\nu = \frac{1}{1 + e^{E_\nu/T}}.$$

(9.83)

10. Discussion

The basic finding in the description of FMBS is the existence of a mean field where fermoins move free of each other, feeling the pullings and pushings of all the other particles only when trying to leave the system. However, nothing is really free, not even an electron moving in vacuum, let alone one inside an atomic cluster or a fullerene. For an electron going from X to Y (Fig. 10.1), the pole of the propagator for a free particle is at $p^2 = m^2$. However, making measurements at X and Y one can not tell if the electron had emitted or absorbed any number of photons. Such processes, the simplest of which is shown in Fig. 10.1(a), cause a shift of the position of the pole. Physically, this means that what one measures (see Figs. 10.1 (b)–(d)), the experimental mass m_{exp}, is not the "bare" mass but something else which includes the effect of virtual processes dressing the fermion.

This discussion shows that the bare mass is in fact not directly an observable. This is particularly true in the case of electrons in a metal cluster or in a fullerene, where the process depicted in Fig. 10.1(a), can either indicate the dressing of an electron through its coupling to a phonon or to a plasmon.

As it is the case in other FMBS, like for example, the atomic nucleus, low- and high–frequency modes lead to equally important renormalization effects like e.g. the effective mass and the effective charge [30, 45, 72]. This is the reason why the electron–phonon and electron–plasmon couplings have to be treated on equal footing. In particular in the description of the optical properties of these systems, where the process shown in Fig. 10.1(a) is also at the basis of processes like those shown in Figs. 10.1(e)–(g) and correspond to e.g. self-energy and vertex corrections of plasmon excitations (carried out to all orders and thus going beyond the Random Phase Approximation).

In Chapters 2–8 we discussed the concepts and present the tools to carry out this program, the example of the optical response of Na_9^+ (Chap. 8) being paradigmatic: the electron–plasmon coupling being essential to account for the position of the centroid of the photoabsorption peaks and the energy range over which these peaks are distributed, the electron–phonon coupling providing the smearing of individual lines.

One should dress electrons by treating on equal footing their coupling to plasmons and phonons not only when they participate in the linear response of the system to an incoming photon (as correlated particle–hole excitations),

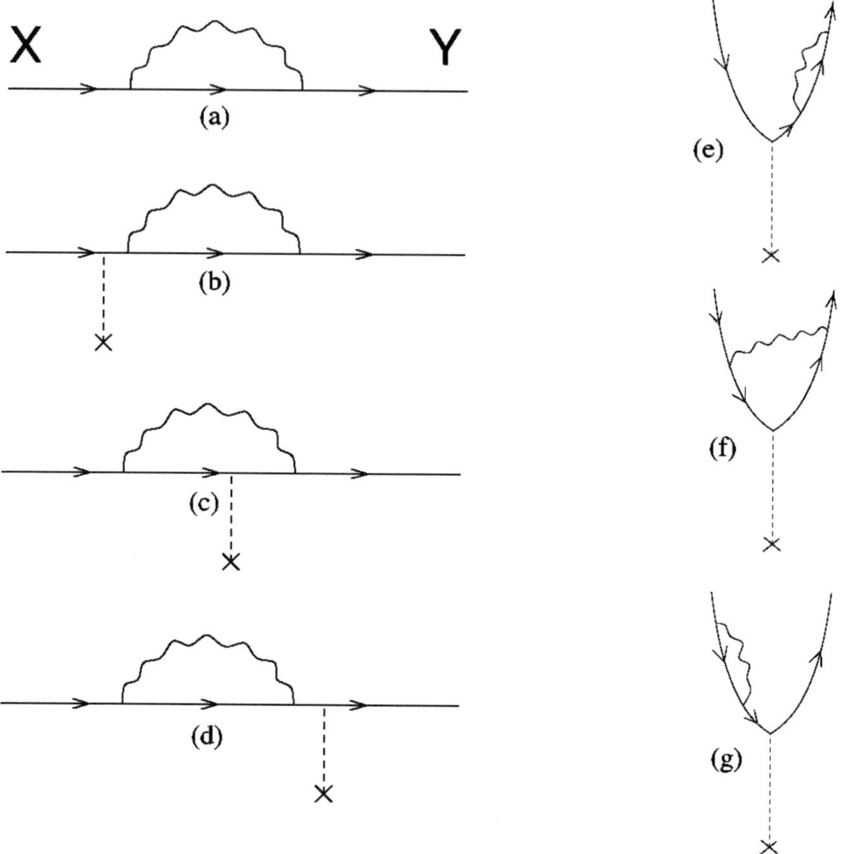

Fig. 10.1. Coupling of a fermion (*arrowed line*) with a boson (*wavy line*). The cross followed by a *dashed line* represents the external probe. In graphs (a)–(d) the time axis is thought to run from left to right (horizontally), while in graphs (e)–(g), obtained from (b)–(d) by deforming the fermion lines upwards leaving unchanged the instant in which the external field acts and the fermion (*black dot*), times run from bottom to top (vertically)

but also when they correlate to form Cooper pairs. Cooper pairs which eventually lead to superconductivity in materials like e.g. doped fullerides (see also [219]), made out making use of atomic aggregates as building blocks. Such a program has not been yet been carried out from first principles and we hope the present lecture notes can be of help in pursuing this goal.

After we had completed the present notes, we became aware of the monography O. Gunnarson was preparing [220]. In it, the need to treat the electron–phonon and electron–electron interactions on equal footings to obtain a consistent description of superconductivity in fullerides is clearly shown, within the framework of a simplified but still realistic model.

References

1. R.P. Feynman: The Caltech Alumni Magazine, February (1960) 22; reprinted 1992, IEEE J. MEMS **1**, 60.
2. H. Ehrenreich and F. Spaepen eds.: *Fullerenes, Solid State Physics*, Vol. 48 (Academic Press, London 1994).
3. M.S. Dresselhaus, G. Dresselhaus, P.C. Ecklund: *Science of Fullerenes and Carbon Nanotubes* (Academic Press, San Diego, 1996).
4. R.A. Broglia: Surf. Sc. **500**, 759 (2002).
5. W. de Heer, Rev. Mod. Phys. **65**, 611 (1993).
6. M. Brack: Rev. Mod. Phys. **65**, 677 (1993).
7. R.A. Broglia: Cont. Phys. **35**, 95 (1994).
8. H. Haberland ed., *Clusters of Atoms and Molecules* (Springer, Berlin Heidelberg New York, 1994).
9. T.P. Martin, in *Elemental and Molecular Clusters* eds. G. Benedek, T.P. Martin and G. Pacchioni (Springer, Berlin Heidelberg New York, 1998), p. 2.
10. G.F. Bertsch, R.A. Broglia: *Oscillations in Finite Quantum Systems* (Cambridge Univ. Press, Cambridge, 1994).
11. M. Barranco, E.S. Hernandez, R.J. Lombard, Ll. Serra: Z. Phys. **D22**, 659 (1992).
12. I. Spagnolatti, M. Bernasconi, G. Benedek: Europhys. Lett. **59**, 572 (2002); ibid., Europhys. Lett. **60**, 329 (2002).
13. R. Saito, G. Dresselhaus, M.S. Dresselhaus: *Physical Properties of Carbon Nanotubes* (Imperial College Press, London, 1998).
14. L.-C. Qin, X. Zhao, K. Hirahara, Y. Miyamoto, Y. Ando, S. Iijima: Nature **408**, 50 (2000).
15. N. Wang, Z. K. Tang, G. D. Li, J. S. Chen: Nature **408**, 50 (2000).
16. M. Tulej, D. A. Kirkwood, M. Pachkov, J. P. Maier: Astrophys. J. **506**, L69 (1998).
17. L. Henrad, A.A. Lucas, Ph. Lambin: Astrophys. J. **406**, 92 (1993).
18. M. Bianchetti, P.F. Buonsante, F. Ginelli, H.E. Roman, R.A. Broglia, F. Alasia: Phys. Rep. **357**, 459 (2002).
19. A.G. Rinzler, J.H. Hafner, P. Nikolaev, L. Lou, S.G. Kim, D. Tomanek, P. Nordlander, D.T. Colbert, R.E. Smalley: Science **269**, 1550 (1995).
20. A. Lorenzoni, H.E. Roman, F. Alasia, R.A. Broglia: Chem. Phys. Lett. **276**, 237 (1997).
21. G.F. Bertsch, K. Yabana, Z. Physik **D 42**, 219 (1997).
22. H.E. Roman, A. Lorenzoni, N. Breda, R.A. Broglia, G. Onida: Il Nuovo Cimento **110**, 1165 (1997).
23. R.A. Broglia, A. Lorenzoni, H.E. Roman, F. Alasia: Il Nuovo Cimento **110**, 1157 (1997).

24. M. Bianchetti, P.F. Buonsante, F. Ginelli, A. Lorenzoni, H.E. Roman, R.A. Broglia: in Electronic Properties of Novel Materials–Science and Technology of Molecular Nanostructures, XIIIth Winter School, no. 486, pp. 448, *AIP Conference Proceedings*, edited by H. Kuzmany, J. Fink, M. Mehring and S. Roth (American Institute of Physics, Melville, New York, 1999).

25. L.Lou, P. Nordlander, R.E. Smalley: Phys. Rev. **B52**, 1429 (1995).

26. E.A. Rohlfing, D.M. Cox, A. Kaldor: J. Chem. Phys. **31**, 3322 (1984).

27. T. Rabenau, A. Simon, R.K. Kremer, E. Sohmen: Z. Phys. **B90**, 69 (1993).

28. L. Hedin, Phys. Rev. **139**, A796 (1965).

29. G. Grosso, G. Pastori-Parravicini: *Solid State Physics*, (Academic Press, San Diego, 2000)

30. C. Mahaux, P.F. Bortignon, R.A. Broglia, C.H. Dasso: Phys. Rep. **120**, 1 (1985).

31. W. Kohn, L.J. Sham: Phys. Rev. **A140**, 1133 (1965).

32. R.O. Jones, O. Gunnarsson: Rev. Mod. Phys. **61**, 689 (1989).

33. G. Onida, L. Reining, A. Rubio: Rev. Mod. Phys. **74**, 601 (2002).

34. N.W. Ashcroft, J.W. Wilkins: Phys. Lett. **14**, 285 (1965).

35. N.W. Ashcroft, N.D. Mermin: *Solid State Physics* (Saunders College Publishing, Philadelphia, 1976).

36. E. Krotscheck: Ann. Phys. (N.Y.) **155**, 1 (1984).

37. J.C. Slater: Phys. Rev. **81**, 385 (1951); ibid. **82**, 538 (1951); ibid. **91**, 528 (1953).

38. L.J. Llanto: Phys. Rev. **B22**, 1380 (1980).

39. J. Lam: Phys. Rev. **B23**, 3243 (1971).

40. P. Eisenberger, J. Lam, P.M. Platzman, P. Schmidt: Phys. Rev. **B6**, 3671 (1972).

41. G.D. Mahan, K.R. Subbaswany: *Local Density Theory of Polarizability* (Plenum Press, New York, 1990).

42. O. Gunnarsson, B.I. Lundqvist: Phys. Rev. **B13**, 4274 (1976).

43. G. Grimvall: *The electron–phonon interaction in metals* (North–Holland, Amsterdam, 1981).

44. J.R. Schrieffer: *Theory of superconductivity* (Benjamin, New York, 1964).

45. D. Brink, R.A. Broglia: *Nuclear Superfluidity: Pairing Correlations in finite Systems* (Cambridge University Press, Cambridge, to be published).

46. E.K.U. Gross, E. Runge, O. Heinonen: *Many Particle Theory* (Adam Hilger, Bristol, United Kingdom, 1991)

47. J. Callaway: *Quantum Theory of the Solid State* (Academic Press, London, 1974)

48. P. Hohenberg, W. Kohn: Phys. Rev. **B136**, 6 (1964).

49. R.G. Parr, W. Yang: *Density Functional Theory of Atoms and Molecules* (Oxford University Press, Inc., New York, 1989).

50. R.M. Dreizler, E.K.U. Gross: *Density Functional Theory: an approach to the quantum many-body problem* (Springer, Berlin Heidelberg New York, 1990)

51. J.P. Perdew, A. Zunger: Phys. Rev. **B23**, 5048 (1981).

52. D.R. Hamann, M. Schlüter, C. Chiang: Phys. Rev. Lett. **43**, 1494 (1979).

53. G.B. Bachelet, D.R. Hamann, M. Schlüter: Phys. Rev. **B26**, 4199 (1982).

54. R. Car, M. Parrinello: Phys. Rev. Lett. **55**, 2471 (1985).

55. J.D. Jackson: *Classical Electrodynamics* (John Wiley & Sons, New York, 1975).

56. G.B. Bachelet: in *Strategies for computer chemistry*, M.P. Tosi, ed., (Kluwer, Dordrecht, 1989), pag. 119-160.

57. L.D. Landau, E.M. Lifshits: *Quantum Mechanics* (Pergamon press, Oxford, 1966)

58. N. Troullier, J.L. Martins: Phys. Rev. **B43**, 1993 (1991); M. Troullier, J.L. Martins: Solid State Comm. **74**, 613 (1990).

59. F. Alasia, R.A. Broglia, H.E. Roman, Ll. Serra, G. Coló, J.M. Pacheco: J. Phys. **B27**, L643 (1994).

60. P.O. Löwdin: Adv. Phys. **5**, 2 (1956).

61. E.K.U. Gross, J.F. Dobson, M. Petersilka: *Density Functional Theory II*, Vol. 181, Topics in Current Chemistry, Ed. R.F. Nalewajski (Springer, Berlin Heidelberg New York, 1996)

62. A.L. Fetter, J.D. Walecka: *Quantum Theory of many Particle Systems* (Mc Graw-Hill, New York, 1971).

63. D. Bohm, D. Pines: Phys. Rev. **82**, 625 (1951).

64. A. Zangwill, P. Soven: Phys. Rev. **A21**, 1561 (1980).

65. E. Lipparini, S. Stringari: Phys. Rep. **175**, 103 (1989).

66. O. Bohigas, A.M. Lane, J. Martorell: Phys. Rep. **51**, 267 (1979).

67. D.J. Thouless: Nucl. Phys. **22**, 78 (1961).

68. C. Ellert, M. Schmidt, C. Schmitt, H. Haberland, C. Guet: Phys. Rev. **B 59**, R7841 (1999).

69. A.J. Read, R.J. Needs: Phys. Rev. **B 44**, 13071-13073 (1991).

70. J.P. Van Dyke: Phys. Rev. **B 5**, 1489-1493 (1972).

71. D.J. Rowe: *Nuclear Collective Motion. Models and Theory* (Methuen, London, 1970)

72. P.F. Bortignon, A. Bracco, R.A. Broglia: *Giant resonances: Nuclear Structure at Finite Temperature* (Harwood Academic Publishers, Amsterdam, 1998).

73. M. Bianchetti: *Ab initio calculations of isolated cylindrical carbon systems: From linear chains to nanotubes* (Ph. D. Thesis, Physics Dept., University of Milan, 2000, unpublished)

74. D.R. Lide: *CRC Handbook of Chemistry and Physics* (CRC Press, Boca Raton, 1998).

75. P. Milani, M. Manfredini, G. Guizzetti, F. Marabelli, M. Patrini: Solid State Commun. **90**, 639 (1994).

76. I.V. Hertel, H. Steger, J. de Vries, B. Weisser, C. Menzel, B. Kamke, W. Kamke: Phys. Rev. Lett. **68**, 784 (1992)

77. G. Mie: Ann. Phys. (Leipzig) **25**, 377 (1908).

78. G.F. Bertsch, A. Bulgac, D. Tomanek, Y. Wang: Phys. Rev. Lett. **67**, 2690 (1991).

79. S. Leach, M. Vervloet, A. Despres, E. Breheret, J.P. Hare, T.J. Dennis, H.W. Kroto, R. Taylor, D.R.M. Walton: Chem. Phys. **160**, 451 (1992).

80. M. Braga, S. Larson, A. Rosén, A. Volosov: Astron. Astrophys. **245**, 232 (1991).

81. J.M. Pacheco, F. Alasia, H.E. Roman, R.A. Broglia: Z. Phys. **D37**, 277 (1996).

82. C. Milani, C. Giambelli, H.E. Roman, F. Alasia, G. Benedek, R.A. Broglia, S. Sanguinetti, K. Yabana: Chem. Phys. Lett. **258**, 554 (1996).

83. C. Milani, C. Giambelli, F. Alasia, R.A. Broglia, G. Coló, H.E. Roman: Chem. Phys. Lett. **258**, 559 (1996).

84. T.W. Ebbesen: Annu. Rev. Mat. Sci. **24**, 235 (1994).

85. E.G. Gamaly, T.W. Ebbesen: Phys. Rev. **B52**, 2083 (1995).

86. H.E. Roman, G. Coló, F. Alasia, R.A. Broglia: Chem. Phys. Lett. **251**, 111 (1996).

87. J.W.G. Wilder, L.C. Venema, A.G. Rinzler, R.E. Smalley, C. Dekker: Nature **391**, 59 (1998).

88. T.W. Odom, J.-L. Huang, P. Kim, C.M. Lieber: Nature **391**, 62 (1998).

89. L.C. Venema, J.W.G. Wilder, H.L.J. Temminck Tuinstra, C. Dekker, A.G. Rinzler, R.E. Smalley: Appl. Phys. Lett. **71**, 2629 (1997).

90. R.A. Broglia, J.M. Pacheco, C. Yannouleas: Phys. Rev. **B44**, 5901 (1991).
91. G. Pacchioni, J. Koutecky: J. Chem. Phys. **88**, 1066 (1988).
92. P. Botschwina: J. Chem. Phys. **101**, 853 (1994).
93. J. Hütter, H.P. Lüthi, F. Diederich: J. Am. Chem. Soc. **116**, 750 (1994).
94. M. Kolbuszewsky: J. Chem. Phys. **102**, 3679 (1995).
95. J.M.L. Martin, J. El-Yazal, J.-P. Francois: Chem. Phys. Lett. **252**, 9 (1996).
96. K. Yabana, G.F. Bertsch: Z. Phys. **D42**, 219 (1997).
97. R.O. Jones, G. Seifert: Phys. Rev. Lett. **79**, 443 (1997).
98. N. Breda, G. Onida, G. Benedek, G. Coló, R.A. Broglia: Phys. Rev. **B58**, 1 (1998).
99. A. Van Orden, R.J. Saykally: Chem. Rev. **98**, 2313 (1998).
100. R.O. Jones: J. Chem. Phys. **110**, 5189 (1999).
101. M. Wyss, M. Grutter, J.P. Maier: Chem. Phys. Lett. **304**, 35 (1999).
102. M. Grutter, M. Wyss, E. Riaplov, J.P. Maier: J. Chem. Phys. **111**, 7397 (1999).
103. D. Forney, P. Freivogel, M. Grutter, J.P. Maier: J. Chem. Phys. **104**, 4954 (1996).
104. K.W. Chang, W.R.M. Graham: J. Chem. Phys. **77**, 4300 (1982).
105. R.A. Broglia, F. Alasia, P. Arcagni, G. Coló, F. Ghielmetti, C. Milani, H.E. Roman: Z. Phys. **D40**, 240 (1997).
106. C. Bohren, D.R. Huffman: *Absorption and Scattering of Light by Small Particles* (John Wiley & Sons, New York, 1983).
107. B.P. Feuston, W. Andreoni, M. Parrinello, E. Clementi: Phys. Rev. **B44**, 4056 (1991).
108. A. Fortini, M. Mazzola, A. Mina, D. Provasi, G. Coló, G. Onida, H.E. Roman, R.A. Broglia: preprint (2002).
109. M. Hamermesh: *Group Theory and its Application to Physical Problems* (Addison-Wesley Publish. Co., Reading (Mass.), 1962).
110. E.B. Wilson, J.C. Decius, P.C. Cross: *Molecular Vibrations: the Theory of Infrared and Raman Vibrational Spectra* (McGraw-Hill, New York, 1955).
111. H. Bethe: Annalen Der Physik **3**, 133 (1929).
112. H. Goldstein: *Classical Mechanics* (Addison-Wesley, Reading (Mass.), 1980).
113. S. Baroni, P. Giannozzi, A. Testa: Phys. Rev. Lett. **58**, 1861 (1987)
114. W. Weber: Phys. Rev. **B15**, 4789 (1977).
115. G. Onida, G. Benedek: Europhys. Lett. **18**, 403 (1992) and **19**, 343 (1992).
116. C.S. Yannoni, P.P. Bernier, D.S. Bethune, G. Meijer, J.R. Salem: J. Am. Chem. Soc. **113**, 3190 (1991).
117. G.B. Adams, J.B. Page, O.F. Sankey, K. Sinha, J. Menendez, D.R. Huffmann: Phys. Rev. **B44**, 4052 (1991).
118. S.J. Cyvin, E. Brendsdal, B.N. Cyvin, J. Brunvoll: Chem. Phys. Lett. **143**, 377 (1988).
119. D. Bakowies, W. Thiel: J. Am. Chem. Soc. **113**, 3704 (1991).
120. F. Negri, G. Orlandi, F. Zerbetto: Chem. Phys. Lett. **144**, 31 (1988); J. Am. Chem. Soc. **113**, 6037 (1991).
121. G. Coló, E. Vigezzi, M. Bernasconi: (unpublished)
122. D.S. Bethune, G. Meijer, W.C. Tang, H.J. Rosen, W.G. Golden, H. Seki, C.A. Brown, M.S. de Vries: Chem. Phys. Lett. **179**, 181 (1991).
123. C.I. Frum, R. Engleman, H.G. Hedderich, R.F. Bernath, L.D. Lamb, D.R. Huffmann: Chem. Phys. Lett. **176**, 504 (1991).
124. D.E. Weeks, W.G. Harter: J. Chem. Phys. **90**, 4744 (1989).
125. Z. Slanina, J.M. Rudzinski, M. Togasi, E. Osawa: J. Mol. Struct. (Theochem), **202**, 169 (1989).

126. D.S. Bethune, G. Meijer, W.C. Tang, H.J. Rosen: Chem. Phys. Lett. **174**, 219 (1990).
127. G. Gensterblum, J.J. Pireaux, P.A. Thiry, R. Caudano, J.P. Vigneron, P. Lambin, A.A. Lucas: Phys. Rev. Lett. **67**, 2171 (1991).
128. K. Prassides, T.J.S. Dennis, J.P. Hare, J. Tomkinson, H.W. Kroto, R. Taylor, D.R.M. Walton: Chem. Phys. Lett. **187**, 455 (1992).
129. R.L. Cappelletti, J.R.D. Copley, W.A. Kamitakahara, F. Li, J.S. Lannin, D. Ramage: Phys. Rev. Lett. **66**, 3261 (1991).
130. K. Raghavachari, C.M. Rholfing: J. Phys. Chem. **95**, 5768 (1991).
131. S. Sanguinetti, G. Benedek, M. Righetti, G. Onida: Phys. Rev. B **50**, 6743 (1994).
132. W. Andreoni, F. Gygi, M. Parrinello: Chem. Phys. Lett. **189**, 241 (1992).
133. F. Negri, G. Orlandi, F. Zerbetto: J. Am. Chem. Soc. 113, 6037 (1991).
134. P. Procacci, G. Cardini, P.R. Salvi, V. Schettino: Chem. Phys. Lett. **195**, 347 (1992).
135. B.N. Cyvin, E. Brendsdal, J. Brunvoll, S.J. Cyvin: J. of Mol. Struct. **352/353**, 481 (1995).
136. K. Raghavachari, D.L. Strout, G.K. Odom, G.E. Scuseria, J.A. Pople, B.G. Johnson, P.M.W. Gill: Chem. Phys. Lett. **214**, 357 (1993).
137. J.M.L. Martin, J. El-Yazal, J.-P. Francois: Chem. Phys. Lett. **248**, 345 (1996).
138. M. Sawtarie, M. Menou, K.R. Subbaswamy: Phys. Rev. B **49**, 7739 (1994); F. Varga, L. Nemes, G.I. Csonka: J. Mol. Struct. **376**, 513 (1996).
139. J.R. Heath, A.L. Cooksy, M.H.W. Gruebele, C.A. Schmuttenmaer, R.J. Saykally: Science **244**, 564 (1989).
140. M. Vala, T.M. Chandrasekhar, J. Szczepanski, R. Van Zee, W. Weltner Jr.: J. Chem. Phys. **90**, 595 (1989).
141. N. Mohazzen-Ahmadi, A.R.W. McKellar, T. Amano: J. Chem. Phys. **91**, 2140 (1989).
142. J.R. Heath, R.A. Sheeks, A.L. Cooksy, R.J. Saykally: Science **249**, 895 (1990).
143. C.A. Schmuttenmaer, R.C. Cohen, N. Pugliano, J.R. Heath, A.L. Cooksy, K.L. Busarow, R.J. Saykally: Science **249**, 897 (1990).
144. D.W. Arnold, S.E. Bradforth, T.N. Kitsopoulos, D.M. Neumark: J. Chem. Phys. **95**, 8753 (1991).
145. J.M.L. Martin, J. P. Francois, R. Gijbels: J. Chem. Phys **93**, 8850 (1990).
146. G. Benedek, G. Onida: Phys. Rev. B **47**, 16471 (1993).
147. K.R. Thompson, R.L. De Kock, W. Weltner jr.: J. Am. Chem. Soc. **93**, 4688 (1971)
148. T.F. Giesen, A. Van Orden, H.J. Hwang, R.S. Fellers, R.A. Provencal, R.J. Saykally: Science **265**, 756 (1994)
149. J.R. Heath, R.J. Saykally: J. Chem. Phys. **94**, 1724 (1991)
150. J.C. Grossman, L. Mitas, K. Raghavachari: Phys. Rev. Lett. **75**, 3870 (1995).
151. V. Parasuk, J. Almlof: Chem. Phys. Lett. **184**, 187 (1991).
152. Z. Wang, P. Dayl, R. Pachter: Chem. Phys. Lett. **248**, 121 (1996).
153. S. Baroni, S. de Gironcoli, A. Dal Corso, P. Giannozzi: Rev. Mod. Phys. **73**, 515 (2001)
154. J. Rickaert, G. Ciccotti, H.J.C. Berendsen: J. Comput. Phys. **23**, 327 (1977).
155. I. Stich, R. Car, M. Parrinello, S. Baroni: Phys. Rev. **B39**, 4997 (1988).
156. D.K. Remler, P.A. Madden: Molecular Physics **70**, 921 (1990).
157. J. Kohanoff, W. Andreoni, M. Parrinello: Phys. Rev. B **46**, 4371 (1992).
158. J. Kohanoff, W. Andreoni, M. Parrinello: Chem. Phys. Lett. **198**, 472 (1992).
159. G. Onida, W. Andreoni, J. Kohanoff, M. Parrinello: Chem. Phys. Lett. 219, 1 (1994)

160. N. Breda, R.A. Broglia, G. Colò, G. Onida, D. Provasi, E. Vigezzi: Phys. Rev. **B62**, 130 (2000).
161. J.M. Ziman: Proc. Camb. Phil. Soc. **51**, (1955) 707.
162. A. Haug: Z. Phys. **146**, 75 (1956).
163. B. Goodman: Phys. Rev. **110**, 888 (1958).
164. M. Lannoo, G.A. Baraff, M. Schlüter, D. Tomanek: Phys. Rev. **B46**, 14264 (1992).
165. N. Breda, R.A. Broglia, G. Colò, H.E. Roman, F. Alasia, G. Onida, V. Yu. Ponomarev, E. Vigezzi: Chem. Phys. Lett. **286**, 350 (1998).
166. N. Breda: M. Sc. Thesis (Physics Dept., University of Milan, 1998, unpublished)
167. O. Gunnarsson, H. Handschuh, P.S. Bechtold, B. Kessler, G. Gantetör, W. Eberhardt: Phys. Rev. Lett. **74**, 1875 (1995).
168. C.M. Varma, J. Zaanen, K. Ragavachari: Science **254**, 989 (1991).
169. M. Schlüter, M. Lannoo, M. Needles, G.A. Baraff, D. Tomanek: Phys. Rev. Lett. **68**, 526 (1992).
170. V.P. Antropov, O. Gunnarsson, A.I. Liechtenstein: Phys. Rev. **B48**, 7651 (1993).
171. J.C.R. Faulhaber, D.Y.K. Kho, P.R. Briddon: Phys. Rev. **B48**, 661 (1993).
172. O. Gunnarsson: Phys. Rev. **B51**, 3493 (1995).
173. O. Gunnarsson: Rev. Mod. Phys. **69**, 575 (1997).
174. R. Taylor, J.P. Hare, A.K. Abdul-Sada, H.W. Kroto: J. Chem. Soc. Chem. Comm. **20**, 1423 (1990).
175. P.W. Fowler, J. Woolnick: Chem. Phys. Lett. **127**, 78 (1986).
176. T. Kato, T. Kodama, T. Schioda, T. Nakagawa, Y. Matsui, S. Suzuki, H. Shiromaru, K. Yamauchi, Y. Achiba: Phys. Lett. **180**, 446 (1991).
177. G.E. Scuseria: Chem. Phys. Lett. **180**, 451 (1991).
178. S. Saito, A. Oshiyama: Phys. Rev. **B44**, 11532 (1991).
179. F. Alasia, R.A. Broglia, G. Coló, H.E. Roman: Chem. Phys. Lett. **247**, 502 (1995).
180. R.E. Haufler, L.S. Wang, L.P.F. Chibante, C. Jin, J. Conceicao, Y. Chai, R.E. Smalley: Chem. Phys. Lett. **179**, 449 (1991).
181. D. Provasi, N. Breda, R.A. Broglia, G. Colò, H.E. Roman, G. Onida: Phys. Rev. **B 61**, 7775 (2000).
182. F. Barranco, R.A. Broglia, G. Gori, E. Vigezzi, P.F. Bortignon, J. Terasaki: Phys. Rev. Lett. **83**, 2147 (2000).
183. F. Barranco, P.F. Bortignon, R.A. Broglia, G. Colò, E. Vigezzi: preprint (2003).
184. F. Barranco, R.A. Broglia, G. Colò, E. Vigezzi: Eur. J. Phys. **A11**, 385 (2001).
185. G. Gori: Ph. Thesis, Univ. of Milano, unpublished (2002).
186. P. Donati, T. Døssing, Y.R. Shimizu, S. Mizutori, P.F. Bortignon, R.A. Broglia: Phys. Rev. Lett. **84**, 4317 (2000).
187. Z.H. Wang, M.S. Dresselhaus, G. Dresselhaus, P.C. Eklund: Phys. Rev. **B48**, 16881 (1993).
188. D.R. Bes, R.A. Broglia, G.G. Dussel, R.J. Liotta, H.M. Sofia: Nucl. Phys. **A260**, 27 (1976).
189. F. Barranco, P.F. Bortignon, R.A. Broglia, G. Colò, E. Vigezzi: Eur. Phys. J. **A11**, 385 (2001).
190. M. Bernath, M.S. Hansen, P.F. Bortignon, R.A. Broglia: Phys. Rev. **A 49**, 1115 (1994).
191. P.-G. Reinhard: Phys. Lett. **A 169**, 281 (1992).
192. G. Colò, N. Van Giai, P.F. Bortignon, R.A. Broglia: Phys. Rev. **C 50**, 1496 (1994).

193. M. Schmidt, C. Ellert, W. Kronmuller, H. Haberland: Phys. Rev. **B 59**, 10970 (1999).
194. W.L. McMillan: Phys. Rev. **167**, 331 (1968).
195. P.B. Allen: Phys. Rev. B6 (1972) 2577.
196. L. Cooper: Phys. Rev. **104**, 1189 (1956).
197. J. Bardeen, L. Cooper, J.R. Schrieffer: Phys. Rev. **106**, 162 (1957); **108**, 1175 (1957).
198. A. Leggett: *The New Physics* (Cambridge University Press, Cambridge, 1989).
199. G.M. Eliashberg: JETP **11**, 6963 (1960); **16**, 78 (1963).
200. H.W. Kroto, J.R. Heath, S.C. O'Brien, R.F. Curl, R.E. Smalley: Nature **318**, 162 (1985).
201. W. Krätschmer, L.D. Lamb, K. Fostiropoulos, D.R. Huffmann: Nature **347**, 354 (1990).
202. S.C. Erwin: in *Buckminsterfullerenes*, edited by W.E. Billungs and M.A. Ciufolini (VCH, New York, 1993) p. 217.
203. C.M. Lieber, Z. Zhang: Solid State Physics **48**, 48 (1994).
204. A.P. Ramirez, Supercond. Rev. **1**, 1 (1994).
205. L.A. Paquette, R.J. Ternansky, D.W. Balogh, G.J. Kentgen: J. Am. Chem. Soc. **105**, 5446 (1983).
206. H. Prinzbach, A. Weller, P. Landenberger, F. Wahl, J. Worth, L.T. Scott, M. Gelmont, D. Olevano, B. van Issendorff: Nature **407**, 60 (2000).
207. Z. Iqbal: in *Nanostructured Carbon for Advanced Applications*, edited by G. Benedek, P. Milani and V. G. Ralchenko (Kluwer Academic Press, Dordrecht, 2001) p. 309.
208. T. Guo, M.D. Diener, Yan Chai, M.J. Alford, R.E. Hauffer, S.M. McClure, T. Ohno, J.H. Weaver, G.E. Scuseria, R.E. Smalley: Science **257**, 1661 (1992).
209. J.C. Grossman, M. Coté, S.G. Louie, M.L. Cohen: Chem. Phys. Lett. **284**, 344 (1998).
210. C. Piskoti, J. Yager, A. Zettl: Nature (London) **373**, 771 (1998).
211. E. Kaxiras, L.M. Zeger, A. Antonelli, Yu-min Juan: Phys. Rev. **B49**, 8446 (1994).
212. N.A. Romero, J. Kim, R.M. Martin: *Electron–Phonon Interactions in C_{28}-derived Molecular Solids*, Cond-mat/0403003 (27 Feb 2004).
213. H.W. Kroto: Nature **329**, 529 (1987).
214. M. Coté, J.C. Grossman, M.L. Cohen, S.G. Louie: Phys. Rev. Lett. **81**, 697 (1998).
215. X. Blase, A. Rubio, S.G. Louie, M.L. Cohen: Phys. Rev. **B52**, R2225 (1995).
216. *The Structure and Evolution of Neutron Stars*, edited by D. Pines, R. Tamagaki and S. Tsuruta (Addison-Wesley, Reading, Massachusetts, 1992)
217. J.H. Schön, Ch. Kloc, B. Batlogg: Nature **408**, 549 (2000); J.H. Schön, Ch. Kloc, B. Batlogg: Nature **406**, 702 (2000).
218. Retractions: Nature **422**, 93 (2003).
219. P.W. Anderson: "*Theories of fullerenes T_c which will not work*", (1991) (unpublished as quoted in [173]).
220. O. Gunnarsson: *Superconductivity in fullerides*, (World Scientific, Singapore, to be published).

Index

$\mathcal{D}_{2d}$ symmetry   142
$\mathcal{D}_{5h}$ symmetry   121

ab initio methods   43, 113
accidental degeneracy   140
adiabatic condition   117
adiabaticity   110
alkali-doped C_{60} compounds   201
average potential   3

bare mass   15
BCS
– gap equation   189
– number equation   189
BCS theory   183
Becke 3-parameter-Lee-Yang-Parr
   method (B3LYP)   127–129
bending force–constant   125
Bond Charge Model (BCM)   105, 114,
   157, 161
bond charges   114
Born–Oppenheimer approximation
   35, 37, 132, 136, 145, 147, 173
bosons   30
bowl isomer   127
breathing mode   117, 122, 127
buckminsterfullerene   115
bulk silicon   139

canonical transformation   165
Car–Parrinello Lagrangian   139
Car–Parrinello method   6, 44, 130, 131,
   157, 163, 172
Car–Parrinello molecular dynamics
   code (CPMD code)   142
carbon structures   79
– C_{20} cluster   122, 197
– C_{28} cluster   141, 197, 199
– C_{36} cluster   197, 199
– C_{60} fullerene   79
– carbon nanotubes   4, 85, 200
– – electronic structure   86
– cyclic chains   126
– fullerenes   4, 128
– fullerite   119
– linear chains   4, 89, 123
– – electronic structure   89
center of mass   110
charge density   132
charge transfer   132
coherence length   183
collective coordinate   108
collective excitation   2, 67
commutation relations   109
computational cost   133
conjugate momentum   109
Cooper pairs   27, 183
creation and annihilation operators
   29, 109
critical temperature   183
curvature effects   197

Debye frequency   28
deformation potential   147
Density Functional Perturbation
   Theory (DFPT)   129
Density Functional Theory (DFT)   6,
   10, 37, 38, 113, 131, 169
density–density correlation function
   63, 72
– unperturbed propagator   74
diatomic molecule   110
diffuse interstellar bands   4
dipole approximation   140
dipole resonance peak   67
Dirac representation   16
displacement field   142
dynamical matrix   105, 137
dynamical minimization   132
Dyson equation   63

effective electron–electron interaction
   165
– in C_{60} 166

effective force–constant matrix   116
effective mass   15
– ω–mass   20
– k–mass   10, 11
– total   21
electromagnetic response
– of Na_8 98
– of C_{60} 82
– of carbon nanotubes   87
– of linear carbon chains   93
electron spill out   197
electron–phonon coupling   141, 147
– diagonal matrix elements   148
– dimensionless coupling constant λ
    155, 157, 161, 163, 175
– in C_{20} and C_{70} 160
– in C_{60} 149, 154
– in Na clusters   163
electron–plasmon coupling   16, 70, 166
electronic degrees of freedom   135
electronic polarizability
– dynamical   65
–– in carbon nanotubes   87
– static   66
–– in C_{60} 82
energetic decoupling   138
energy cutoff   142
energy transfer   138
equations of motion   134
equilibrium condition   113, 125
equilibrium distance   113
equilibrium geometry   132
exchange potential   10
exchange-correlation energy   39, 41
exchange-correlation potential   40–42,
    51, 64, 169
external potential   134

fermions   28
fictitious kinetic energy   133
fictitious mass   133
first-neighbors interaction   116
force constant matrix   141
force constants   105, 111
force exerted on atoms   141
force–constant matrix   106, 115
frozen phonon   129, 141, 157, 163, 172
functional derivatives   39

Generalized Gradient Approximation
    (GGA)   127
golden rule   67
Grahm–Schmidt   135

graphite   117

harmonic approximation   105
– beyond the   137
harmonic regime   137
harmonicity   110
Hartree equations   8
Hartree potential   8, 40, 50
Hartree–Fock equations   9
Hartree–Fock theory   37, 38, 41, 61
Hessian matrix   106
highest-occupied molecular orbital   6
– for C_{60} 80
Hohenberg–Kohn theorem   38, 40
hole doped fullerites   202
homogeneous electron gas   41
Hybrid Orbital Model (HOM)   151,
    157, 161

icosahedral symmetry   117
– C_{60} 79
– group theory   100
independent particle response   63
induced interaction   21
infrared active modes   140
infrared and Raman active modes
    113, 120
interatomic forces   132
invariance, translational and rotational
    106
ion diffusion   137
ionic kinetic energy   139
ionicity   132
irreducible representations   111, 122
isotope effect   190

Jahn–Teller   123
jellium model   12, 13

Keating potential   116, 123
Kohn–Sham
– approach to density functional theory
    10
– eigenvalues   136
– equations   38, 62, 131
–– cylindrical basis   51
–– spherical basis   45
– Hamiltonian   41, 132
– orbitals   39, 40, 45, 46, 51, 57, 133
– states for C_{60} 79

Löwdin transformation   47, 79
Lagrange equations   107
Lagrange multipliers   40

Lagrangian   107
linear molecules   108
linear response   61, 70
linear response theory   113
linear transformation   185
linear triatomic molecule   111
local density approximation   6, 41, 61
longitudinal vibrations   111, 124
lowest-unoccupied molecular orbital   6
– for C_{60} 81

mass abundance   5
mass enhancement factor   27, 175
matrix-vector multiplications   136
mean field   7
mean field theory   4
metal clusters   4, 79
– Na_8 98
metastable state   138
microcanonical   138
Modified Neglect of Diatomic Overlap
   (MNDO)   122, 157
Molecular Dynamics (MD)   131, 135,
   138
– parameter-free   132
multiplet   141
multipolar expansion   114

Na_8 142
nearest-neighbors force constants   112
neutron scattering   119
Newton equation   135
non-central interactions   114
normal modes   105, 108

odd-numbered rings   127
on-the-energy-shell transition   19
optical absorption
– of C_{60} 81
optical response   1
optically active modes   140
orbital kinetic energy   139
orthogonality condition   139
orthogonality constraint   134
oscillator strength   67

particle–hole transitions   63, 68
particle–vibration coupling
– in a spherical system   16, 151
pentagonal pinch mode   117
perturbation theory
– first order   173, 178
– second order   178
phase transition   183

phonon frequencies   132
phonons   26, 105
– Hamiltonian for   146
photoabsorbtion   171
photoabsorption
– cross–section   67
photoemission   7, 155, 157
– in C_{60}^- 155
plane-wave basis   132, 133
plasmon resonance
– in C_{60} 83
plasmon wavefunction
– separable interaction   73
plasmons   2, 16, 23, 65, 68, 79
polyacetylenic   127
– ring   129
pseudo-wavefunction   55
pseudopotentials   44
– basics on   54
– for carbon   56
– local part   47
– non-local
– – C_{60} 81
– non-local part   48
– norm-conserving (NCPP)   44, 131
– sum rules   66

Quantum Chemical Force Field
   (QCFF)   121
quantum Monte Carlo   37, 42
quasiparticles   7

radial modes   119
Raman active modes   140
Random Phase Approximation (RPA)
   23, 61, 64
– density–density correlation function
   63
reduced electron–phonon coupling
   200
reduced mass   110
Reflection Electron Energy Loss
   Spectroscopy (REELS)   119
residual interaction   64, 68, 73, 75
restoring forces   142
ring C_{20} 127

screened Coulomb interaction   128
screened Coulomb potential   22, 28,
   173
second quantization   28
self-consistent potential   131, 167
self-energy   17, 18, 32

shell model   114
simulated annealing   133, 135
single-particle potential   5
small oscillations   137
sodium clusters   142
specific heat   26
spherical harmonics   46, 47
squashing mode   117, 141
stationary point   135
strength function   65, 70
stretching force–constant   118
strong coupling limit   190
sum rules   65
– energy weighted   66, 70

thermodynamic equilibrium   138
Thomas–Fermi approximation   22
Thomas–Reiche–Kuhn sum rule   67
time-dependent
– density   71
– Density Functional Theory   61
– external field   70
– local density approximation   6, 61,
  64, 172
– – cylindrical basis   76
– – in configuration space   72
– – including electron–phonon and
  electron–plasmon couplings   169,
  171
– – propagator   64

– – spherical basis   74
– perturbations   61
timescale   138
total energy   36, 41, 132
trajectory   139
transition density   68, 167
transition potential   168
translational invariance   111
transversal modes   125
two-body contributions   112
two-body interaction   137

unitary transformation   136

van der Waals   3
van der Waals crystals   113
van der Waals solid   194
Verlet algorithm   131, 135, 137
vibrational eigenstates   110, 112
vibrational eigenvectors   143
vibrational properties   139
vibrational spectra   141
virtual excitation   18

weak coupling limit   190
Wigner symbols   50
Wigner–Eckart theorem   150

zero point energy   111
zero point motion   109

Printing: Mercedes-Druck, Berlin
Binding: Stein+Lehmann, Berlin